Eukaryotic Gene Transcription

Frontiers in Molecular Biology

SERIES EDITORS

B. D. Hames

Department of Biochemistry and Molecular Biology
University of Leeds, Leeds LS2 9JT, UK

AND

D. M. Glover

Cancer Research Campaign Laboratories
Department of Anatomy and Physiology
University of Dundee, Dundee DD1 4HN, UK

TITLES IN THE SERIES

---

1. Human Retroviruses
Bryan R. Cullen

2. Steroid Hormone Action
Malcolm R. Parker

3. Mechanisms of Protein Folding
Roger H. Pain

4. Molecular Glycobiology
Minoru Fukuda and Ole Hindsgaul

5. Protein Kinases
Jim Woodgett

6. RNA–Protein Interactions
Kyoshi Nagai and Iain Mattaj

7. DNA–Protein: Structural Interactions
David Lilley

8. Mobile Genetic Elements
David Sherratt

9. Chromatin Structure and Gene Expression
Sarah C. R. Elgin

10. Cell Cycle Control
Chris Hutchinson and D. M. Glover

11. Molecular Immunology (Second Edition)
B. David Hames and David M. Glover

12. Eukaryotic Gene Transcription
Stephen Goodbourn

13. Molecular Biology of Parasitic Protozoa
Deborah F. Smith and Marilyn Parsons

14. Molecular Genetics of Photosynthesis
Bertil Andersson, A. Hugh Salter, and Jim Barber

15. Eukaryotic DNA Replication
J. Julian Blow

16. Protein Targeting
Stella M. Hurtley

Eukaryotic Gene Transcription

EDITED BY

Stephen Goodbourn

Division of Biochemistry
Department of Cellular and Molecular Sciences
St George's Hospital Medical School, London, UK

IRL PRESS
at
OXFORD UNIVERSITY PRESS
Oxford New York Tokyo

Oxford University Press, Walton Street, Oxford OX2 6DP

Oxford New York
Athens Auckland Bangkok Bombay
Calcutta Cape Town Dar es Salaam Delhi
Florence Hong Kong Istanbul Karachi
Kuala Lumpur Madras Madrid Melbourne
Mexico City Nairobi Paris Singapore
Taipei Tokyo Toronto

and associated companies in
Berlin Ibadan

Oxford is a trade mark of Oxford University Press

Published in the United States
by Oxford University Press Inc., New York

A catalogue record for this book is available from the British Library

Library of Congress Cataloging in Publication Data
Eukaryotic gene transcription / edited by Stephen Goodbourn. — 1st ed.
(Frontiers in molecular biology ; 12)
Includes bibliographical references.
1. Genetic transcription–Regulation. I. Goodbourn, Stephen. II Series.
[DNLM: 1. Transcription, Genetic. 2. Eukaryotic Cells. 3. Transcription Factors. 4. Gene Expression Regulation. QH 450.2 E86 1996]
QH450.2.E84 1996 574.87'3223–dc20 95–31163
ISBN 0 19 963487 4 (Hbk)
ISBN 0 19 963486 6 (Pbk)

Typeset by Footnote Graphics, Warminster, Wilts
Printed in Great Britain by The Bath Press, Avon

Preface

In recent years significant advances have been made in our understanding of gene regulation, of which the control of transcription appears to be the most important component. While it has long been realized that the recognition of promoter and enhancer regions by sequence-specific DNA binding proteins lies at the heart of transcriptional regulation, the means by which the activities of these proteins are modulated, and the consequences of these changes are only now beginning to be understood. The objectives of this book are to review the underlying principles of transcriptional control and to examine some of the aims and directions of current research. The nine essays presented here have been chosen as representative of particularly active areas of investigation, and the authors have been encouraged to speculate about the future directions of their respective fields.

One of the major goals of recent research has been the elucidation of the mechanisms by which regulatory transcription factors influence the rate of transcription. Our current views on this are discussed by Jyrki Eloranta and Stephen Goodbourn. Alan Wolffe takes the discussion further by indicating the extent to which histones play a more than structural role in transcriptional activation. Perhaps the simplest way to regulate transcription is to control the access of sequence-specific activators or repressors to their DNA target sites. In what is probably the best studied eukaryotic example of this, Tom Gilmore discusses the activation of NF-κB and related factors from latent cytoplasmic stores in response to extracellular stimuli. Although the biochemical details of the intracellular events that lead to NF-κB activation are not yet understood, this is not the case for signal transduction through the second messenger cyclic AMP. Nicholas Foulkes and Paolo Sassone-Corsi discuss the properties of DNA binding proteins that mediate transcriptional responses to cyclic AMP. These proteins can bind to DNA in the absence of stimulation, indicating that it is their ability to stimulate transcription that is regulated. The cyclic AMP response proteins are members of a large family of transcription factors, referred to as leucine zipper proteins, that form dimers, and tissue-specific expression of a variety of members of this family can have profound effects on the ability to respond to cyclic AMP. A similar situation is observed during myogenesis, where members of the helix–loop–helix family of DNA binding proteins play a key role; Craig Neville and Nadia Rosenthal discuss how the activities of these factors are regulated by their choice of interacting partner, and indicate how subsequent downstream decisions also involve specific transcription factors. Cornelius Gross and Bill McGinnis discuss *Drosophila* development where the developmentally-restricted availability of homeobox factors and their differential interactions with other factors determines patterns of gene expression. The transcriptional stimulatory properties of DNA binding proteins are also regulated during the cell cycle, and Nic Jones discusses our knowledge of these events in both yeast and higher eukaryotes. Despite the advances in our knowledge of specific transcription factors, activation often involves the combined actions of several factors, or the integration of several signals. Peter King and Stephen Goodbourn review models of how this is achieved. Finally, Jon Karn and colleagues discuss the mechanism of anti-termination of HIV transcription by the Tat protein and remind us that transcriptional regulation need not operate solely at the level of initiation.

I hope that the importance of transcriptional regulation in so many aspects of biology is reflected by the range of topics covered in this book, and that it will correspondingly be of interest to a wide spectrum of readers. I would like to thank the authors for their contributions and patience during the production of this book, and Jyrki Eloranta, Mark Fisher, Nic Jones, Zoia Larin, and Peter King for their comments.

London S.G.
November 1995

Contents

Contributors

MARK J. CHURCHER
MRC Laboratory of Molecular Biology, Hills Road, Cambridge CB2 2QH.

JYRKI J. ELORANTA
Division of Biochemistry, Department of Cellular and Molecular Sciences, St. George's Hospital Medical School, London SW17 0RE, UK.

NICHOLAS S. FOULKES
Institut de Génétique et de Biologie Moléculaire et Cellulaire CNRS-INSERM, BP163, 67404 Illkirch, CU de Strasbourg, France.

MICHAEL J. GAIT
MRC Laboratory of Molecular Biology, Hills Road, Cambridge CB2 2QH.

THOMAS D. GILMORE
Biology Department, Boston University, 5 Cummington Street, Boston MA 02215, USA.

STEPHEN GOODBOURN
Division of Biochemistry, Department of Cellular and Molecular Sciences, St. George's Hospital Medical School, London SW17 0RE, UK.

CORNELIUS T. GROSS
Department of Molecular Biophysics and Biochemistry, Yale University, New Haven, CT 06511, USA.

NIC JONES
Gene Regulation Laboratory, Imperial Cancer Research Fund, 44 Lincoln's Inn Fields, London WC2A 3PX, UK.

JONATHAN KARN
MRC Laboratory of Molecular Biology, Hills Road, Cambridge CB2 2QH.

NICHOLAS J. KEEN
MRC Laboratory of Molecular Biology, Hills Road, Cambridge CB2 2QH.

PETER KING
Division of Biochemistry, Department of Cellular and Molecular Sciences, St. George's Hospital Medical School, London SW17 0RE, UK.

WILLIAM J. McGINNIS
Department of Molecular Biophysics and Biochemistry, Yale University, New Haven, CT 06511, USA.

CRAIG NEVILLE
Cardiovascular Research Center, Harvard University, Massachusetts General Hospital—East 149 13th Street, Charlestown, MA 02129, USA.

NADIA ROSENTHAL
Cardiovascular Research Center, Harvard University, Massachusetts General Hospital—East 149 13th Street, Charlestown, MA 02129, USA.

KAROLA RITTNER
MRC Laboratory of Molecular Biology, Hills Road, Cambridge CB2 2QH.

PAOLO SASSONE-CORSI
Institut de Génétique et de Biologie Moléculaire et Cellulaire CNRS-INSERM, BP163, 67404 Illkirch, CU de Strasbourg, France.

ALAN P. WOLFFE
Laboratory of Molecular Embryology, National Institute of Child Health and Human Development, National Institutes of Health, Bethesda, MD 20892, USA.

Abbreviations

AChR	acetylcholine receptor
Ad-MLP	adenovirus major late promoter
ADI	ATP-dependent inhibitor
AE	autoregulatory element
AER	apical ectodermal ridge
ARIA	acetylcholine receptor-inducing activity
ATF	activating transcription factor
bHLH	basic–helix–loop–helix
bZip	basic–leucine zipper
cAMP	3′, 5′-cyclic adenosine monophosphate
CARE	cAMP autoregulatory responsive element
CBP	CREB-binding protein
cdk	cyclin-dependent kinase
CEF	chicken embryo fibroblasts
CGRP	calcitonin gene-related peptide
CK	casein kinase
CRE	cyclic AMP responsive element
CREB	cyclic AMP response element binding protein
CREM	cyclic AMP response element modulator
CsA	cyclosporin A
CTD	carboxy-terminal domain
DAG	diacylglycerol
DBD	DNA-binding domain
Dr1	TFIID repressor 1
dsRNA	double-stranded RNA
EBP	enhancer-binding protein
EC	embryonal carcinoma
EDL	extensor digitorum longus
EGF	epidermal growth factor
FGF	fibroblast growth factor
FSH	follicle-stimulating hormone
GAF	gamma interferon activated factor
GAS	gamma interferon activation site
GBP	guanylate binding protein
HCF	host-cell factor
HIV	human immunodeficiency virus
HLH	helix–loop–helix
HMG	high mobility group protein
HP1	heterochromatin protein 1

HTH	helix–turn–helix
ICER	induced cAMP early repressor
IFN	interferon
IGF	insulin-like growth factor
IgH	immunoglobulin heavy chain
IL	interleukin
Inr	initiator
IPTG	isopropyl-thio-β-D-galactoside
ISRE	IFN stimulatable response element
IST	inducer of short transcripts
KID	kinase-inducible domain
LH	luteinizing hormone
LP	long-photoperiod
LPS	lipopolysaccharide
LRF	liver-regenerating factor
LTR	long terminal repeat
MAT	mating type locus
MBF	Mlu I cell cycle box binding factor
MCB	Mlu I cell cycle box
MCK	muscle creatine kinase
MDF	myogenic determination factor
MEF	myocyte enhancer factor
MHC	major histocompatiblity compex
MHC	myosin heavy chain
MLC	myosin light chain
MMTV	mouse mammary tumour virus
mot1	modifier of transcription 1 gene
NLS	nuclear localization sequences
NMJ	neuromuscular junction
NRD	negative regulatory domain
PDGF	platelet-derived growth factor
PIC	pre-initiation complex
PKA	protein kinase A
PKC	protein kinase C
PP-1	protein phosphatase-1
PRD	positive regulatory domain
Py	pyrimidine
RACE-PCR	rapid amplification of cDNA ends-polymerase chain reaction
RE	response element
RH	Rel homology (domain)
RRE	rev response element
RT-PCR	reverse transcriptase-polymerase chain reaction
SBF	Swi4/6 cell-cycle box binding factor
SCB	Swi4/6 cell-cycle box
SCG	superior cervical ganglia
SCN	suprachiasmatic nucleus

SIE(F)	sis-inducible element(factor)
sin	switch independent
SL1	RNA polymerase I-specific complex
SNAPc	snRNA-activating protein complex
snf	sucrose non-fermenting
snRNA	small nuclear RNA
SP	short-photoperiod
SRE	serum-responsive element
SRF	serum response factor
SSR	start site region
STAT	signal transducing activators of transcription
SWI	switch
TAF	TBP-associated factor
TAR	*trans*-activation response region
TBP	TATA-box binding protein
TCF	ternary complex factor
TdT	terminal deoxynucleotidyltransferase
TNF	tumour necrosis factor
TPA	12-O-tetradecanoyl phorbol-13-acetate
TR	thyroid hormone receptor
TRE	TPA-responsive element
TREB	*tax*-responsive element binding protein
UAS	upstream activator sequence

1 | Positive and negative regulation of RNA polymerase II transcription

JYRKI J. ELORANTA and STEPHEN GOODBOURN

1. Introduction

The correct temporal and spatial expression of specific genes governs the identity of eukaryotic cells. For most genes, the primary control point is the regulation of transcriptional initiation. The DNA sequences responsible for determining the exact start site and the level of mRNA synthesis are generically referred to as promoters. The core promoter elements (TATA boxes and initiator elements) determine, and reside immediately 5′ to and overlap with, the mRNA start points. The minimal basal level of transcription can be supported by basal transcription machinery that assembles on the core promoter elements. In addition, promoters contain recognition sites for regulatory transcription factors, which are sequence-specific DNA-binding proteins. Regulatory factors can influence transcription levels by either enhancing (activators) or antagonizing (repressors) the assembly or activity of the basal transcription machinery. Typical regulatory factors are modular, so that the DNA-binding domain is usually separable from the one mediating activation or repression (1). The functional analysis of various activator proteins has revealed that activation domains often appear to fall into distinct classes, such as acidic, glutamine-rich, and proline-rich, according to the predominance of particular amino acid residues. Transcriptional repression domains have not been as well defined, although alanine-rich or basic regions found in some repressor proteins may serve as such negative effector domains (2–4).

One of the major goals of recent research has been the elucidation of the mechanisms by which regulatory transcription factors influence the rate of initiation by the basal transcription machinery. Influence on both the rate and extent of assembly of basal transcription machinery on core promoters has been suggested. As discussed in this chapter, regulatory factors are believed to achieve their effect by protein–protein interactions with specific components of the basal transcription machinery.

2. The RNA polymerase II basal transcriptional machinery

The basal transcription factors (5–7) required for initiation by RNA polymerase II from most promoters were first identified in nuclear extracts as activities capable of supporting basal levels of transcription, and have been separated into biochemically defined fractions referred to as TFIIA, -B, -D, -E, -F, and -H. Subsequent cloning and biochemical analyses have shown that many of these factors are composed of multiple subunits. The basal factors can be assembled in a defined order on core promoters to form a pre-initiation complex (5–8) (Fig. 1). In most RNA polymerase II promoters, an A+T-rich sequence element known as the TATA box (consensus T A T A $^{T}/_{A}$ A $^{T}/_{A}$ A) is the critical determinant for the multi-step assembly. Since TFIID is the only basal factor known to possess sequence-specific DNA-binding activity, it is responsible for the initial template commitment. The TATA box is specifically recognized by a particular component of the TFIID complex, referred to as the TATA box binding protein (TBP). However, several footprinting studies have indicated that the DNA contacts made by the TFIID complex are more extensive than those by TBP alone (9–11). TBP alone typically generates a footprint of approximately 20 base pairs, centred around the TATA box, whereas the extended region protected by TFIID encompasses the transcriptional start site, as well as sequences further downstream. Other subunits of TFIID are thus likely to make additional DNA contacts. Indeed, a recently cloned subunit of the TFIID complex, a 150-kDa TBP-associated factor (dTAF150), has been shown to bind specifically to DNA sequences overlapping the start site of transcription and also to a further 30 bp downstream (12). Together, TBP and dTAF150 seem to be largely responsible for the extended TFIID footprint pattern.

Several RNA polymerase II genes do not contain discernible TATA boxes, yet they appear to be transcribed using the same basal machinery. The pre-initiation complex assembly on TATA-less promoters is typically specified by initiator (Inr) elements that overlap the precise transcription start sites (13, 14). The functional consensus sequence for Inr elements is rather loose (Py-Py-A(+1)-N-$^{T}/_{A}$-Py-Py, where Py is pyrimidine and N is any nucleotide; 15). The initial recognition of TATA-less promoters through these Inr elements probably utilizes the dTAF150 component of TFIID (12, 16, 17). The specific binding of dTAF150 to the core promoter may explain how TFIID can become recruited and then orchestrate the transcriptional initiation events on TATA-less RNA polymerase II promoters that do not allow TBP binding. It should be stressed that TBP has been shown to be required for TATA-less promoters, even if it does not function as the initial template-recognizing factor (18). Interestingly, even when recruited to certain TATA-less promoters by protein–protein interactions, TBP makes DNA contacts in the –30 region, regardless of the exact DNA sequence (19). While the template recognition events between TATA-containing and TATA-less promoters are clearly distinct, the subsequent DNA-independent steps in the initiation process are likely to be highly similar for both promoter types (20).

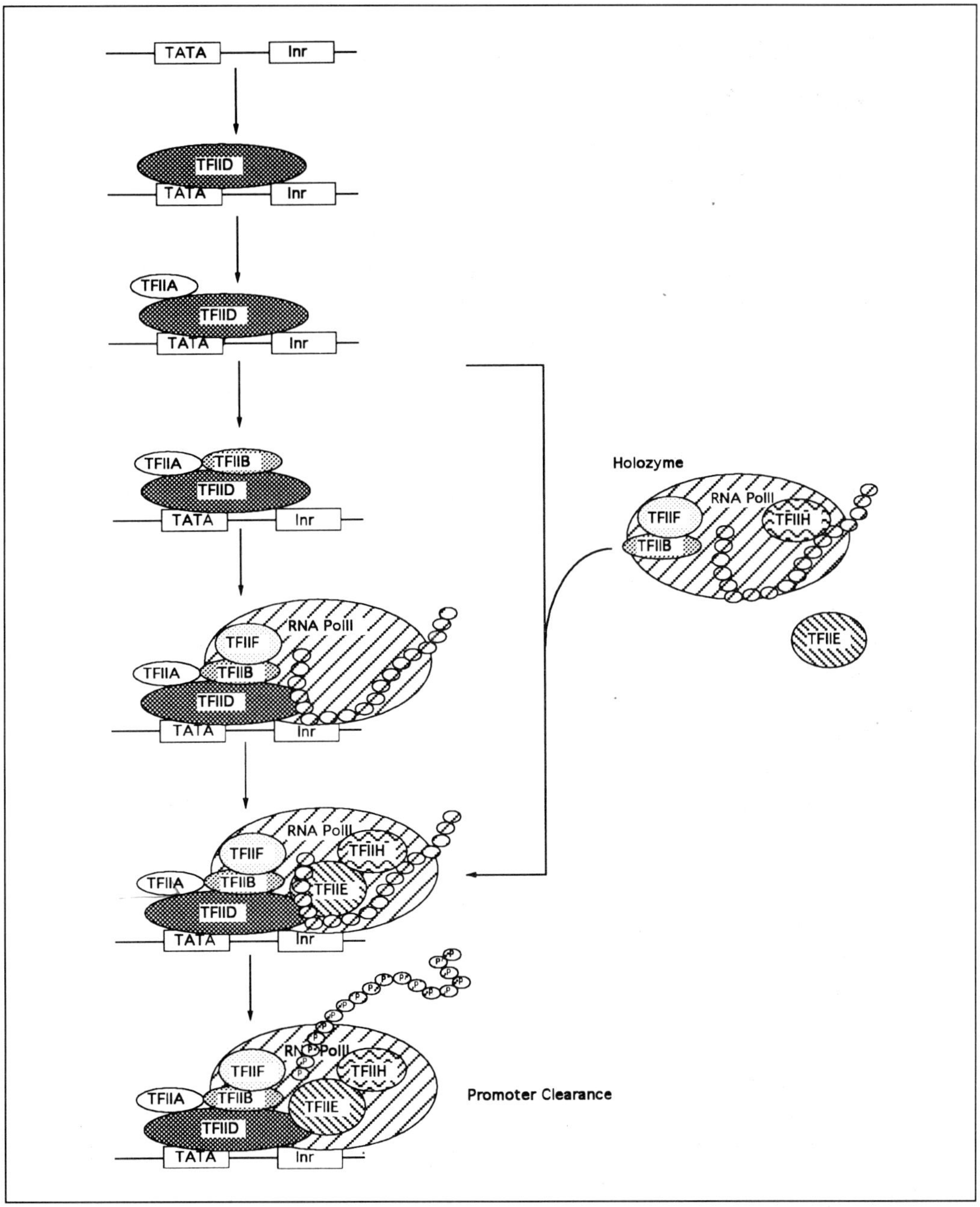

Fig. 1 Stepwise assembly of the pre-initiation complex on an RNA polymerase II-dependent promoter. The template commitment step is illustrated as taking place through either TATA box or Inr element recognition by TFIID. The subsequent individual assembly steps are described in the text. Once TFIIE and TFIIH are recruited to the complex, the carboxy-terminal domain of the large subunit of RNA polymerase II (CTD) becomes hyperphosphorylated, which leads to release of the RNA polymerase II from the promoter (an event referred to as promoter clearance) and initiation of transcription. The holozyme indicated to the right of the figure is a partially pre-assembled complex that has been identified in yeast; it is probable that this can be recruited to the template in a simplified assembly process.

Following template commitment, pre-initiation complex assembly continues with the association of TFIIA with TBP, an event that stabilizes the TFIID–template interaction on TATA-containing promoters. TFIIA is believed to achieve this effect by counteracting the negative activities that are inhibitory to the TBP function (see below). It should be noted that TFIIA can also enter the assembly pathway at any later stage (21). TFIIB is the next basal factor to enter the pre-initiation complex, and also interacts directly with TBP. Whereas TFIIA is capable of stabilizing the interaction of TFIID with a TATA box, TFIIB appears to have a similar effect on the TFIID–Inr interaction. The pre-initiation complex intermediate TFIID–TFIIA–TFIIB serves as a nucleation site for entry of the remaining basal transcription factors and RNA polymerase II. TFIIF mediates the entry of RNA polymerase II into the complex, and markedly decreases non-specific DNA binding of the polymerase. On eukaryotic RNA polymerase II-dependent promoters the assembly continues with the further association of TFIIE and TFIIH. In the presence of the four ribonucleoside triphosphates, the fully assembled complex is capable of initiating RNA synthesis from specific start sites. On TATA-containing promoters these are typically located 25 to 30 nucleotides downstream of the TATA box.

The six basal transcription factors consist of more than 30 polypeptides, and the RNA polymerase itself of 10±2 subunits; thus, the potential for multiple protein–protein interactions within an assembled pre-initiation complex is vast. For example, RNA polymerase II interacts directly with TBP (22), TFIIB (23, 24), TFIIF (25, 26), TFIIE (26, 27), and TFIIH (28). Recent studies suggest that many of the interactions involving basal factors are inductive, causing the interaction partners to assume new conformations or activities necessary to allow the subsequent steps in the assembly process (27, 29–31).

The stepwise assembly model for the pre-initiation complex has recently been challenged by the purification of a largely pre-assembled complex, termed holozyme, from crude homogenates of the yeast *Saccharomyces cerevisiae* (32; see Fig.1). The holozyme contains TFIIB, TFIIF, and TFIIH, together with RNA polymerase II, but no significant amounts of TFIID, TFIIA, or TFIIE. It will be interesting to learn whether similar partial pre-initiation complexes assemble in solution in the nuclei of higher eukaryotic cells.

All promoters do not appear to require the full set of basal transcription factors. For example, it has been shown that the immunoglobulin heavy chain (IgH) gene promoter can be accurately transcribed *in vitro* by RNA polymerase II in the presence of only TBP and TFIIB (33). This minimal requirement is dependent on the template DNA being negatively supercoiled, since transcription from the IgH promoter requires the entire array of basal factors when the template is linearized. This suggests that the free energy stored in DNA supercoils promotes the formation of an open complex by the minimal set of basal factors. In contrast to the IgH promoter, transcription from the adenovirus major late promoter requires the complete set of basal transcription factors, independent of the superhelical state of the template. The determinants for the differential requirements of different promoters for basal transcription factors remain somewhat unclear. The reason why TFIIE and

TFIIH appear dispensable for transcription of the immunoglobulin heavy chain (IgH) gene promoter may be that, instead of being required for the formation or function of an initiation complex, their role is to convert a transcriptional initiation complex into an elongation complex, an event referred to as promoter clearance (27, 34). Also, TFIIH has a helicase activity, believed to be responsible for the local unwinding of DNA around the initiation site; supercoiling may abrogate the requirement for TFIIH. While *in vitro* studies have been highly informative, the factor requirement *in vivo* might be different, owing to the influence of chromatin components (see Chapter 2).

3. RNA polymerase I and III promoters

In addition to its requirement for basal transcription from both TATA-containing and TATA-less RNA polymerase II promoters, TBP is also required for transcription by RNA polymerases I and III (19, 35, 36). None of the RNA polymerase I promoters have TATA-like sequences, and the human RNA polymerase I-specific TBP-containing complex, SL1, cannot bind independently to them, but needs to be tethered via protein–protein interactions with an upstream binding factor (37, 38). Most RNA polymerase III promoters also have no TATA box and TBP is recruited to them via other proteins (39). Some RNA polymerase III promoters, such as that of the U6 snRNA gene, do contain a functional TATA box, which directs pre-initiation complex assembly to these promoters by binding TBP. Paradoxically, when the U6 TATA box is mutated, the promoter switches to be transcribed by RNA polymerase II (40, 41). This provides an exceptional case, where the presence of a TATA box determines the selection of RNA polymerase III over RNA polymerase II.

4. TATA box binding protein

TBP is necessary for transcription by all three eukaryotic nuclear RNA polymerases, whether the promoters have identifiable TATA box sequences or not (19, 35, 36). TBP is thus an integral component of the transcription initiation complex in all three systems, despite the differences in the modes of promoter recognition and biochemical properties of the three RNA polymerases. Its universal presence in transcription complexes suggests that TBP is likely to have been a member of an early initiation complex, which preceded the evolutionary divergence of the three eukaryotic RNA polymerases. In fact, the recent cloning of a TBP homologue from nuclei-less Archaea suggests that the role of TBP in transcription was established before the divergence of eukaryotic and Archaean lineages (42, 43).

Little or no free TBP exists in solution in human or *Drosophila* cells, instead it is found in large multi-component complexes that contain additional polypeptides that are tightly bound to TBP, and are consequently referred to as TBP-associated factors, or TAFs (10, 44, 45; see below). The ability of TBP to support transcription by each of the three RNA polymerases appears to be determined by its interactions with different arrays of TAFs. One of the future challenges is to learn the basis of channelling the newly synthesized TBP pool into RNA polymerase I-, II-, and III-specific transcription complexes, and to determine the degree of conservation in protein–protein interactions involving TBP in the different complexes. Three point mutations have been identified within the conserved carboxy-terminal region of TBP (see below) that abolish *in vivo* transcription by all three nuclear RNA polymerases, suggesting that the different classes of transcription machinery contact the same regions of the TBP polypeptide (46). However, two other point mutations produce specific effects: one prevents transcription by RNA polymerases II and III but allows transcription by RNA polymerase I; another abolishes TATA binding, but still functions on TATA-less RNA polymerase I- and III-dependent genes but not on a TATA-containing RNA polymerase II-dependent promoter (47). Therefore, components of the three different transcription systems are likely to utilize overlapping but distinct regions of TBP. Temperature-sensitive yeast strains that contain TBP specifically defective in RNA polymerase III-dependent transcription accumulate increased amounts of RNA polymerase II-dependent transcripts when grown at the non-permissive temperature; conversely, an RNA polymerase II-specific TBP mutant shows increased RNA polymerase III transcription (48). This suggests that factors specific for the RNA polymerase II or RNA polymerase III system compete for limiting amounts of TBP *in vivo*; the competition could reflect overlapping recognition sites on the TBP surface for class-specific factors. Consistent with this, studies on the RNA polymerase I-specific complex SL1 indicate that the protein–protein interactions that associate TBP with either RNA polymerase I- or II-specific TAFs are mutually exclusive (49).

The carboxy-terminal 180 amino acids of TBP are highly conserved (19); for example, this domain shows 85% identity between human and yeast. Even the carboxy terminus of TBP from Archaea is approximately 40% identical to human TBP, despite the large evolutionary distance (43). The conserved core domain of TBP has been shown to be sufficient to bind to TATA boxes, and to allow basal transcription (10, 50, 51). Furthermore, gene replacement studies indicate that the conserved domain of TBP is sufficient for supporting growth in yeast (52–55). The high conservation of sequence translates into functional conservation, as human and yeast TBPs are interchangeable for both basal and activated transcription by RNA polymerase II *in vitro* (56); furthermore, human and yeast TBPs can support transcription in a cell-free Archaean transcription system (57). Surprisingly, while the *S. pombe* TBP can complement and support cell growth of a *S. cerevisiae* TBP null mutation (55, 58), the human or *Drosophila* TBPs cannot (52, 59). Nevertheless, it has been shown that the yeast TBP can mediate transcriptional activation by retinoic acid or GAL4-VP16 in some mammalian cell lines as efficiently as the human TBP (60). A study on the *in vivo* transcriptional activity of human TBP at different classes of yeast promoters indicates that the failure of human TBP to support yeast growth is not a consequence of its inability to generally support RNA polymerase II-dependent transcription (61).

TBP binds DNA as a monomer (62). Unlike most sequence-specific DNA-binding proteins, TBP interacts primarily with the minor groove of DNA (63, 64). The crystal structures of TBP from *Arabidopsis thaliana* (65) and the conserved carboxy-terminal domain of yeast TBP (66) have been solved. These analyses reveal a highly symmetrical, evolutionarily conserved structure, in which the carboxy-terminal region of TBP forms a novel DNA-binding fold positioned on DNA in a saddle-like manner. The concave inner surface mediates contacts with DNA, whereas the convex outer surface provides an interaction interface for several proteins, such as TFIIA, certain TAFs, and various activator proteins. The structure is in complete agreement with previous mutational analyses that identified regions in TBP as being involved in either DNA binding or protein interactions (19). Co-crystallization of TBP with DNA has confirmed that TBP induces an extremely strong bend in DNA (67, 68), as previously suggested by gel mobility shift analysis using permuted binding sites as probes (69). TBP-induced DNA bending brings separate DNA sequences on linear DNA closer together in three-dimensional space, which may facilitate the interactions between the pre-initiation complex on a core promoter and regulatory factors on distant binding sites, and may also permit closer association between RNA polymerase II and the basal transcription factors. Interestingly, TBP bends the TATA element in an opposite direction to that seen in nucleosome-bound DNA, suggesting a mechanism for the observed competition between nucleosomes and the binding of TBP.

TFIID is likely to form a similar structure to TBP on the TATA box, although a high-resolution structure has not yet been obtained.

5. TBP-associated factors (TAFs)

TFIID can be purified from HeLa or *Drosophila* cells as a multi-protein complex composed of TBP and several stably associated proteins (eight in *Drosophila*), called TAFs, that range in size from 250 to 30 kDa (10, 44, 45, 70–73; see Fig.2). Several of the *Drosophila* and human TAFs have been shown to be structural homologues; furthermore, conservation of protein–protein interactions suggests conservation of function (Table 1). It is therefore not surprising that the conserved TBP core domain is sufficient to allow association with TAFs (72). Whether the assembly of the multiprotein TFIID complex follows a precisely ordered pathway within cells is not known. However, the dependence on partially reconstituted complexes for the incorporation of some TAFs into TFIID (74), and the vast number of mutual protein–protein interactions between the TFIID components, would suggest that there are some assembly restrictions. For example, dTAF250 appears to be a 'scaffold' protein that is needed for many other TAFs to be incorporated in TFIID. Similarly, dTAF40 incorporation requires dTAF60 to be pre-associated with the complex (see Fig. 2). It is not known whether the assembly of the TFIID complex is a regulated process. However, in the absence of the other TAFs, dTAF250 specifically inhibits the binding activity of TBP to the TATA box (75, 76); neutralization of this inhibitory activity may be an important mechanism for transcriptional regulation.

Until recently, yeast TBP was believed to exist as an uncomplexed 27-kDa

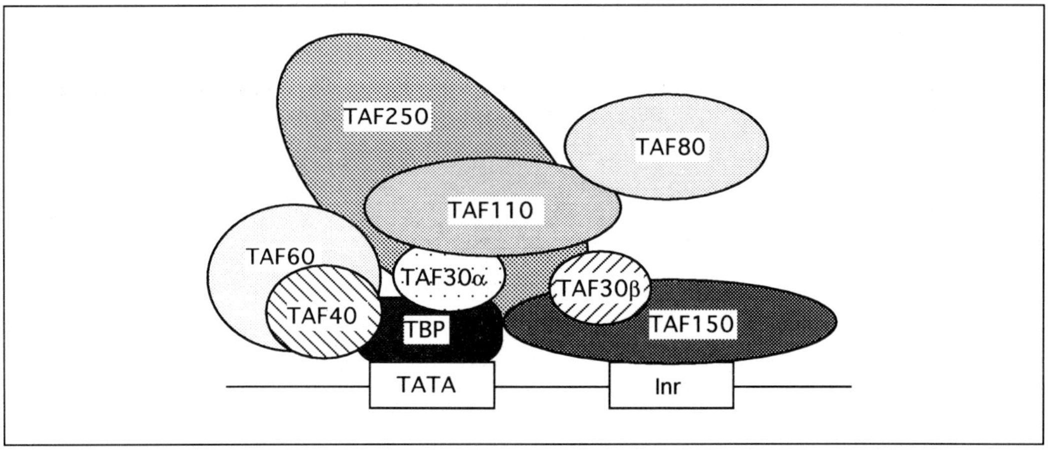

Fig. 2 Components of TFIID. TATA box recognition is mediated by the TATA box binding protein (TBP), which also serves as a template for the assembly of other factors (TBP-associated factors, or TAFs). Some of these factors are indicated as interacting directly with TBP (TAF30α, TAF40, TAF60, TAF150, and TAF250), consistent with published data (see Table 1), whereas others (TAF30β, TAF80, and TAF110) appear to require contacts with other TAFs for their incorporation into TFIID. TAF150 is thought to be required for Inr element recognition.

monomer (77–79). However, given the conservation of sequence and function of yeast, human, and *Drosophila* TBPs, it has always seemed likely that yeast TBP would also be part of larger multi-protein complexes. Indeed, it has recently been reported that approximately 70% of yeast TBP is complexed with at least seven TAFs (80). Such a multi-subunit complex associated with the yeast TBP has been shown to be specifically required for activated RNA polymerase II transcription (81). Preliminary analysis of the cloned yeast TAFs has indicated that many of them represent homologues of the *Drosophila* and human TAFs (Table 1). The existence of a multi-subunit TFIID complex thus appears to be a universal phenomenon in eukaryotes. The observations that the yeast TBP core domain can assemble *in vitro* into a complete and functional TFIID complex together with the human TAFs (82), and that yeast TBP can mediate transcriptional activation in mammalian cells (60), suggest that the functionally relevant protein–protein interactions are conserved. For example, the previously identified essential yeast gene product TSM-1 is strikingly similar to dTAF150, suggesting that TSM-1 is a yeast TAF homologue (12). As expected for a true TAF, TSM-1 directly associates with the yeast TBP both *in vivo* and *in vitro*. Furthermore, like dTAF150, TSM-1 can interact with dTAF250, which further implies that yeast is also likely to contain a dTAF250 counterpart. Even if the *in vitro* interaction specificity between TSM-1 and dTAF150 thus appears to be conserved, the *Drosophila* protein is, perhaps surprisingly, unable to rescue the TSM-1 mutant *in vivo*. It is also interesting to note that while the activation domain of Sp1 can activate transcription with yeast TBP in human cells (83), it cannot function in yeast (84). Thus, functional conservation is not absolute.

It is not clear whether there is only one type of TFIID complex. The observation that a single, major TFIID species is capable of supporting activation by a diverse

Table 1 Properties of RNA polymerase II TAFs

Drosophila TAF	Homologues (human/yeast)	Contacts with other TAFs	Contacts with other components of the basal transcription machinery	References
dTAF250	hTAF250(CCG1)/ yTAF145	dTAF150 dTAF110 dTAF60 hTAF70 dTAF30α dTAF30β	TBP	(81, 106, 107 174–176)
dTAF150	?/yTAF150 (TSM-1)	dTAF250 dTAF30α dTAF30β	TBP	(12)
dTAF110	?/?	dTAF250 dTAF80 dTAF30α dTAF30β	TFIIA	(71, 174–176)
dTAF80	?/yTAF90	dTAF110		(81, 177, 178)
dTAF60	hTAF70/ yTAF60	dTAF250 hTAF250 dTAF40	TBP	(73, 151. 179)
dTAF40	hTAF32/?	dTAF60 hTAF70	TBP dTFIIB hTFIIB	(73, 104, 179)
dTAF30α	?/?	dTAF250 dTAF110	TBP	(73, 175)
dTAF30β	?/?	dTAF250 dTAF150		(175)
?	hTAF55/?	hTAF250		(109)
?	?/yTAF170 (ADI/MOT1)		TBP	(151)

class of activation domains suggests that this is the case (10). However, the existence of distinct low-abundance TFIID complexes specifically involved in transcription of certain promoters cannot be excluded. The known TAFs listed in Table 1 are present in stoichiometric ratios in isolated TFIID complexes. However, there may be TAFs that are only present in certain tissues, or during certain developmental periods, that have escaped detection in studies utilizing only a limited variety of extract sources.

As discussed above, TAFs are not restricted to the RNA polymerase II system. Transcription of vertebrate ribosomal RNA (rRNA) by RNA polymerase I requires the selectivity factor SL1, which is composed of TBP and three RNA polymerase I-specific TAFs, all of which are essential for transcription (85). Reconstitution of transcription from the TATA-less RNA polymerase III promoters utilizes yet another distinct protein complex, TFIIIB, which contains TBP together with specific TAFs (86, 87). SNAPc (snRNA-activating protein complex) consists of TBP and at least three specific TAFs, and is specifically engaged in vertebrate snRNA (small nuclear RNA) transcription (88). Depending on the presence of TATA sequences, some snRNA promoters are transcribed by RNA polymerase II, while others are transcribed by RNA polymerase III (see above); SNAPc is involved in both cases. Unlike TFIID, SL1, and TFIIIB, which recognize either the TATA box or Inr, or have little or no affinity for specific DNA sequences, SNAPc binds specifically to a distinct DNA element located upstream of transcription initiation sites on snRNA promoters. Thus, the TAFs in SNAPc seem to reprogramme the binding specificity of the TBP-containing complex, directing it to target a sequence that is unrelated to a TATA box. Reconstitution of *in vitro* transcription from a TATA-containing RNA polymerase III promoter requires TBP in addition to SNAPc, suggesting that this initiation complex contains at least two forms of TBP, one bound to the TATA box, and another as part of SNAPc.

6. How do regulatory transcription factors work?

Transcriptional activators are thought to contact targets in the basal transcriptional machinery directly or indirectly via intermediary proteins. Once activator proteins are bound to specific DNA sequences, they interact with the basal transcription machinery, and either recruit it to DNA, stabilize components already recruited to DNA, or cause a productive conformational change or some other kind of modification (e.g. phosphorylation) in one or more components of the basal transcription machinery, and thereby initiate a cascade of events that lead to the increased initiation of transcription.

6.1 TBP as a target for activators

Because of the central role played by the TATA box binding protein (TBP) in the initiation of transcription, it is an obvious candidate for one of the targets of upstream activator proteins. However, while TBP together with other RNA polymerase II basal transcription factors can perform basal transcription, other components of the TFIID complex (the TAFs) are thought to be required for efficient transcriptional enhancement by all classes of activator proteins (9, 44, 45, 50, 89, 90). Since TBP alone is incapable of responding to activators, activator–TBP associations cannot completely explain the function of activation domains. Nevertheless, these interactions are believed to be important contributors to the efficacy of transcriptional stimulation.

Various groups have demonstrated direct physical interactions *in vitro* between human TBP and the activation domains of several viral transcriptional activators, including the acidic activation region of herpes simplex virus VP16 (Table 2). Subsequently, it has been shown that activation domains of several cellular activators are also capable of directly interacting with TBP (Table 2). It is clear that the activation domains do not fall into any single, defined structural class. Activator proteins have also been identified that interact with TBP through regions other than their activa-

Table 2 Transcription factor interactions with TBP

Activator protein	Region interacting with TBP	References
VP16 (herpes virus)	Activation domain	(91,180)
Zta (Epstein–Barr virus)	Activation domain	(93)
E1A (adenovirus)	Activation domain	(95,181)
Tat (HIV)	Activation domain	(182)
IE2 (cytomegalovirus)	Activation domain	(183)
Tax1 (HTLV-1)	Activation domain	(184)
pX (hepatitis B virus)	Activation domain	(185)
Rta (Epstein–Barr virus)	Activation domain	(186)
p53 (mammalian)	Activation domain	(92,101,187)
Sp1 (mammalian)	Activation domain	(96)
PU.1 (mammalian)	Activation domain	(188)
E2F (mammalian)	Activation domain	(189)
Antennapedia (*Drosophila*)	Activation domain	(96)
c-Rel (chicken)	Activation domain	(190)
c-Fos (mammalian)	DNA binding and/or dimerization domain	(191)
c-Jun (mammalian)	DNA binding and/or dimerization domain	(191)
c-Rel (mammalian)	DNA binding and/or dimerization domain	(100)
Oct-1 (mammalian)	DNA binding and/or dimerization domain	(98)
Oct-2 (mammalian)	DNA binding and/or dimerization domain	(98)

tion domains (Table 2). This implies that at least some regulatory proteins are capable of hitting more than one target in the basal transcription machinery, since by definition their activation domains also generate a signal to activate transcription.

Evidence to support the importance of interactions between activation domains and TBP is provided by studies showing that mutations in these activation domains that decrease affinity for TBP *in vitro* also reduce transcriptional activation *in vivo* (91, 92). The viral transactivator Zta stabilizes the DNA binding of TFIID *in vitro* by decreasing the dissociation rate of DNA-bound TBP (93). As expected, the domain in Zta protein responsible for transactivation *in vivo* is required for the association with TBP. The stabilization effect by Zta appears to be especially marked in the context of weak TATA boxes; thus association with TBP is more likely to be rate limiting in promoters with a poor TATA consensus. It has been recently reported that Zta also increases the stability of the TFIIA–TFIID-promoter complex *in vitro* in a manner dependent on TAFs (94), although it is not clear that this effect is directly attributable to the activation domain–TBP interaction, since there may be additional contacts between Zta and other polypeptides induced by the formation of a ternary complex.

Some transcription factors, such as E1A (95), Sp1 (96), and p53 (97), show a preference for interaction with human TBP over the yeast homologue, despite interacting with the conserved core domain of TBP. This effect may depend on the activator-specific contact residues on the surface of TBP, and is reflected in the transactivation potency of these factors. Other activators, like Oct-2, appear to interact equally strongly with the TBPs from either species (98). E1A, p53, and c-Rel are all capable of interacting with the TFIID complex, as well as with isolated TBP, which indicates that the surface on TBP recognized by these activators must be available for interaction, even though TBP is tightly associated with several TAFs (99–101).

6.2 TAFs as targets for activators

TAFs are required in addition to TBP for activator proteins to stimulate transcription, and they have also been postulated to provide protein surfaces for activator proteins to target.

Although the glutamine-rich activation domains of Sp1 can bind directly to TBP (see above), they also appear to interact with the TFIID complex by binding to dTAF110 (71). Interestingly, the interaction seems to occur between two glutamine-rich regions, since the amino terminus of dTAF110 has a high content of glutamines. A positive correlation has been observed between the level of transcriptional activation by various GAL4-Sp1 mutants in *Drosophila* Schneider cells and the degree of binding of the GAL4-Sp1 derivatives to dTAF110 in the yeast two-hybrid assay (102). Sp1 can mediate efficient transcriptional activation in *in vitro* reconstitution experiments with a partial TFIID complex containing TBP, dTAF250, dTAF150, and dTAF110, although it does not interact with the other two TAFs present (74). Another glutamine-rich activation domain, namely that of the cAMP-regulated transcription factor CREB (cyclic AMP response element binding protein) also interacts with dTAF110 (103). Interestingly, however, several other activation domains classified as glutamine-rich fail to associate with dTAF110 (71), implying that predominance of glutamines is not the only determinant for activation within this class of activators.

In *in vitro* protein-binding assays, dTAF40 binds directly and specifically to the VP16 acidic activation domain (104). Supporting the importance of this interaction, specific antibodies raised against dTAF40 inhibit VP16-mediated activation, while leaving basal transcription unaffected. TBP and dTAF40 appear to associate with different subdomains within the VP16 activation region, thus it is possible that VP16 interacts with the two components of TFIID simultaneously. Another acidic activation domain, that of the p53 protein, can interact with two TAFs, dTAF40 and dTAF60; furthermore, the ability of this domain to bind either TAF correlates with its transcription activation potential (105). It is not known whether the p53 activation region can contact the two TAFs alternatively or simultaneously. A reconstituted TFIID complex containing dTBP, dTAF250, and dTAF60, with or without dTAF40, can support activation by the p53 activation domain; whether dTAF40 can function without dTAF60 cannot be tested, since it cannot be incorporated into the TFIID complex without the presence of dTAF60. The isoleucine-rich activation domain of the activator protein NTF-1 can similarly contact two TAFs, dTAF150 and dTAF60 (74). In this case, two distinct, partial TFIID ternary complexes, TBP–dTAF250–dTAF150 and TBP–dTAF250–dTAF60, can support activation by NTF-1.

Sequence analysis of the largest human TAF, hTAF250 has revealed its identity to CCG1, previously cloned as a DNA-binding protein that overcomes the cell cycle arrest in late G1 phase in the temperature-sensitive rodent cell line ts13 (106, 107). It is intriguing to speculate that hTAF250 would somehow link transcriptional control to cell cycle, by being specifically involved in the regulation of genes important for cell cycle progression. The ts13 cells do not manifest a global defect in transcription at the non-permissive temperature, thus the cell cycle block may result from altered expression of a specific set of genes. In agreement with the hypothesis that TAFs are required for activated transcription, only activated, but not basal, transcription is temperature sensitive in ts13 extracts *in vitro*; as expected, the recombinant hTAF250 can restore transcriptional activation *in vitro* (108). While the connection between cell cycle-linked transcription and hTAF250 is intriguing, a more universal role for this TAF has also been proposed. In mutational analysis of TBP, its general ability to respond to activators correlates with the nature of its direct interactions with hTAF250 (83).

Individual TAFs appear to serve as targets for distinct types of activators; however, in some cases there does not seem to be a simple one-to-one relationship. From the discussion above it is clear that a single activator can contact more than one TAF; conversely, a human TAF, hTAF55, appears to be capable of interacting with multiple activators (Sp1, CTF, E1A, YY1, and USF) with different overall structures (109). Interestingly, this particular TAF is the only component of the human TFIID, for which a *Drosophila* homologue has not been identified.

6.3 TFIIB as a target for activators

Interactions have also been demonstrated between TFIIB and the activation domains of VP16, as well as other transactivation regions (110–114). Variant TFIIB proteins that cannot associate with the VP16 activation region support basal transcription at 80–100% of wild-type levels; however, these mutants are severely defective in supporting transcription directed by an acidic activation region (115). Conversely, a single amino acid substitution mutant in VP16 that renders it defective in transactivation also fails to interact with TFIIB. The stable entry of TFIIB can indeed be a rate-limiting step in pre-initiation complex assembly *in vitro* (110). Since TFIIB dissociates from the pre-initiation complex following initiation, the activators could stimulate transcription by re-recruiting TFIIB to the promoter. In addition to quantitative recruitment to the pre-initiation complex, activators also affect TFIIB qualitatively, by inducing conformational changes such that pre-initiation complex assembly is driven forward (116). Whilst the enhancement of TFIIB recruitment by acidic activators requires the presence of TBP, TAFs are dispensable (112).

6.4 TFIIH as a target for activators

The carboxy-terminal domain (CTD) of the largest subunit of RNA polymerase II is highly conserved and consists of several repeats of the consensus heptapeptide Tyr-Ser-Pro-Thr-Ser-Pro-Ser (117). Due to variable phosphorylation of these heptapeptide repeats, RNA polymerase II exists as two forms: the extensively phosphorylated and the hypophosphorylated forms. Changes in CTD phosphorylation occur during the transcription cycle, so that the hypophosphorylated enzyme preferentially associates with the pre-initiation complex, while actively elongating complexes mainly contain the phosphorylated RNA polymerase II (118–120; see Fig. 1). Phosphorylation of CTD has thus been believed to be important in enhancing the escape of RNA polymerase II from the promoter, in order to shift from the initiation to the elongation phase of transcription. The observations that the promoter-bound TBP and the putative promoter clearance factor TFIIE can both selectively interact only with the hypophosphorylated RNA polymerase II are consistent with this model (22, 27).

Although various other kinases in many organisms have been suggested to be capable of CTD phosphorylation *in vitro*, the strongest candidate for a true regulatory CTD kinase activity is contained in the multi-subunit basal transcription factor TFIIH (121). The TFIIH kinase activity is enhanced in the context of the assembled pre-initiation complex. In particular, the presence of TFIIE, which associates before TFIIH with the pre-initiation complex, strongly stimulates the kinase activity of the purified TFIIH *in vitro* (31). Interestingly, the TFIIH subunit possessing the kinase activity appears to be a previously identified cyclin-dependent kinase (122, 123).

Two acidic activation domains, namely those of VP16 and p53, have recently been shown to interact directly with the 62-kDa subunit of TFIIH (124). Furthermore, point mutations in VP16 that reduce its activation potential also weaken its interaction with both the human and yeast TFIIH. Thus, transactivator proteins could directly affect the kinase activity of TFIIH to enhance transcriptional initiation.

6.5 TFIIF as a target for activators

TFIIF may function in activated as well as basal transcription. It has recently been shown that TFIIF can relieve squelching (see below) by the activator protein SRF (125), and TFIIF can directly interact with SRF (126). Specifically, the larger of the two TFIIF subunits, RAP74, can associate with the activation domain of SRF. There appears to be a good correlation between binding of SRF to RAP74 and the ability of the RAP74 protein to mediate transcriptional activation by SRF; however, the effect of SRF on the properties of TFIIF remains unclear.

6.6 Adaptors as targets for activators

Adaptors are proteins that can transmit a signal from activators to the basal transcription machinery without possessing independent DNA-binding activity or being stably associated with TBP (although they may well interact with TBP in a more transient manner than do TAFs). The distinction between a TAF and an adaptor is sometimes vague, and the classification of a novel factor as one or the other may partly depend on whether it has been identified by biochemical or genetic means.

Excessive amounts of transcriptional activators suppress the level of transcription, a phenomenon referred to as squelching. This is thought to be due to sequestration of a limiting component of the transcriptional machinery into a non-productive form. Over-expression of a GAL4-VP16 fusion protein in yeast squelches expression and eventually proves toxic. The growth inhibition by high levels of GAL4-VP16 requires the integrity of both the VP16 acidic activation domain and the GAL4 DNA-binding domain, implying that the effect is a consequence of the recruitment of basal transcription factors into non-productive DNA-bound complexes to genomic GAL4 binding sites (127).

Yeast selection screens have been carried out to isolate mutants that can resist the toxic effect of GAL4-VP16. Such yeast strains might be expected to have alterations (alteration/deficiency in activation) in limiting components of the transcriptional machinery. Three genes, *ada2*, *ada3*, and *gcn5*, have been isolated by this approach (128–130). The ADA2, ADA3, and GCN5 products appear to act as transcriptional adaptors for certain acidic activators, such as GAL4-VP16, but not for some others. Physical interaction has been demonstrated between ADA2 and the VP16 activation domain (131). ADA2, ADA3, and GCN5 form a trimeric complex (132), providing an explanation of why double mutants in any two of the three genes do not affect yeast growth any more severely than any single mutant. The precise target within the pre-initiation complex for the adaptor function of the ADA2–ADA3–GCN5 complex has not been determined.

Adaptors for activator function also exist in higher eukaryotes. Squelching experiments have identified groups of activation domains that cross-compete in transfection assays. For example, ectopic expression of the acidic transactivation domains of c-Jun itself, JunB, GAL4, and VP16 can repress c-Jun activity, whereas the activation regions belonging to different classes cannot (133). Three proteins could be co-immunoprecipitated with c-Jun in a transactivation domain-dependent manner; this association is sensitive to competition with the excess of VP16 protein, suggesting that these proteins are adaptors.

A mammalian protein termed CBP (CREB-binding protein) has been identified that bears a 50 amino acid region of homology with the ADA2 gene product (134). CBP is a large protein of 250 kDa and also contains a bromodomain (a conserved structural motif found in *Drosophila* and yeast proteins involved in signal-dependent but not basal transcription and thought to be important for protein–protein interactions). CBP can specifically interact with the protein kinase A-phosphorylated, activated form of CREB, and in transfection experiments enhances the CREB-mediated activation of reporter constructs *in vivo* in a manner that depends on the integrity and function of the protein kinase A phosphorylation site (135). Furthermore, microinjection of specific anti-CBP antibodies blocked the activity of a cAMP-regulated promoter in fibroblasts (136). Strikingly, anti-CBP antibodies also blocked activated transcription from other signal-responsive promoters containing TREs (12-*O*-tetradecanoylphorbol-13-acetate (TPA)-responsive elements) or SREs (serum-responsive elements), whilst leaving basal promoter activities unaffected. Like cAMP, the stimulating agents of TREs and SREs (phorbol esters and serum growth factors, respectively) also induce activatory phosphorylations in specific transcription factors. CBP has since been shown to specifically interact with c-Jun that is phosphorylated on sites that are known to stimulate transcription *in vitro*, and which become phosphorylated upon mitogen induction *in vivo* (136). Perhaps CBP, or a family of related proteins, has a role as an end-point or integrator of various signal transduction pathways initiated at the cell surface, providing a link between signalling cascades and transcriptional effect.

As expected for an adaptor protein, CBP activates transcription in a sequence-dependent manner when fused to a heterologous DNA-binding domain (134, 135). CBP interacts with TFIIB *in vitro* (135). The CBP-interacting region in TFIIB partially overlaps the domain contacting VP16. CBP shares extensive sequence similarity with p300, a protein capable of associating with the adenovirus E1A protein (137, 138). p300 appears to be another transcriptional adaptor, whose function is negatively regulated by its association with E1A. The DNA-binding factors that target p300 to specific promoters have yet to be identified; however, p300 has a cell cycle inhibitory function, and is likely to regulate the promoters for genes required for arrest in G0/G1. The similarity between CBP and p300 suggests that there may be a family of related, but perhaps functionally distinct, adaptor proteins.

The recently cloned PC4 is a small (14 kDa) human protein that also has characteristics of an adaptor protein (139). PC4 markedly enhances transcriptional stimulation *in vitro* by various types of activation domain, including acidic (VP16), proline-rich (CTF), and glutamine-rich (Sp1), but has only minor effects on basal transcription. TAFs need to be present for PC4 to exert its stimulatory function. PC4 is capable of directly interacting with both the VP16 activation domain and TFIIA, either in solution or as DNA-bound VP16–TFIIA–TBP complexes. The transcriptional potency of VP16 activation domain mutants correlates strongly with their binding affinity to PC4. The adaptor activity of PC4 is regulated by the phosphorylation status of its amino terminus, so that only the non-phosphorylated form is able to interact with VP16 and potentiate transcriptional activation (140, 141).

6.7 Repressors within the pre-initiation complex: a role for TFIIA

Several negative activities specifically associated with TBP have been identified. These may also serve as targets for activator proteins, which may promote the release of such negative factors from a repressed pre-initiation complex, thus 'anti-repressing' the promoter in question. The net effect of the presence of negative activities capable of repressing the basal core promoter activity is potentiation of the stimulatory effect of gene-specific activators. Some transcriptional repressors may stabilize the association of the inhibitory factors with the pre-initiation complexes, or enhance their incorporation, although no such examples have been reported yet.

Human factor Dr1 (TFIID repressor 1) is a 19-kDa nuclear phosphoprotein that can associate with TBP and repress both basal and activated transcription (142, 143). In a co-transfection assay, Dr1 can repress transcription directed by both TATA-containing and TATA-less promoters. When Dr1 associates with TBP as a monomer, it precludes the association of TFIIB with TBP, presumably because both Dr1 and TFIIB compete for binding to the same, or an overlapping region, in TBP. Dr1 can also associate with TBP as a dimer, trimer, or tetramer, and as a tetramer Dr1 inhibits both TBP–TFIIB and TBP–TFIIA interactions. In addition to RNA polymerase II transcription, Dr1 can also repress *in vitro* transcription from RNA polymerase III promoters, while leaving RNA polymerase I-mediated transcription unaffected (144). In addition to its TBP-binding domain, Dr1 requires an alanine-rich region for transcriptional inhibition. Interestingly, this type of domain is also required for repression in many repressor proteins that bind DNA in a sequence-specific manner. However, when the TBP-binding domain in Dr1 is replaced by a heterologous DNA-binding domain, the repressive activity cannot be restored. A subset of transcriptional activators, such as E1A, appear to be able to overcome Dr1-mediated repression in co-transfection analyses. The adenoviral transactivator E1A can disrupt the interaction of Dr1 with TBP, thus allowing the productive association of TBP with TFIIA (145). The domain of E1A required for TATA box-dependent transcriptional activation is also needed for the dissociation of Dr1 from the pre-initiation complex.

Dr2 activity has been purified to homogeneity and shown to be identical to human topoisomerase I (TopoI) (146, 147). Dr2/TopoI associates with TBP and prevents TFIID–TFIIA–DNA ternary complex formation. In the absence of transcriptional activators, Dr2/TopoI represses basal transcription specifically from TATA box-containing promoters by associating with TBP. Topoisomerases isolated from sources other than human are unable to repress the human *in vitro* transcription system, and the repression mechanism of Dr2/TopoI remains unclear, since a TopoI mutant lacking the DNA unwinding activity is still capable of repressing transcription. In addition to repressing basal transcription in the absence of an activator, Dr2/TopoI can also effect a net increase in activation by a chimeric activator GAL4-AH. Together, these events acting on both the basal and activated level lead to a high activator-mediated transcriptional induction. The transcriptional activation by Dr2/TopoI requires TAFs, whereas the repression effect can be seen with recombinant TBP alone.

In the absence of an activator, TFIIA is capable of overcoming the repression imposed by TBP-associated Dr2/TopoI (147). It has thus been hypothesized that the function of TFIIA is to remove negative components present in the TFIID complex. This could commit the complex to a productive pathway, and render it further responsive to activators. The observation that TFIIA is not required for those *in vitro* transcription systems that use recombinant TBP (148), which—unlike the TFIID complex purified from cell extracts — is devoid of the negatively acting factors, is in complete agreement with this model.

A TBP-inhibitory factor, ADI (ATP-dependent inhibitor), has also been identified in yeast that prevents the DNA binding of TBP in an ATP-dependent manner (149). ATP hydrolysis seems to be required for the inhibition, since non-hydrolysable ATP analogues cannot substitute for ATP. TFIIA can counteract the effect of ADI by stabilizing the binding of TBP to DNA; TFIIB also has the ability to prevent dissociation of TBP from DNA, but to a much lesser degree than TFIIA. Sequence analysis has revealed ADI to be the product of the *mot1* (modifier of transcription 1) gene, which was originally isolated in a genetic yeast screen for mutants causing transcriptional derepression of several genes (150). The repression of transcription by ADI appears specific to the RNA polymerase II system, and the presence of a TATA box is not sufficient to allow a promoter to be regulated by ADI. Recently, ADI has been demonstrated to be a yeast TAF, yTAF170, but does not appear to be part of the major yeast TFIID complex (151).

6.8 Action of sequence-specific transcriptional repressors

Perhaps the simplest means to repress transcription is to prevent promoter access by transcriptional activators, by steric occlusion or by cytoplasmic partitioning of the activator (152–154). However, it is becoming increasingly clear that repressors can also function by binding to promoters in a sequence-specific manner, without impeding the access of other components of transcription machinery. By analogy with transcriptional activators, these factors presumably mediate repression by interfering with the activity or assembly of the basal transcription machinery, or perhaps by enhancing the incorporation of inhibitory factors, such as Dr1, Dr2, and ADI. Accordingly, such factors possess specific repression domains (3, 155, 156).

The Even-skipped (Eve) protein belongs to the large family of homeodomain proteins that control the embryonic development of *Drosophila* (see Chapter 6). Eve can repress its target promoters, such as the *Ubx* promoter, by specifically binding to upstream sites and interfering with the assembly of a functional pre-initiation complex (157). Eve does not affect the kinetics of the assembly reaction but rather reduces the number of functional pre-initiation complexes that form on the promoters. Pre-initiation complexes become resistant to Eve repression early in the assembly pathway, which indicates that Eve affects one of the first steps in their formation. The immediate physical target of Eve repression remains to be determined.

Utilizing the *in vitro* transcription assay, it has been shown that basal transcription machinery is also the target for repression mediated by unliganded thyroid hormone receptor (TR) (158). Like Eve, TR affects an early step in pre-initiation complex formation, and fully formed pre-initiation complexes are refractory to repression. *In vitro*, the transcriptional repression by TR appears independent of TAFs. A specific physical interaction between TFIIB and TR has been observed

(158, 159), the result being in agreement with the earlier observations that TFIIB is capable of contacting several other members of the steroid/nuclear receptor superfamily (160). Two distinct regions of TFIIB are targeted by two distinct regions of TR, one of the two regions in TR overlaps with its ligand-binding domain. Accordingly, the appropriate ligand thyroid hormone, which converts TR into an activator, significantly decreases the interaction between TR and TFIIB, perhaps by inducing a conformational change in the ligand-binding domain.

In haploid α-mating-type *S. cerevisiae* cells **a**-specific genes are repressed by the α2-MCM1 complex that binds to the conserved operator sequences upstream of each **a**-specific gene (161; see Chapter 5). The occupancy of the operators by α2-MCM1 *in vivo* is not sufficient for efficient repression, but rather requires two additional proteins known as SSN6 and TUP1. SSN6 and TUP1 associate to form a protein complex, involved in transcriptional repression of a diverse set of genes, including **a**-specific, haploid-specific, and glucose-repressible genes (162). SSN6–TUP1 is thus proposed to function as a general repressor, targeted to specific promoters by interaction with DNA-bound complexes such as α2-MCM1. This recruitment to DNA appears to be carried out by TUP1, which contains seven β-transducin repeats, capable of interacting with the α2 (163). The SSN6–TUP1 complex appears to interfere directly with either the assembly or function of the basal transcription machinery, rather than inhibit the function of transactivator proteins (164). All the genes naturally repressed by α2-MCM1 are transcribed by RNA polymerase II. Nevertheless, it was recently shown that when an α2-MCM1 operator is positioned upstream of a transcriptional initiation site in the context of RNA polymerase I promoters, α2-MCM1 could repress these in an SSN6–TUP1-dependent fashion (165). In contrast, no effect was seen on RNA polymerase III promoters. These results imply that the SSN6–TUP1 repressor complex may target, and perhaps interact directly with, a component of the basal transcription machinery common to both RNA polymerase I and II systems. The putative SSN6–TUP1 target is either not a component of the RNA polymerase III machinery, or not accessible to the repression effect in RNA polymerase III transcription complexes. Intriguingly, SSN6–TUP1 may provide an example of a negative adaptor activity transferring a repressive signal to the basal transcription machinery. The SSN6–TUP1 complex lacks any detectable independent DNA-binding activity, but, as expected for a negatively acting adaptor, SSN6 can act as a LexA-operator-dependent and TUP1-dependent transcriptional repressor when brought to DNA as a fusion with the heterologous LexA DNA-binding domain (161). In contrast, the repression by a LexA–TUP1 fusion protein is not dependent on the presence of SSN6 (166); thus, in addition to linking the repressor complex to DNA-binding proteins, TUP1 exhibits partial independent repressor function. Indeed, TUP1 contains two alanine-rich regions, implicated in repressor function in many regulatory proteins. By use of a reconstituted oestrogen-responsive transcription system in yeast, it has been shown that SSN6 can repress transcriptional stimulation by the oestrogen receptor, suggesting that a mammalian counterpart to the SSN6 protein may exist (167).

7. Conclusions and implications for transcriptional synergy

Transcriptional activators appear to enhance transcriptional initiation by interacting favourably with adaptors and/or members of the basal machinery, displacing general or specific negative regulators, or altering chromatin structure, thus enabling themselves or other regulatory factors to engage certain critical promoter regions. In a similar manner, transcriptional repressors are thought to inhibit either the formation or the function of a pre-initiation complex. The nature of the biochemical consequences of interactions between regulatory factors and basal transcriptional machinery remains largely unknown. An interaction may cause a conformational change or other type of modification in a target protein. This could lead to the stabilization of a pathway intermediate, or perhaps to novel or altered interactions between members of basal transcription machinery. Eventually signals from activators have to catalyse the switch to increased transcription; the phosphorylation of RNA polymerase II CTD, which facilitates its escape from the core promoter, might be the late event where signals finally merge.

It is not yet clear how limited the possible varieties of targets are for regulatory factors bearing a certain type of activation domain, but the list of interactions between certain members of the basal transcription machinery and specific activators is rapidly growing. For example, the well-studied acidic activation region of VP16, consisting of two subdomains, has six reported potential targets in mammalian cells: TBP, TFIIB, dTAF40, TFIIH, PC4, and a possible mammalian homologue of the yeast ADA complex (see Fig. 3). The p53 protein can interact with TBP, dTAF60, dTAF40, and TFIIH, whereas CREB can contact both dTAF110 and an adaptor protein CBP. The fact that activator proteins function by multiple steps is evidenced by the observation that raising the concentration of a single basal transcription factor does not overcome the requirement for activators (112). It remains to be seen whether the necessary steps towards the formation of a functional pre-initiation complex and transcription initiation are the same for every promoter, how many separable steps are necessary, and whether these steps have to take place in a strict temporal order.

Typical eukaryotic promoters contain a mosaic arrangement of DNA-binding elements for several activators and/or repressors. The effect of multiple activators acting on the same promoter may lead to a synergistic effect on the stimulation of transcriptional initiation; in other words, two or more activators produce transcription levels greater than the sum of the levels resulting from the action of individual activators (168). Synergistic activation may be a consequence of co-operative DNA binding of transcription factors (169, 170); however, certain activators can also work co-operatively under conditions in which their DNA-binding sites are saturated (171). The finding that several components of the transcriptional machinery can serve as targets for activators offers an alternative explanation for synergistic activation. In this model (see Fig. 4) activators bound to adjacent sites could contact dif-

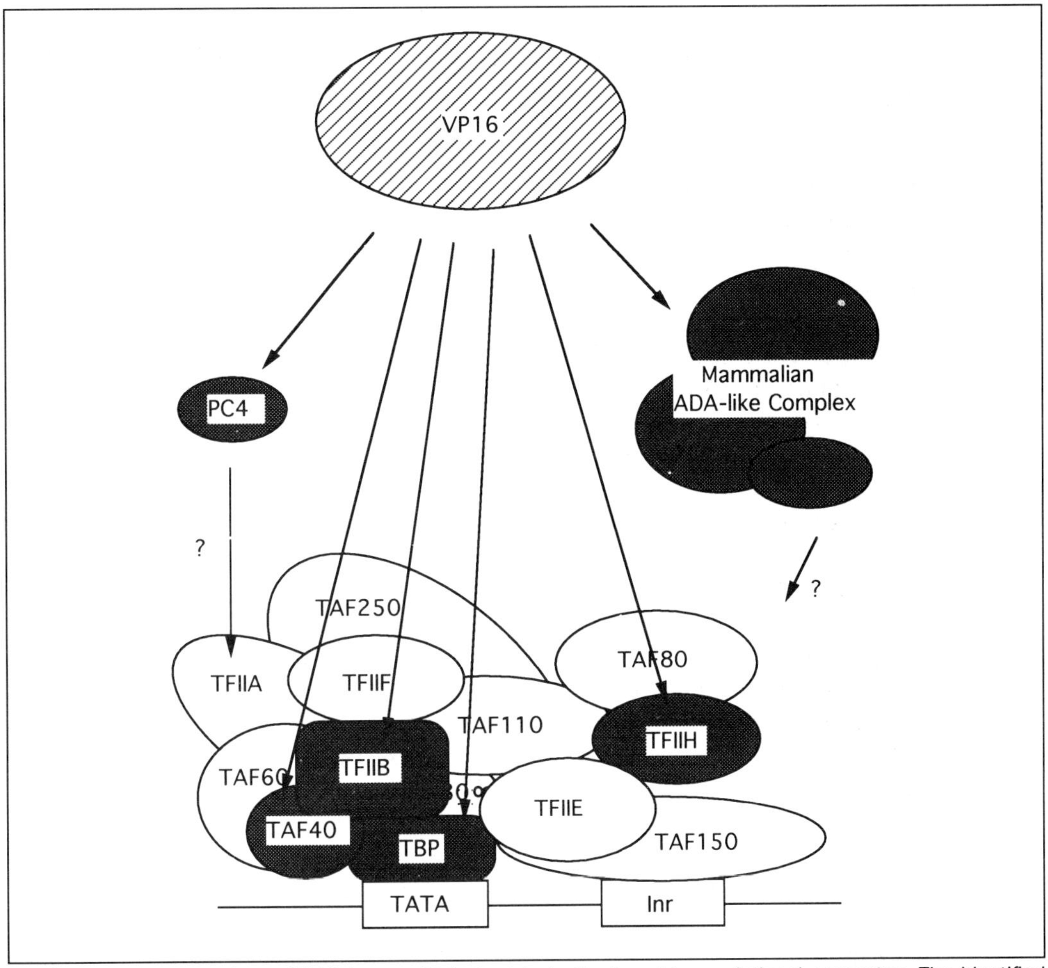

Fig. 3 The viral transactivator VP16 has multiple targets in the basal transcriptional apparatus. The identified interactions between VP16 and specific proteins in TFIID or other basal transcriptional components are indicated by connecting arrows. The TFIID components are represented in the same format as in Fig. 2; the other TFII components have been superimposed on to this structure such that TAF30α and TAF30β are obscured. Only the identified targets are highlighted. In addition to TFII components, VP16 can interact with adaptor proteins, and these are indicated by the PC4 and putative mammalian ADA-like complex interactions. The TFII targets for interactions of the adaptors are unknown, although PC4 has been shown to interact with TFIIA (see text).

ferent targets in the transcriptional machinery, thus enhancing the rate of assembly or stability of the complex.

Artificial or natural test promoters containing multiple binding sites for a single activator can sometimes be synergistically activated; for example, Sp1 is capable of synergistic activation of the promoters containing multiple Sp1 sites, even if the DNA binding does not appear co-operative (172). This may be explained by the activator under study having the capacity to contact multiple targets. The *in vivo* synergy on promoters consisting of binding sites for several distinct activators

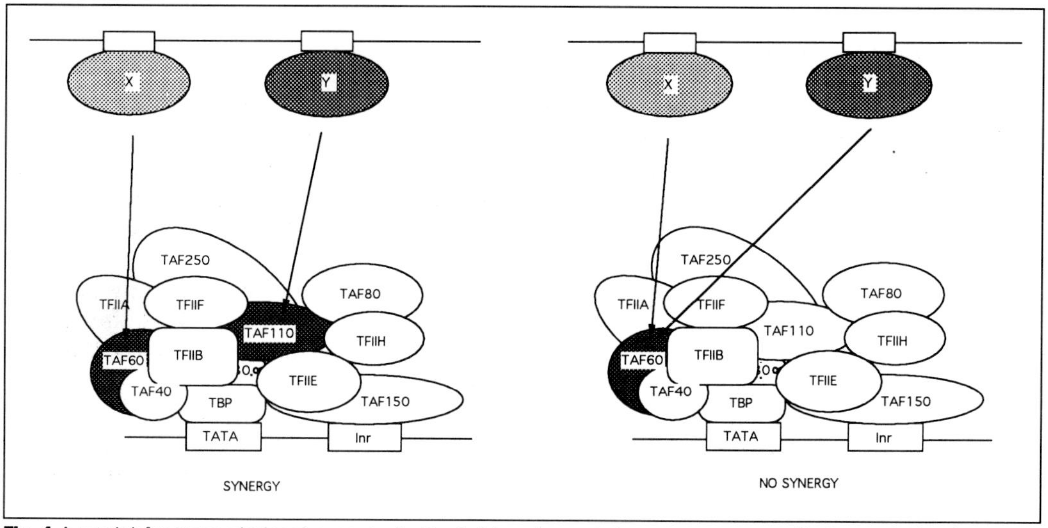

Fig. 4 A model for transcriptional synergy. In the left-hand panel, two sequence-specific transcription factors, X and Y, are shown binding to upstream promoter elements. Each factor is capable of interacting with a different component of the basal transcription apparatus. In this hypothetical example, factor X interacts with TAF60 while factor Y interacts with TAF110; this representation is not meant to imply that interactions with other TFII components cannot participate in synergy. The consequence of this two-site interaction is the synergistic enhancement of recruitment, stabilization, or modification of the properties of the basal transcriptional machinery. In the right-hand panel, the sequence-specific transcription factors, X and Y, are only capable of interacting with the same component of the basal transcription apparatus (shown here as TAF60) with a consequent inability to show co-operativity in transcriptional enhancement.

appears to depend on the extent of 'co-operation compatibility' between factors, and some response elements seem to co-operate nearly universally, whereas others exhibit strong selectivity (173). It will be of great interest to determine whether this in any way correlates with the variety of contacts that the regulatory factors under examination can make with the adaptors and/or the basal transcription machinery.

Finally, the recently reported existence of partially pre-assembled pre-initiation complexes, holozymes (32), in yeast could simplify models for synergistic transcriptional activation. In the classical stepwise pathway for pre-initiation complex assembly, any of the successive stages is a potential target for activators. The pre-assembled factors could present a contiguous target surface for multiple activators and, as a consequence, the simultaneous action of these activators could recruit the complex to the TFIID-bound promoter DNA in a co-operative manner.

Acknowledgements

We thank Cath Lindon for her critical reading of the manuscript. J. J. E. acknowledges the Academy of Finland for support.

References

1. Mitchell, P. J. and Tjian, R. (1989) Transcriptional regulation in mammalian cells by sequence-specific DNA binding proteins. *Science*, **245**, 371.
2. Licht, J. D., Hanna-Rose, W., Reddy, J. C., English, M. A., Ro, M., Grossel, M., Shaknovic, R., and Hansen, U. (1994) Mapping and mutagenesis of the amino-terminal transcriptional repression domain of the Drosophila Krüppel protein. *Mol. Cell. Biol.*, **14**, 4057.
3. Han, K. and Manley, J. L. (1993) Transcriptional repression by the Drosophila even-skipped protein: definition of a minimal repression domain. *Genes Dev.*, **7**, 491.
4. Saha, S., Brickman, J. M., Lehming, N., and Ptashne, M. (1993) New eukaryotic transcriptional repressors. *Nature*, **363**, 648.
5. Weinmann, R. (1992) The basic RNA polymerase II transcriptional machinery. *Gene Expression*, **2**, 81.
6. Conaway, R. C. and Conaway, J. W. (1993) General initiation factors for RNA polymerase II. *Annu. Rev. Biochem.*, **62**, 161.
7. Zawel, L. and Reinberg, D. (1993) Initiation of transcription by RNA polymerase II: a multi-step process. *Prog. Nucl. Acid Res. Mol. Biol.*, **44**, 67.
8. Buratowski, S. (1994) The basics of basal transcription by RNA polymerase II. *Cell*, **77**, 1.
9. Peterson, M. G., Tanese, N., Pugh, B. F., and Tjian, R. (1990) Functional domains and upstream activation properties of cloned human TATA binding protein. *Science*, **248**, 1625.
10. Zhou, Q., Lieberman, P. M., Boyer, T. G., and Berk, A. J. (1992) Holo-TFIID supports transcriptional stimulation by diverse activators and from a TATA-less promoter. *Genes Dev.*, **6**, 1964.
11. Chiang, C.-M., Ge, H., Wang, Z., Hoffman, A., and Roeder, R. G. (1993) Unique TATA-binding protein-containing complexes and cofactors involved in transcription by RNA polymerases II and III. *EMBO J.*, **12**, 2749.
12. Verrijzer, C. P., Yokomori, K., Chen, J.-L., and Tjian, R. (1994) Drosophila TAF150: similarity to yeast gene TSM-1 and specific binding to core promoter DNA. *Science*, **264**, 933.
13. Weis, L. and Reinberg, D. (1992) Transcription by RNA polymerase II: initiator-directed formation of transcription-competent complexes. *FASEB J.*, **6**, 3300.
14. Gill, G. (1994) Taking the initiative. *Curr. Biol.*, **4**, 374.
15. Javahery, R., Khachi, A., Lo, K., Zenzie-Gregory, B., and Smale, S. T. (1994) DNA sequence requirements for transcriptional initiator activity in mammalian cells. *Mol. Cell. Biol.*, **14**, 116.
16. Kaufmann, J. and Smale, S. T. (1994) Direct recognition of initiator elements by a component of the transcription factor IID complex. *Genes Dev.*, **8**, 821.
17. Purnell, B. A., Emanuel, P. A., and Gilmour, D. S. (1994) TFIID sequence recognition of the initiator and sequences farther downstream in Drosophila class II genes. *Genes Dev.*, **8**, 830.
18. Hernandez, N. (1993) TBP, a universal eukaryotic transcription factor. *Genes Dev.*, **7**, 1291.
19. Zenzie-Gregory, B., Khachi, A., Garraway, I. P., and Smale, S. T. (1993) Mechanism of initiator-mediated transcription: evidence for a functional interaction between the

TATA-binding protein and DNA in the absence of a specific recognition sequence. *Mol. Cell. Biol.*, **13**, 3841.

20. Zenzie-Gregory, B., O'Shea-Greenfeld, A., and Smale, S. T. (1992) Similar mechanisms for transcription initiation mediated through a TATA box or an initiator element. *J. Biol. Chem.*, **267**, 2823.
21. Cortes, P., Flores, O., and Reinberg, D. (1992) Factors involved in specific transcription by mammalian polymerase II: purification and analysis of transcription factor IIA and identification of transcription factor IIJ. *Mol. Cell. Biol.*, **12**, 413.
22. Usheva, A., Maldonado, E., Goldring, A., Lu, H., Houbavi, C., Reinberg, D., and Aloni, Y. (1992) Specific interaction between the nonphosphorylated form of RNA polymerase II and the TATA-binding protein. *Cell*, **69**, 871.
23. Tschochner, H., Sayre, M. H., Flanagan, P. M., Feaver, W. J., and Kornberg, R. D. (1992) Yeast RNA polymerase II initiation factor e: isolation and identification as the functional coumterpart of human transcription factor IIB. *Proc. Natl Acad. Sci. USA*, **89**, 11292.
24. Ha, I., Roberts, S., Maldonado, E., Sun, X., Green, M. R., and Reinberg, D. (1993) Multiple functional domains of human transcription factor IIB: distinct interactions with two general transcription factors and RNA polymerase II. *Genes Dev.*, **7**,1021.
25. McCracken, S. and Greenblatt, J. (1991) Related RNA polymerase-binding regions in human RAP30/74 and Escherichia coli σ^{70}. *Science*, **253**, 900.
26. Flores, O., Maldonado, E., and Reinberg, D. (1989) Factors involved in specific transcription by mammalian RNA polymerase II. Factors IIE and IIF independently interact with RNA polymerase II. *J. Biol. Chem.*, **264**, 8913.
27. Maxon, M. E., Goodrich, J. A., and Tjian, R. (1994) Transcription factor IIE binds preferentially to RNA polymerase IIa and recruits TFIIH: a model for promoter clearance. *Genes Dev.*, **8**, 515.
28. Gerard, M., Fischer, L., Moncollin, V., Chipoulet, J. M., Chambon, P., and Egly, J. M. (1991) Purification and interaction properties of the human RNA polymerase B(II) general transcription factor BTF2. *J. Biol. Chem.*, **266**, 20940.
29. Lee, D. K., Dejong, J., Hashimoto, S., Horikoshi, M., and Roeder, R. G. (1992) TFIIA induces conformational changes in TFIID via interactions with the basic repeat. *Mol. Cell. Biol.*, **12**, 5189.
30. Malik, S., Lee, D. K., and Roeder, R. G. (1993) Potential RNA polymerase II-induced interactions of transcription factor TFIIB. *Mol. Cell. Biol.*, **13**, 10.
31. Ohkuma, Y. and Roeder, R. G. (1994) Regulation of TFIIH ATPase and kinase activities by TFIIE during active initiation complex formation. *Nature*, **368**, 160.
32. Koleske, A. J. and Young, R. A. (1994) An RNA polymerase II holoenzyme responsive to activators. *Nature*, **368**, 466.
33. Parvin, D. P. and Sharp, P. A. (1993) DNA topology and a minimal set of basal factors for transcription by RNA polymerase II. *Cell*, **73**, 533.
34. Goodrich, J. A. and Tjian, R. T. (1994) Transcription factors IIE and IIH and ATP hydrolysis direct promoter clearance by polymerase II. *Cell*, **77**, 145.
35. White, R. J. and Jackson S. P. (1992) The TATA-binding protein: a central role in transcription by RNA polymerases I, II and III. *Trends Genet.*, **8**, 284.
36. Rigby, P. W. J. (1993) Three in one and one in three: it all depends on TBP. *Cell*, **72**, 7.
37. Bell, S. P., Learned, R. M., Jantzen, H.-M., and Tjian, R. (1988) Functional cooperativity between transcription factors UBF1 and SL1 mediates human ribosomal RNA synthesis. *Science*, **241**, 1192.

38. Bell, S. P., Jantzen, H.-M., and Tjian, R. (1990) Assembly of alternative multiprotein complexes directs rRNA promoter selectivity. *Genes Dev.*, **4**, 943.
39. White, R. J. and Jackson, S. P. (1992) Mechanism of TATA-binding protein recruitment to TATA-less class III promoter. *Cell*, **71**, 1041.
40. Mattaj, I. W., Dalhan, N. A., Parry, H. D., Carbon, P., and Krol, A. (1988) Changing the RNA polymerase specificity of U snRNA gene promoters. *Cell*, **55**, 435.
41. Lobo, S. M. and Hernandez, N. (1989) A 7bp mutation converts a human RNA polymerase II snRNA promoter into an RNA polymerase III promoter. *Cell*, **58**, 55.
42. Marsh, T. L., Reich, C. I., Whitelock, R. B., and Olsen, G. J. (1994) Transcription factor IID in the Archaea: sequences in the Thermococcus celer genome would encode a product closely related to the TATA-binding protein of eukaryotes. *Proc. Natl Acad. Sci. USA*, **91**, 4180.
43. Rowlands, T., Baumann, P., and Jackson, S. P. (1994) The TATA-binding protein: a general transcription factor in eukaryotes and archaebacteria. *Science*, **264**, 1326.
44. Dynlacht, B. D., Hoey, T., and Tjian, R. (1991) Isolation of coactivators associated with the TATA-binding protein that mediate transcriptional activation. *Cell*, **66**, 563.
45. Tanese, N., Pugh, B. F., and Tjian, R. (1991) Coactivators for a proline-rich activator purified from the multisubunit human TFIID complex. *Genes Dev.*, **5**, 2212.
46. Cormack, B. P. and Struhl, K. (1992) The TATA-binding protein is required for transcription by all three nuclear RNA polymerases in yeast cells. *Cell*, **69**, 685.
47. Schultz, M. C., Reeder, R. H., and Hahn, S. (1992) Variants of the TATA-binding protein can distinguish subsets of RNA polymerase I, II, and III promoters. *Cell*, **69**, 697.
48. Cormack, B. P. and Struhl, K. (1993) Regional coding randomization: defining a TATA-binding protein surface required for RNA polymerase III transcription. *Science*, **262**, 244.
49. Comai, L., Zemendijk, J. C., Beckmann, H., Zhou, S., Admon, A., and Tjian, R. (1994) Reconstitution of transcription factor SL1: exclusive binding of TBP by SL1 or TFIID subunits. *Science*, **266**, 1966.
50. Hoey, T., Dynlacht, B. D., Peterson, M. G., Pugh, B. F., and Tjian, R. (1990) Isolation and characterization of the Drosophila gene encoding the TATA box binding protein, TFIID. *Cell*, **61**, 1179.
51. Lieberman, P. M., Schmidt, M. C., Kao, C. C., and Berk, A. J. (1991) Two distinct domains in the yeast transcription factor IID and evidence for a TATA box-induced conformational change. *Mol. Cell. Biol.*, **11**, 63.
52. Cormack, B. P., Strubin, M., Ponticelli, A. S., and Struhl, K. (1991) Functional differences between yeast and human TFIID are localized to the highly conserved region. *Cell*, **65**, 341.
53. Reddy, P. and Hahn, S. (1991) Dominant negative mutations in yeast TFIID define a bipartite DNA-binding region. *Cell*, **65**, 349.
54. Zhou, Q., Schmidt, M. C., and Berk, A. J. (1991) Requirement of acidic amino acid residues immediately amino-terminal to the conserved domain of Saccharomyces cerevisiae TFIID. *EMBO J.*, **10**, 1843.
55. Poon, D., Schroeder, S., Wang, C. K., Yamamoto, T., Horikoshi, M., Roeder, R. G. and Weil, P. A. (1991) The conserved carboxy-terminal domain of Saccharomyces cerevisiae is sufficient to support normal cell growth. *Mol. Cell. Biol.*, **11**, 4809.
56. Kelleher, III, R. J., Flanagan, P. M., Chasman, D. I., Ponticelli, A. S., Struhl, K., and Kornberg, R. D. (1992) Yeast and human TFIIDs are interchangeable for the response to acidic transcriptional activators in vitro. *Genes Dev.*, **6**, 296.

57. Wettach, J., Gohl, H. P., Tschochner, H., and Thomm, M. (1995) Functional interaction of yeast and human TATA-binding proteins with an archaeal RNA polymerase and promoter. *Proc. Natl Acad. Sci. USA*, **92**, 472.
58. Fikes, J., Becker, D. M., Winston, F., and Guarente, L. (1990) Striking conservation of TFIID in Schizosaccharomyces pombe and Saccharomyces cerevisiae. *Nature*, **346**, 291.
59. Gill, G. and Tjian, R. (1991) A highly conserved domain of TFIID displays species specificity in vivo. *Cell*, **65**, 333.
60. Keaveney, M., Berkenstam, A., Feigenbutz, M., Vriend, G., and Stunnenberg, H. G. (1993) Residues in the TATA-binding protein required to mediate a transcriptional response to retinoic acid in EC cells. *Nature*, **365**, 562.
61. Cormack, B. P., Strubin, M., Stargell, L. A., and Struhl, K. (1994) Conserved and nonconserved functions of the yeast and human TATA-binding proteins. *Genes Dev.*, **8**, 1335.
62. Horikoshi, M., Yamamoto, T., Ohkuma, Y., Weil, P. A., and Roeder, R. G. (1990) Analysis of structure-function relationships of yeast TATA box binding factor TFIID. *Cell*, **61**, 1171.
63. Lee, D. K., Horikoshi, M., and Roeder, R. G. (1991) Interaction of TFIID in the minor groove of the TATA element. *Cell*, **67**, 1241.
64. Starr, D. B. and Hawley, D. K. (1991) TFIID binds in the minor groove of the TATA box. *Cell*, **67**, 1231.
65. Nikolov, D. B., Hu, S.-H., Lin, J., Gasch, A., Hoffman, A., Horikoshi, M., Chua, N.-H., Roeder, R. G., and Burley, S. K. (1992) Crystal structure of TFIID TATA box binding protein. *Nature*, **360**, 40.
66. Chasman, D. I., Flaherty, K. M., Sharp, P. A., and Kornberg, R. D. (1993) Crystal structure of yeast TATA-binding protein and model for interaction with DNA. *Proc. Natl Acad. Sci. USA*, **90**, 8174.
67. Kim, Y., Geiger, J. H., Hahn, S., and Sigler, P. B. (1993) Crystal structure of a yeast TBP/TATA box complex. *Nature*, **365**, 512.
68. Kim, J. L., Nikolov, D. B., and Burley, S. K. (1993) Co-crystal structure of TBP recognizing the minor groove of a TATA element. *Nature*, **365**, 520.
69. Horikoshi, M., Bertuccioli, C., Takada, R., Wang, J., Yamamoto, T., and Roeder, R. G. (1992) Transcription factor TFIID induces DNA bending upon binding to the TATA element. *Proc. Natl Acad. Sci. USA*, **89**, 1060.
70. Takada, R., Nakatani, Y., Hoffman, A., Kokubo, T., Hasegawa, S., Roeder, R. G., and Horikoshi, M. (1992) Identification of human TFIID components and direct interaction between a 250-kDa polypeptide and the TATA box-binding protein (TFIIDτ). *Proc. Natl Acad. Sci. USA*, **89**, 11809.
71. Hoey, T., Weinzierl, R. O. J., Gill, G., Chen, J.-L., Dynlacht, B. D., and Tjian, R. (1993) Molecular cloning and functional analysis of Drosophila TAF110 reveal properties expected of coactivators. *Cell*, **72**, 247.
72. Zhou, Q., Boyer, T., and Berk, A. J. (1993) Factors (TAFs) required for activated transcription interact with TATA box-binding protein conserved core domain. *Genes Dev.*, **7**, 180.
73. Kokubo, T., Gong, D.-W., Wootton, J. C., Horikoshi, M., Roeder, R. G., and Nakatani, Y. (1994) Molecular cloning of Drosophila TFIID subunits. *Nature*, **367**, 484.
74. Chen, J.-L., Attardi, L. D., Verrijzer, C. P., Yokomori, K., and Tjian, R. (1994) Assembly of recombinant TFIID reveals differential coactivator requirements for distinct transcriptional activators. *Cell*, **79**, 93.

75. Kokubo, T., Gong, D.-W., Yamashita, S., Horikoshi, M., Roeder, R. G., and Nakatani, Y. (1993) Drosophila 230-kD TFIID subunit, a functional homolog of the human cell cycle gene product, negatively regulates DNA binding of the TATA box-binding subunit of TFIID. *Genes Dev.*, **7**, 1033.
76. Kokubo, T., Yamashita, S., Horikoshi, M., Roeder, R., and Nakatani, Y. (1994) Interaction between the N-terminal domain of the 230-kDa subunit and the TATA box-binding subunit of TFIID negatively regulates TATA box binding. *Proc. Natl Acad. Sci. USA*, **91**, 3520.
77. Buratowski, S. P., Hahn, S., Sharp, P. A., and Guarente, L. (1988) Function of a yeast TATA element-binding protein in a mammalian transcription system. *Nature*, **334**, 37.
78. Horikoshi, M., Wang, C. K., Fujii, H., Cromlish, J. A., Weil, P. A., and Roeder, R. G. (1989) Cloning and structure of a yeast gene encoding a general transcription initiation factor TFIID that binds to the TATA box. *Proc. Natl Acad. Sci. USA*, **86**, 4843.
79. Schmidt, M. C., Kao, C. C., Pei, R., and Berk, A. J. (1989) Yeast TATA box transcription factor gene. *Proc. Natl Acad. Sci. USA*, **86**, 7785.
80. Poon, D. and Weil, P. A. (1993) Immunopurification of yeast TATA-binding protein and associated factors. *J. Biol. Chem.*, **268**, 15325.
81. Reese, J. C., Apone, L., Walker, S. S., Griffin, L. A., and Green, M. R. (1994) Yeast TAF_{II}s in a multisubunit complex required for activated transcription. *Nature*, **371**, 523.
82. Zhou, Q. and Berk, A. J. (1995) The yeast TATA-binding protein (TBP) core domain assembles with human TBP-associated factors into a functional TFIID complex. *Mol. Cell. Biol.*, **15**, 534.
83. Tansey, W., Ruppert, S., Tjian, R., and Herr, W. (1994) Multiple regions of TBP participate in the response to transciptional activators in vivo. *Genes Dev.*, **8**, 2756.
84. Künzler, M., Braus, G. H., Georgiev, O., Seipel, K., and Schaffner, W. (1994) Functional differences between mammalian transcription activation domains at the yeast GAL1 promoter. *EMBO J.*, **13**, 641.
85. Comai, L., Tanese, N., and Tjian, R. (1992) The TATA-binding protein and associated factors are integral components of the RNA polymerase I transcription factor. *Cell*, **68**, 965.
86. Lobo, S. M., Tanaka, M., Sullivan, M. L., and Hernandez, N. (1992) A TBP complex essential for transcription from TATA-less but not TATA-containing RNA polymerase III promoters is part of the TFIIIB fraction. *Cell*, **71**, 1029.
87. Taggart, A. K. P., Fisher, T. S., and Pugh, B. F. (1992) The TATA-binding protein and associated factors are components of polIII transcription factor TFIIIB. *Cell*, **71**, 1015.
88. Sadowski, C. L., Henry, R. W., Lobo, S. M., and Hernandez, N. (1993) Targeting to a non-TATA box cis-regulatory element: a TBP-containing complex activates transcription from snRNA promoters through the PSE. *Genes Dev.*, **7**, 1535.
89. Hoffman, A., Sinn, E., Yamamoto, T., Wang, J., Roy, A., Horikoshi, M., and Roeder, R. G. (1990) Highly conserved core domain and unique N-terminus with presumptive regulatory motifs in a human TATA factor (TFIID). *Nature*, **346**, 387.
90. Pugh, B. F. and Tjian, R. (1990) Mechanism for transcriptional activation by Sp1: evidence for coactivators. *Cell*, **61**, 1187.
91. Ingles, C. J., Shales, M., Cress, W. D., Triezenberg, S. J., and Greenblatt, J. (1991) Reduced binding of TFIID to transcriptionally compromised mutants of VP16. *Nature*, **351**, 588.
92. Truant, R., Xiao, H., Ingles, J., and Greenblatt, J. (1993) Direct interaction between the transcriptional activation domain of human p53 and the TATA box-binding protein. *J. Biol. Chem.*, **268**, 2284.

93. Lieberman, P. M. and Berk, A. J. (1991) The Zta trans-activator protein stabilizes TFIID association with promoter DNA by direct protein-protein-interaction. *Genes Dev.*, **5**, 2441.
94. Lieberman, P. M. and Berk, A. J. (1994) A mechanism for TAFs in transcriptional activation: activation domain enhancement of TFIID-TFIIA-promoter DNA complex formation. *Genes Dev.*, **8**, 995.
95. Horikoshi, M., Maguire, K., Kralli, A., Maldonado, E., Reinberg, D., and Weinmann, R. (1991) Direct interaction between adenovirus E1A protein and the TATA box binding transcription factor IID. *Proc. Natl Acad. Sci. USA*, **88**, 5124.
96. Emili, A., Greenblatt, J., and Ingles, C. J. (1994) Species-specific interaction of the glutamine-rich activation domains of Sp1 with the TATA box binding protein. *Mol. Cell. Biol.*, **14**, 1582.
97. Seto, E., Usheva, A., Zambetti, G., Momand, J., Horikoshi, N., Weinmann, R., Levine, A. J., and Shenk, T. (1992) Wild-type p53 binds to the TATA-binding protein and represses transcription. *Proc. Natl Acad. Sci. USA*, **89**, 12028.
98. Zwilling, S., Annweiler, A., and Wirth, T. (1994) The POU domains of the Oct1 and Oct2 transcription factors mediate specific interaction with TBP. *Nucl. Acids Res.*, **22**, 1655.
99. Boyer, T. G. and Berk, A. J. (1993) Functional interaction of adenovirus E1A with holo-TFIID. *Genes Dev.*, **7**, 1810.
100. Kerr, L. D., Ransone, L. J., Wamsley, P., Schmitt, M. J., Boyer, T. G., Zhou, Q., Berk, A. J., and Verma, I. M. (1993) Association between proto-oncoprotein Rel and TATA-binding protein mediates transcriptional activation by NF-κB. *Nature*, **365**, 412.
101. Liu, X., Miller, C. W., Koeffler, P. H., and Berk, A. J. (1993) The p53 activation domain binds the TATA box-binding polypeptide in holo-TFIID, and a neighboring p53 domain inhibits transcription. *Mol. Cell. Biol.*, **13**, 3291.
102. Gill, G., Pascal, E., Tseng, Z. H., and Tjian, R. (1994) A glutamine-rich hydrophobic patch in transcription factor Sp1 contacts the dTAF110 component of the Drosophila TFIID complex and mediates transcriptional activation. *Proc. Natl Acad. Sci. USA*, **91**, 192.
103. Ferreri, K., Gill, G., and Montminy, M. (1994) The cAMP-regulated transcription factor CREB interacts with a component of the TFIID complex. *Proc. Natl Acad. Sci. USA*, **91**, 1210.
104. Goodrich, J. A., Hoey, T., Thut, C. J., Admon, A., and Tjian, R. (1993) Drosophila TAF40 interacts with both a VP16 activation domain and the basal transcription factor TFIIB. *Cell*, **75**, 519.
105. Thut, C. J., Chen, J.-L., Klemm, R., and Tjian, R. (1995) p53 transcriptional activation mediated by coactivators $TAF_{II}40$ and $TAF_{II}60$. *Science*, **267**, 100.
106. Hisatake, K., Hasegawa, S., Takada, R., Nakatani, Y., Horikoshi, M., and Roeder, R. G. (1993) The p250 subunit of native TATA box-binding factor TFIID is the cell-cycle regulatory protein CCG1. *Nature*, **362**, 179.
107. Ruppert S., Wang, E. H., and Tjian, R. (1993) Cloning and expression of human TAF250: a TBP-associated factor implicated in cell cycle regulation. *Nature*, **362**, 175.
108. Wang, E. H. and Tjian, R. (1994) Promoter-selective transcriptional defect in cell cycle mutant ts13 rescued by hTAF250. *Science*, **263**, 811.
109. Chiang, C.-M. and Roeder, R. G. (1995) Cloning of an intrinsic human TFIID subunit that interacts with multiple transcriptional activators. *Science*, **267**, 531.
110. Lin, Y. S. and Green, M. R. (1991) Mechanism of action of an acidic transcriptional activator in vitro. *Cell*, **64**, 971.

111. Lin, Y.-S., Ha, I., Maldonado, E., Reinberg, D., and Green, M. R. (1991) Binding of general transcription factor TFIIB to an acidic activating region. *Nature*, **345**, 359.
112. Choy, B. and Green, M. R. (1993) Eukaryotic activators function during multiple steps of preinitiation complex assembly. *Nature*, **366**, 531.
113. Colgan, J., Wampler, S. and Manley, J. L. (1993) Interaction between a transcriptional activator and transcription factor IIB in vivo. *Nature*, **362**, 549.
114. Kim, T. K. and Roeder, R. G. (1994) Proline-rich activator CTF1 targets the TFIIB assembly step during transcriptional activation. *Proc. Natl Acad. Sci. USA*, **91**, 4170.
115. Roberts, S. G. E., Ha, I., Maldonado, E., Reinberg, D., and Green, M. R. (1993) Interaction between an acidic activator and transcription factor TFIIB is required for transcriptional activation. *Nature*, **363**, 741.
116. Roberts, S. G. E. and Green, M. R. (1994) Activator-induced conformational change in general transcription factor TFIIB. *Nature*, **371**, 717.
117. Young, R. A. (1991) RNA polymerase II. *Annu. Rev. Biochem.*, **60**, 689.
118. Laybourn, P. J. and Dahmus, M. E. (1990) Phosphorylation of RNA polymerase IIA occurs subsequent to interaction with the promoter and before the initiation of transcription. *J. Biol. Chem.*, **265**, 13165.
119. Lu, H., Flores, O., Weinman, R., and Reinberg, D. (1991) The nonphosphorylated form of RNA polymerase II preferentially associates with the preinitiation complex. *Proc. Natl Acad. Sci. USA*, **88**, 10004.
120. Weeks, J. R., Hardin, S. E., Shen, J., Lee, J. M., and Greenleaf, A. (1993) Locus-specific variation in phosphorylation state of RNA polymerase II in vivo: correlations with gene activity and transcript processing. *Genes Dev.*, **7**, 2329.
121. Lu, H., Zawel, L., Fisher, L., Egly, J.-M., and Reinberg, D. (1992) Human general transcription factor IIH phosphorylates the C-terminal domain of RNA polymerase II. *Nature*, **358**, 641.
122. Feaver, W. J., Svejstrup, J. Q., Henry, N. L., and Kornberg, R. D. (1994) Relationship of CDK-activating kinase and RNA polymerase II CTD kinase TFIIH/TFIIK. *Cell*, **79**, 1103.
123. Roy, R., Adamczewski, J. P., Seroz, T., Vermeulen, W., Tassan, J.-P., Schaeffer, L., Nigg, E. A., Hoeijemakers, J. H. J., and Egly, J.-M. (1994) The MO15 cell cycle kinase is associated with the TFIIH transcription-DNA repair factor. *Cell*, **79**, 1093.
124. Xiao, H., Pearson, A., Coulombe, B., Truant, R., Zhang, S., Regier, J. L., Triezenberg, S. J., Reinberg, D., Flores, O., Ingles, C. J., and Greenblatt, J. (1994) Binding of basal transcription factor TFIIH to the acidic activation domains of VP16 and p53. *Mol. Cell. Biol.*, **14**, 7013.
125. Zhu, H., Joliot, V., and Prywes, R. (1994) Role of transcription factor TFIIF in serum response factor-activated transcription. *J. Biol. Chem.*, **269**, 3489.
126. Joliot, V., Demma, M., and Prywes, R. (1995) Interaction with RAP74 subunit of TFIIF is required for transcriptional activation by serum response factor. *Nature*, **373**, 632.
127. Berger, S. L., Cress, W. D., Cress, A., Triezenberg, S. J., and Guarente, L. (1990) Selective inhibition of activated but not basal transcription by the acidic activation domain of VP16: evidence for transcriptional adaptors. *Cell*, **61**, 1199.
128. Berger, S. L., Pina, B., Silverman, N., Marcus, G. A., Agapite, J., Regier, J. L., Triezenberg, S. J., and Guarente, L. (1992) Genetic isolation of ADA2: a potential transcriptional adaptor required for function of certain acidic activation domains. *Cell*, **70**, 251.
129. Pina, B., Berger, S., Marcus, G. A., Silverman, N., Agapite, J., and Guarente, L. (1993) ADA3: a gene, identified by resistance to GAL4-VP16, with properties similar to and different from those of ADA2. *Mol. Cell. Biol.*, **13**, 5981.
130. Marcus, G. A., Silverman, N., Berger, S. L., Horiuchi, J., and Guarente, L. (1994) Func-

tional similarity and physical association between GCN5 and ADA2: putative transcriptional adaptors. *EMBO J.*, **13**, 4807.

131. Silverman, N., Agapite, J., and Guarente, L. (1994) Yeast ADA2 protein binds to the VP16 protein activation domain and activates transcription. *Proc. Natl Acad. Sci. USA*, **91**, 11665.
132. Horiuchi, J., Silverman, N., Marcus, G. A., and Guarente, L. (1995) ADA3, a putative transcriptional adaptor, consists of two separable domains and interacts with ADA2 and GCN5 in a trimeric complex. *Mol. Cell. Biol.*, **15**, 1203.
133. Oehler, T. and Angel, P. (1992) A common intermediary factor (p52/p54) recognizing "acidic blob"-type domains is required for transcriptional activation by the proteins. *Mol. Cell. Biol.*, **12**, 5508.
134. Chrivia, J. C., Kwok, R. P. S., Lamb, N., Hagiwara, M., Montminy, M. R., and Goodman, R. H. (1993) Phosphorylated CREB binds specifically to the nuclear protein CBP. *Nature*, **365**, 855.
135. Kwok, R. P. S., Lundblad, J. R., Chrivia, J. C., Richards, J. P., Bächinger, H. P., Brennan, R. G., Roberts, S. G. E., Green, M. R., and Goodman, R. H. (1994) Nuclear protein CBP is a coactivator for the transcription factor CREB. *Nature*, **370**, 223.
136. Arias, J., Alberts, A. S., Brindle, P., Claret, F. X., Smeal, T., Karin, M., Feramisco, J., and Montminy, M. (1994) Activation of cAMP and mitogen responsive genes relies on a common nuclear factor. *Nature*, **370**, 226.
137. Arany, Z., Sellers, W. R., Livingston, D. M., and Eckner, R. (1994) E1A-associated p300 and CREB-associated CBP belong to a conserved family of coactivators. *Cell*, **77**, 799.
138. Eckner, R., Ewen, M. E., Newsome, D., Gerdes, M., DeCaprio, J. A., Lawrence, J. B., and Livingston, D. M. (1994) Molecular cloning and functional analysis of the adenovirus E1A-associated 300kD protein (p300) reveals a protein with properties of a transcriptional adaptor. *Genes Dev.*, **8**, 869.
139. Ge, H. and Roeder, R. (1994) Purification, cloning, and characterization of a human coactivator, PC4, that mediates transcriptional activation of class II genes. *Cell*, **78**, 513.
140. Kretzschmar, M., Kaiser, K., Lottspeich, F., and Meisterernst, M. (1994) A novel mediator of class II gene transcription with homology to viral immediate-early transcriptional regulators. *Cell*, **78**, 525.
141. Ge, H., Zhao, Y., Chait, B. T., and Roeder, R. G. (1994) Phosphorylation negatively regulates the function of coactivator PC4. *Proc. Natl Acad. Sci. USA*, **91**, 12691.
142. Inostroza, J. A., Mermelstein, F. H., Ha, I., Lane, W. S., and Reinberg, D. (1992) Dr1, a TATA-binding protein-associated phosphoprotein and inhibitor of class II gene transcription. *Cell*, **70**, 477.
143. Yeung, K. C., Inostroza, J. A., Mermelstein, F. H., Kannabiran, C., and Reinberg, D. (1994) Structure-function analysis of the TBP-binding protein Dr1 reveals a mechanism for repression of class II gene transcription. *Genes Dev.*, **8**, 2097.
144. White, R. J., Khoo, B. C.-E., Inostroza, J. A., Reinberg, D., and Jackson, S. P. (1994) Differential regulation of RNA polymerases I, II, and III by the TBP-binding repressor Dr1. *Science*, **266**, 448.
145. Kraus, V. B., Inostroza, J. A., Yeung, K., Reinberg, D., and Nevins, J. R. (1994) Interaction of the Dr1 inhibitory factor with the TATA binding protein is disrupted by adenovirus E1A. *Proc. Natl Acad. Sci. USA*, **91**, 6279.
146. Kretzschmar, M., Meisterernst, M., and Roeder, R. G. (1993) Identification of human DNA topoisomerase I as a cofactor for activator-dependent transcription by RNA polymerase II. *Proc. Natl Acad. Sci. USA*, **90**, 11508.

147. Merino, A., Madden, K. R., Lane, W. S., Champoux, J. J., and Reinberg, D. (1993) DNA topoisomerase I is involved in both repression and activation of transcription. *Nature*, **365**, 227.
148. Ma, D., Watanabe, H., Mermelstein, F., Admon, A., Oguri, K., Sun, X., Wada, T., Imai, T., Shiroya, T., Reinberg, D., and Handa, H. (1993) Isolation of a cDNA encoding the largest subunit of TFIIA reveals functions important for activated transcription. *Genes Dev.*, **7**, 2246,
149. Auble, D. T. and Hahn, S. (1993) An ATP-dependent inhibitor of TBP binding to DNA. *Genes Dev.*, **7**, 844.
150. Auble, D. T., Hansen, K., Mueller, C. G. F., Lane, W. S., Thorner, J., and Hahn, S. (1994) Mot1, a global repressor of RNA polymerase II transcription, inhibits TBP binding to DNA by an ATP-dependent mechanism. *Genes Dev.*, **8**, 1920.
151. Poon, D., Campbell, A. M., Bai, Y., and Weil, P. A. (1994) Yeast Taf170 is encoded by MOT1 and exists in a TATA box-binding protein (TBP)-associated factor complex distinct from transcription factor IID. *J. Biol. Chem.*, **269**, 23135.
152. Goodbourn, S. (1990) Negative regulation of transcriptional initiation in eukaryotes. *Biochim. Biophys. Acta*, **1032**, 53.
153. Clark, A. R. and Docherty, K. (1993) Negative regulation of transcription in eukaryotes. *Biochem. J.*, **296**, 521.
154. Herschbach, B. M. and Johnson, A. D. (1993) Transcriptional repression in eukaryotes. *Annu. Rev. Cell. Biol.*, **9**, 479.
155. Jaynes, J. B. and O'Farrell, P. H. (1991) Active repression of transcription by the Engrailed homeodomain protein. *EMBO J.*, **10**, 1427.
156. Baniahmad, A., Kohne, A. C., and Renkawitz, R. (1992) A transferable silencing domain is present in the thyroid hormone receptor, in the v-erbA oncogene product and in the retinoic acid receptor. *EMBO J.*, **11**, 1015.
157. Johnson, F. B. and Krasnow, M. A. (1992) Differential regulation of transcription preinitiation complex assembly by activator and repressor homeo domain proteins. *Genes Dev.*, **6**, 2177.
158. Fondell, J. F., Roy, A. L., and Roeder, R. G. (1993) Unliganded thyroid hormone receptor inhibits formation of a functional preinitiation complex: implications for active repression. *Genes Dev.*, **8**, 1400.
159. Baniahmad, A., Ha, I., Reinberg, D., Tsai, S., Tsai, M.-J., and O'Malley, B. W. (1993) Interaction of human thyroid hormone receptor ß with transcription factor TFIIB may mediate target gene derepression and activation by thyroid hormone. *Proc. Natl Acad. Sci. USA*, **90**, 8832.
160. Ing, N. H., Beekman, J. M., Tsai, S. Y., Tsai, M.-J., and O'Malley, B. W. (1992) Members of the steroid receptor superfamily interact with TFIIB (S300-II). *J. Biol. Chem.*, **267**, 17617
161. Keleher, C. A., Redd, M. J., Schultz, J., Carlson, M., and Johnson, A. D. (1992) Ssn6-Tup1 is a general repressor of transcription in yeast. *Cell*, **68**, 709.
162. Williams, F. E., Varanasi, U., and Trumbly, R. J. (1991) The CYC8 and TUP1 proteins involved in glucose repression in Saccharomyces cerevisiae are associated in a protein complex. *Mol. Cell. Biol.*, **11**, 3307.
163. Komachi, K., Redd, M. J., and Johnson, A. D. (1994) The WD repeats of Tup1 interact with the homeo domain protein α2. *Genes Dev.*, **8**, 2857.
164. Herschbach, B. M., Arnaud, M. B., and Johnson, A. D. (1994) Transcriptional repression directed by the yeast α2 protein in vitro. *Nature*, **370**, 309.

165. Herschbach, B. M. and Johnson, A. D. (1993) The yeast α2 protein can repress transcription by RNA polymerases I and II but not III. *Mol. Cell. Biol.*, **13**, 4029.
166. Tzamarias, D. and Struhl, K. (1994) Functional dissection of the yeast Cyc8-Tup1 transcriptional corepressor complex. *Nature*, **369**, 758.
167. McDonnell, D. P., Vegeto, E., and O'Malley, B. W. (1992) Identification of a negative regulatory function for steroid receptors. *Proc. Natl Acad. Sci. USA*, **89**, 10563.
168. Herschlag, D. and Johnson, F. B. (1993) Synergism in transcriptional activation: a kinetic view. *Genes Dev.*, **7**, 173.
169. Poellinger, L., Yoza, B. K., and Roeder, R. G. (1989) Functional cooperativity between protein molecules bound at two distinct sequence elements of the immunoglobulin heavy-chain promoter. *Nature*, **337**, 573.
170. Baniahmad, C., Muller, M., Altscmied, J., and Renkawitz, R. (1991) Co-operative binding of the glucocorticoid receptor DNA binding domain is one of at least two mechanisms for synergism. *J. Mol. Biol.*, **222**, 155.
171. Carey, M., Lin, Y.-S., Green, M. R., and Ptashne, M. (1990) A mechanism for synergistic activation of a mammalian gene by GAL4 derivatives. *Nature*, **345**, 361.
172. Pascal, E. and Tjian, R. (1991) Different activation domains of Sp1 govern formation of multimers and mediate transcriptional synergism. *Genes Dev.*, **5**, 1646.
173. Wang, W. and Gralla, J. D. (1991) Differential ability of proximal and remote element pairs to cooperate in activating RNA polymerase II transcription. *Mol. Cell. Biol.*, **11**, 4561.
174. Weinzierl, R. O. J., Dynlacht, B. D., and Tjian, R. (1993) Largest subunit of Drosophila transcription factor IID directs assembly of a complex containing TBP and a coactivator. *Nature*, **362**, 511.
175. Yokomori, K., Chen, J.-L., Admon, A., Zhou, S., and Tjian, R. (1994) Molecular cloning and characterization of dTAF30α and dTAF30β: two small subunits of Drosophila TFIID. *Genes Dev.*, **7**, 2587.
176. Kokubo, T., Gong, D.-W., Roeder, R. G., Horikoshi, M., and Nakatani, Y. (1993) The Drosophila 110-kDa transcription factor TFIID subunit directly interacts with the N-terminal region of the 230-kDa subunit. *Proc. Natl Acad. Sci. USA*, **90**, 5896.
177. Dynlacht, B. D., Weinzierl, R. O. J., Admon, A., and Tjian, R. (1993) The dTAF80 subunit of Drosophila TFIID contains ß-transducin repeats. *Nature*, **363**, 176.
178. Kokubo, T., Gong, D.-W., Yamashita, S., Takada, R., Roeder, R. G., Horikoshi, M., and Nakatani, Y. (1993) Molecular cloning, expression, and characterization of the Drosophila 85-kilodalton TFIID subunit. *Mol. Cell. Biol.*, **13**, 7859.
179. Weinzierl, R. O. J., Ruppert, S., Dynlacht, B. D., Tanese, N., and Tjian, R. (1993) Cloning and expression of Drosophila TAF60 and human TAF70 reveal conserved interactions with other subunits of TFIID. *EMBO J.*, **12**, 5303.
180. Stringer, K. F., Ingles, C. J., and Greenblatt J. (1990) Direct and selective binding of an acidic transcriptional activator domain to the TATA box factor TFIID. *Nature*, **345**, 783.
181. Lee, W. S., Kao, C. C., Bryant, G. O., Liu, X., and Berk, A. J. (1991) Adenovirus E1A activation domain binds the basic repeat in the TATA box transcription factor. *Cell*, **67**, 365.
182. Kashanchi, F., Piras, G., Radonovich, M. F., Duvall, J. F., Fattaey, A., Chiang, C.-M., Roeder, R. G., and Brady, J. N. (1994) Direct interaction of human TFIID with the HIV-1 transactivator Tat. *Nature*, **367**, 295.
183. Hagemeier, C., Walker, S. M., Caswell, R. C., Kouzarides, T., and Sinclair, J. (1992) The human cytomegalovirus 80-kilodalton but not the 72-kilodalton immediate-early protein transactivates heterologous promoters in a TATA box-dependent mechanism and interacts directly with TFIID. *J. Virol.*, **66**, 4452.

184. Caron, C., Rousset, R., Beraud, C., Moncollin, V., Egly, J.-M., and Jalinot, P. (1993) Functional and biochemical interaction of the HTLV-I Tax1 transactivator with TBP. *EMBO J.*, **12**, 4269.
185. Qadri, I., Maguire, H. F., and Siddiqui, A. (1995) Hepatitis B virus transactivator protein X interacts with the TATA-binding protein. *Proc. Natl Acad. Sci. USA*, **92**, 1003.
186. Manet, E., Allera, C., Gruffat, H., Mikaelian, I., Rigolet, A., and Sergeant, A. (1993) The acidic activation domain of the Epstein-Barr virus transcription factor R interacts in vitro with both TBP and TFIIB and is cell-specifically potentiated by a proline rich region. *Gene Expression*, **3**, 49.
187. Martin, D. W., Munoz, R. M., Subler, M. A., and Deb, S. (1993) p53 binds to the TATA-binding protein-TATA complex. *J. Biol. Chem.*, **268**, 13062.
188. Hagemeier, C., Bannister, A. J., Cook, A., and Kouzarides, T. (1993) The activation domain of transcription factor PU.1 binds the retinoblastoma (RB) protein and the transcription factor TFIID *in vitro*: RB shows sequence similarity to TFIID and TFIIB. *Proc. Natl Acad. Sci. USA*, **90**, 1580.
189. Hagemeier, C., Cook, A., and Kouzarides, T. (1993) The retinoblastoma protein binds E2F residues required for activation in vivo and TBP binding in vitro. *Nucl. Acids Res.*, **21**, 4998.
190. Xu, X., Prorock, C., Ishikawa, H., Maldonado, E., Ito, Y., and Gelinas, C. (1993) Functional interaction of the v-Rel and c-Rel oncoproteins with the TATA-binding protein and association with transcription factor IIB. *Mol. Cell. Biol.*, **13**, 6733.
191. Ransone, L. J ., Kerr, L. D., Schmitt, M. J., Wamsley, P., and Verma, I. M. (1993) The bZIP domains of Fos and Jun mediate a physical association with the TATA box-binding protein. *Gene Expression*, **3**, 37.

2 | Histones and transcriptional control

ALAN P. WOLFFE

1. Introduction

The eukaryotic transcriptional machinery functions in a chromatin environment. Substantial experimental evidence supports an intimate interrelationship between transcription factors and the histone proteins in regulating gene expression (1–6). We know from genetic experiments that mutations in the histone proteins can have major effects on both transcriptional activation and repression (7–10). In many instances these effects can be related to a precise positioning of the histone proteins along the DNA molecule (11, 12). Importantly, biochemical fractionation and reconstruction experiments have allowed the structural role of individual histones within chromatin to be determined (13–17). With this information it is possible to provide a meaningful structural interpretation of the results emerging from the intense effort to reconstruct the true *in vivo* range of transcriptional regulation for a particular gene within chromatin (18–26).

The purpose of this chapter is to introduce this exciting research area. I will first describe the structure of the natural chromatin template and the role of the histone proteins in organizing the DNA. I will then discuss genetic evidence for a role for this organization in both gene repression and activation, and finally I will illustrate how transcription factors and histones interact together to assist in the transcriptional regulation of specific genes.

2. Chromatin structure

2.1 Histones

The primary proteins whose properties mediate the folding of DNA into chromatin are the histones. Aside from the compaction of DNA, the histone proteins undertake protein–protein interactions between themselves and other distinct chromosomal proteins. These interactions lead to several constraints on the properties of histones, contributing to their high degree of primary sequence conservation through evolution (1, 2, 13). All of the core histones are small, basic proteins (11–16 kDa) containing relatively large amounts of lysine and arginine. All four core his-

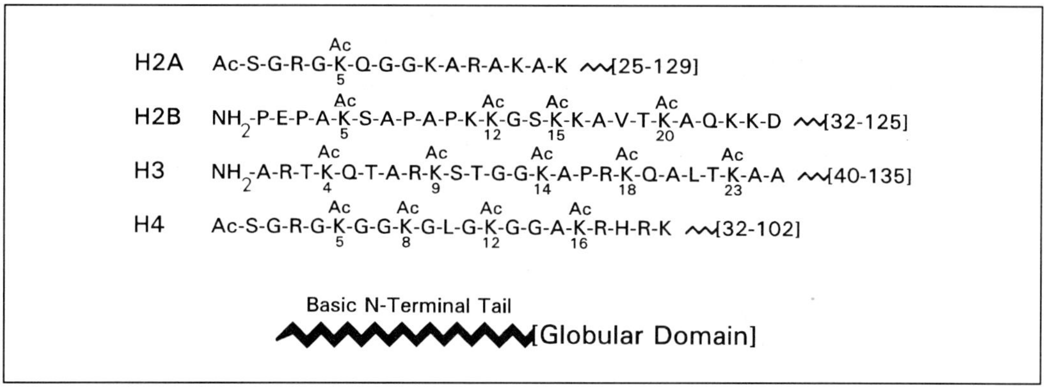

Fig. 1 The core histones have a highly conserved basic N-terminal tail and a highly conserved C-terminal globular domain. Sites of acetylation (Ac) are indicated.

tones contain a globular domain through which histone–histone and histone–DNA interactions occur, and charged amino (N)-terminal tails that contain the bulk of the lysine and arginine residues (see Fig. 1). Like the globular regions of the histones, which might be expected to be conserved due to their central structural role in organizing DNA, the amino acid sequence of the charged tails is also conserved. These charged tails are the sites of many post-translational modifications of the histone proteins. Acetylation of the N-terminal tails (see Fig. 1) of the histones has important consequences for the interaction of transcription factors with their recognition sites in chromatin (27).

Most eukaryotic cells contain a fifth histone called the linker histone, of which the most common is histone H1. Linker histones are highly basic, being particularly rich in lysine, and are slightly larger than the core histones (> 20 kDa). They have a central globular domain and highly charged tails at both the N- and carboxyl (C)-terminus (see Section 2.4). Linker histones are not found within the histone core around which DNA is wrapped, instead they interact with DNA with high affinity only after it has been wrapped around the core histones (1, 13, 15).

2.2 The nucleosome

The fundamental repeating unit of metazoan chromatin is the nucleosome. This structure consists of a protein core containing the globular domains of two molecules of each of the four core histones (H2A, H2B, H3, and H4). Approximately 160 bp of DNA is wrapped in two left-handed, superhelical turns around the histone core (see Fig. 2A). The single molecule of linker histone associates with the exterior of this complex and with the linker DNA (generally 20–30 bp) that is not in contact with the globular domains of the core histones (see Fig. 2B). Many early structural studies made use of nucleosome core particles that contain only the core histones and the 146 bp of DNA that make the strongest histone DNA contacts (see Fig. 2C). The core particle has a disc-like shape, 11 nm in diameter and 5.6 nm in width. The structure has a pseudodyad axis of symmetry that passes through the

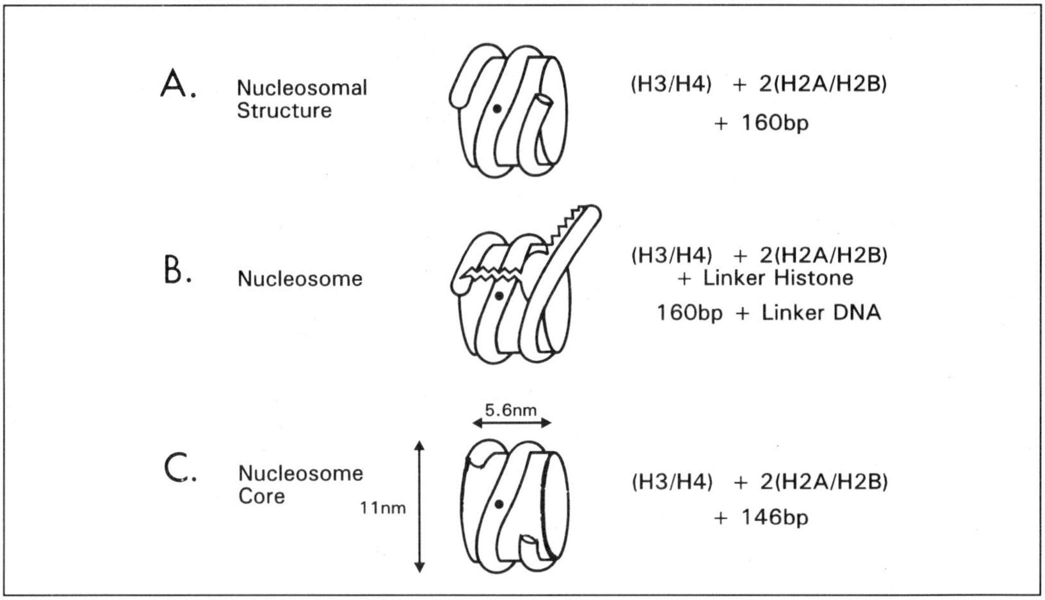

Fig. 2 Three types of histone-DNA complexes. A. A nucleosomal structure consists of two molecules each of all four core histones that organize two turns (160 bp) of DNA around their globular domains. B. A nucleosome contains an additional 20–30 bp of 'linker' DNA to the nucleosomal structure and also has one molecule of a linker histone. C. A nucleosome core consists of two molecules of each of all the core histones and the 146 bp of DNA that make the strongest contacts with the histones. The dimensions of the nucleosome core are indicated. The axis of dyad symmetry is indicated by the filled circle on the DNA.

centre of the nucleosomal DNA, at this position the minor groove of the double helix faces away from the core histones (28; see Fig. 4 below).

The histone core has a tripartite organization (see Fig. 3). A tetramer of histones $(H3/H4)_2$ organizes the central 120 bp of DNA within the nucleosome identically to the intact histone octamer; only when DNA is wrapped around the histone tetramer can two dimers of histones H2A/H2B associate to either side, thereby extending core histone–DNA contacts to the full 160 bp (14). This interaction depends on both protein–protein contacts and the DNA conformation induced by the histone tetramer. The secondary structure of core histones and their relative arrangement in space results in an extensive molecular contact interface between individual histones within dimers of both H3/H4 and H2A/H2B (see Fig. 4; 16). The individual polypeptide chains are folded in a similar manner, described as the histone fold. This consists of a long central helix flanked on either side by a loop segment and a shorter helix (see Fig. 4). The area of the interaction between the H2A/H2B dimer and the $(H3/H4)_2$ tetramer is surprisingly more extensive than the interface between the two H3/H4 dimers that make up the $(H3/H4)_2$ tetramer (see Fig. 4). However, the contacts within the interface between the H2A/H2B dimer and the $(H3/H4)_2$ tetramer are more potentially accessible to solvent, and this interface is the first to be disrupted in the disassembly of the histone octamer (29).

The structure of DNA in the nucleosome and the contributions of the individual

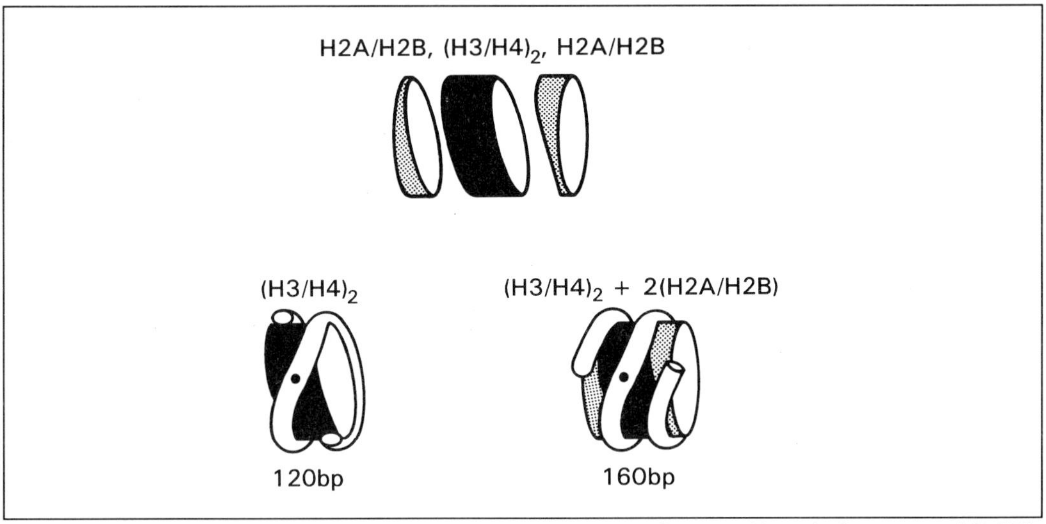

Fig. 3 The histone octamer has a tripartite organization. It consists of a central kernel of a tetramer $(H3/H4)_2$ surrounded by two dimers (H2A/H2B). Each tetramer $(H3/H4)_2$ will organize 120 bp of DNA, each dimer organizes an additional 20 bp to either side of the tetramer–DNA complex.

histones and their domains have been explored in detail using the hydroxyl radical cleavage reagent (14, 15, 30, 31). Three turns of DNA at the dyad axis of the nucleosome have a helical periodicity (10.7 bp per turn) different from the remainder of DNA in the structure (10.0 bp per turn). The junctions between the two DNA structures correspond to sites of sharp DNA bending in the nucleosome (30). The necessity of forming these structural features within nucleosomal DNA is likely to make significant contributions to directing where the histone octamer positions itself along the DNA molecule (14, 15). Remarkably, the histone octamer exerts a dominant constraint on DNA in the nucleosome, such that even unique sequence DNA adopts the same structure as the average of all DNA sequences associated with histone octamers in the nucleus (31).

The N-terminal tails of the core histones (see Fig. 1) make no direct contribution to maintaining the wrapping of DNA around the globular domains of the core histones (14). However, these positively charged domains are in contact with DNA, probably forming electrostatic interactions with the phosphodiester backbone and lying in the major groove (32–35). Modifications of the N-terminal tails, such as acetylation, may weaken their interaction with DNA (35).

2.3 Transcription factor access to DNA in the nucleosome

Formation of nucleosomes presents many impediments to the recognition of DNA by transcription factors (see also Chapter 1). One side of the double helix is occluded because it faces the globular domains of the core histones. The adjacent superhelical turn of DNA will further restrict free access to one side of the DNA molecule. For transcription factors that interact with more than ~ 5 bp at each site of

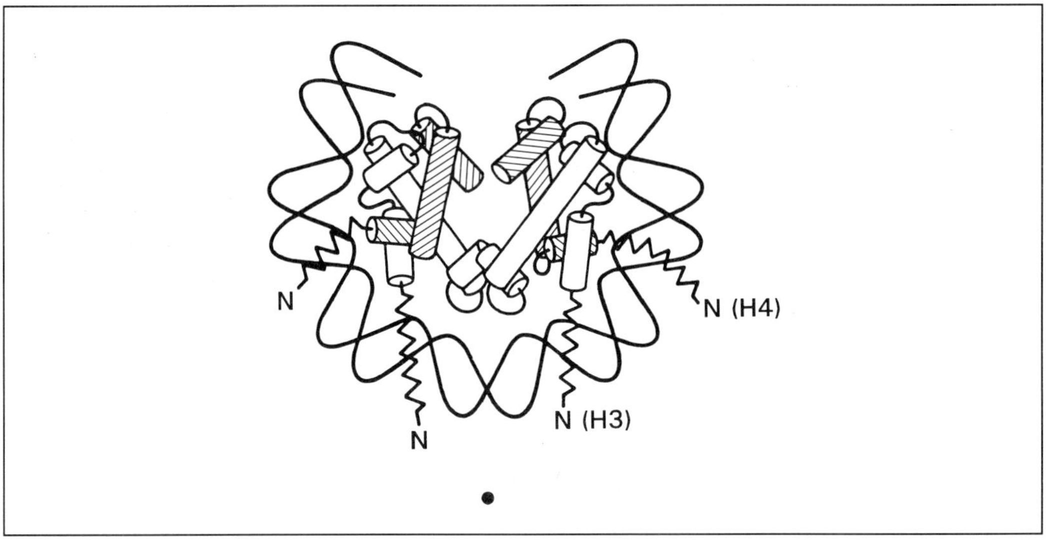

Fig. 4 The association of the histone tetramer $(H3/H4)_2$ with DNA at the dyad axis of the nucleosome. The dyad axis is indicated by the filled circle below the structure; here, the minor groove of DNA faces towards solution. The individual H3/H4 dimers are shown, with the histone fold. H4 has the hatched cylinders, H3 has the open cylinders. A suggested position for the N-terminal tails of histones H3 and H4 is shown.

contact with DNA (i.e. half a helical turn), the distortion of nucleosomal DNA into two 80-bp circles will require the transcription factor to also wrap itself around the histone core. This would demand considerable flexibility between each distinct DNA-binding domain in the transcription factor, if the DNA retains the same path around the histones. The overwinding of DNA to 10.0 bp/turn within most of the nucleosome will also impose constraints on the multiple base pair-specific contacts made by transcription factors that normally recognize DNA in solution that has a helical periodicity of 10.5 bp/turn. Furthermore, the positively charged N-terminal core histone tails that interact with the phosphodiester backbone, and which may lie partially within the major groove of DNA, are also likely to restrict transcription factor access.

2.4 Histone H1 and the chromatin fibre

The problem that transcription factors encounter in recognizing DNA within chromatin is further compounded by the presence of specialized linker histones such as H1. Histone H1 association stabilizes the interaction of the core histones with the 160 bp of DNA wrapped around them, forming the chromatosome (see Fig. 5). This stabilization probably occurs through additional interactions with this DNA mediated by both the globular and tail domains of the linker histone (see Fig. 5A). The globular domain of the linker histones contains two binding sites for DNA (36) and the juxtaposition of DNA segments, following their wrapping around the histone octamer, promotes linker histone association with core histone-bound DNA over binding to free DNA in solution (15). Histone H1 also has close contacts with

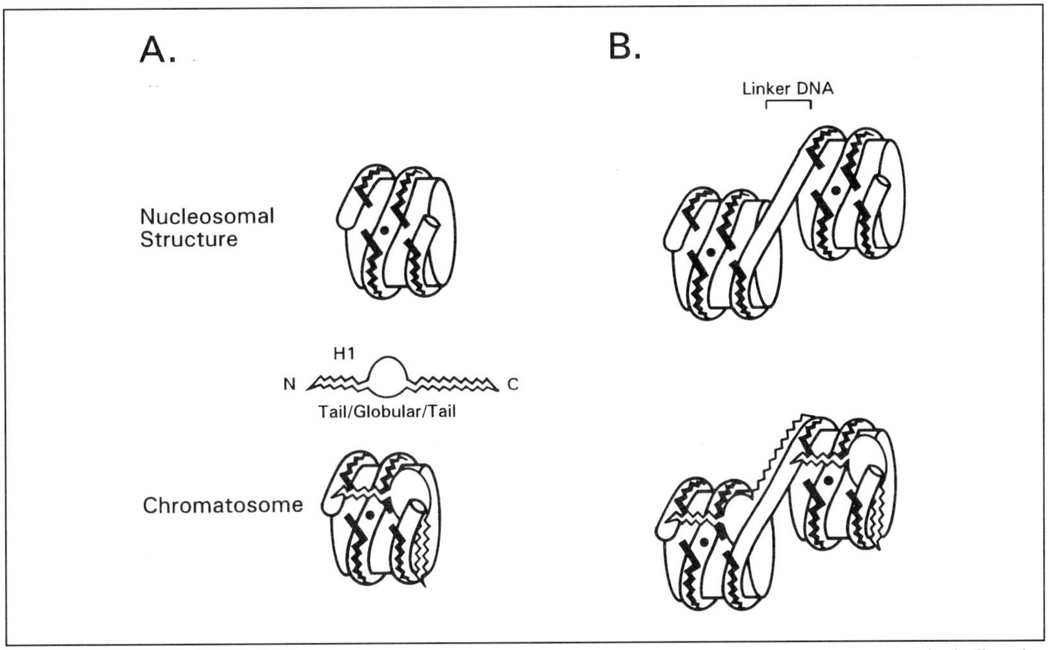

Fig. 5 Linker histone association with DNA in chromatin. A. A cartoon of the nucleosomal structure including the core histone tails (back zigzags), the linker histone H1, and a nucleosomal structure associated with H1 (chromatosome) are shown. B. A dinucleosomal structure and a dinucleosome are shown. The positions of linker DNA and of histone H1 are indicated. In each case, the axis of dyad symmetry is indicated by a filled circle.

histone H2A suggesting that protein–protein interactions may also have a role in nucleosome formation (37). It should also be noted that the true substrate for association of linker histones is a structure in which two octamers of core histones (each associated with 160 bp of DNA) are actually separated by 20–30 bp of linker DNA (see Fig. 5B, and Section 4.2, Fig.9).

The role of linker histones in chromatin structure is to stabilize the folding of nucleosomal arrays into the chromatin fibre (38). The chromatin fibre has a solenoidal structure, each turn of which requires six or more nucleosomes, including more than 1000 base pairs of DNA. The stable formation of a solenoid will rapidly render large segments of DNA invisible to other DNA-binding proteins. Chromatin fibres will form with the core histones alone; however, under these conditions they are more likely to be dynamic, folding and unfolding more frequently than in the presence of linker histones (39). The key requirement for the folding of nucleosomal arrays is charge neutralization of the phosphodiester backbone of linker DNA through interaction with basic proteins (40). The basic N-terminal tails of the core histones can interact with this linker DNA and possibly with DNA around adjacent nucleosomes to stabilize the fibre (41, 42). Thus, modifications of the N-terminal tails, such as acetylation, might also impair these interactions and thus destabilize the chromatin fibre. Acetylation of the histone tails may also reduce the association of histone H1 with DNA in individual nucleosomes (43).

2.5 Chromosomal domains

Beyond the folding of DNA into the nucleosome and of nucleosomal arrays into the chromatin fibre, a long-range organization of the chromatin fibre into domains within the chromosome occurs. Little is known about the determinants of this organization. Specialized chromatin structures exist at the centromeres and telomeres of chromosomes (44, 45). This chromatin is predominantly transcriptionally inert and is described as heterochromatin (46). An extreme example of a heterochromatin domain is the inactive X chromosome in female mammals (47). Heterochromatin, such as the inactive X chromosome, is normally late replicating during S phase, enriched in methylcytosine, and lacks acetylation of the N-terminal tail of histone H4 (47–49). What leads to the generation of these features is currently unknown.

Within the transcriptionally active chromosomal regions there also exist local domain-specific variations in chromatin composition. Immunofluorescence microscopy using specific antibodies has revealed individual chromosomal domains to contain core histones that are differentially acetylated (50) or that contain distinct types of linker histone (51). Most of these studies have employed the large polytene chromosomes of *Drosophila* and *Chironomus*. However, there is increasing evidence for local domains of differential histone acetylation or linker histone content in the normal somatic nucleus (52, 53).

Substantial evidence now suggests that the nucleus is highly organized into organelles, or factories, that are defined by particular functions such as replication, transcription, RNA processing, and export (54). The nucleolus is the classical example of a nuclear organelle devoted to the synthesis of rRNA (55); however, it is clear that similar centres (~ 100–500) exist that are the sites of mRNA synthesis (56, 57). The organization of such factories and the contribution of chromatin structure to their assembly and function is unknown.

3. Genetic evidence of a role for chromatin structure in transcriptional control

3.1 Genetic evidence for histone function in transcription in *Saccharomyces cerevisiae*

Genetic techniques have been established in yeast whereby most genes can be disrupted or replaced by exogenous DNA. The individual chromosomal copies of histone genes have been replaced with normal or mutant histone genes that are under inducible control (58–61). When core histone H4 or H2B synthesis is repressed, the yeast cells arrest in the G2 stage of the cell cycle. Associated with this arrest, the normally silent *PHO5*, *CUP1*, and *HIS3* genes become transcriptionally activated. Since histone H4 and H2B synthesis has not occurred, the chromosomal DNA is only partially assembled into nucleosomes, and it is probable that the promoter elements of these genes have become accessible to *trans*-acting factors and RNA polymerase (59–61). Thus, transcriptional activation can be directly attributed to the disruption of chromatin structure. Under normal circumstances molecular mechanisms must exist to disrupt locally repressive histone–promoter interactions.

Having established that histones are an essential component of regulated transcription in yeast, the roles of the N-terminal tail and C-terminal globular domains of the individual histones in gene regulation were determined. The N-terminal tail of histone H4 is required for the repression of the silent mating loci. Two regions within the tail involved in silencing have been identified: a very basic segment (amino acids 16–19) and an adjacent, relatively uncharged sequence (amino acids 21–29) (62–67). The N-terminal tail of H4 also has a role in gene activation, since amino acids 4 to 23 are required for activation of the GAL1 promoter and several other genes (67). In surprising contrast, the N-terminal tail of histone H3 is required for the repression of many of the same genes that require the N-terminal tail of histone H4 to be activated (68). The N-terminal tails of histones H3 and H4 are almost completely conserved between *S. cerevisae* and human, providing evolutionary evidence for their essential biological roles. The N-terminal tails of histones H2A and H2B are much less conserved, and their deletion has little effect on yeast mating (7).

Genetic analysis of the transcription process in yeast has identified several genes that have pleiotropic effects. Studies on expression of both the *HO* endonuclease gene involved in mating and the *SUC2* invertase gene involved in sucrose metabolism led to the identification of common genes required for both processes (8, 9, 69–71). The important products are encoded by *snf* (sucrose non-fermenting) *2*, *-5*, and *-6* and *SWI* (switch) *1*, *-2*, and *-3* genes, and biochemical evidence suggests that they function together in a single large 'activator' complex. Subsequently, *snf2* was shown to be identical to *SWI2*. The Swi1, -2, -3, snf5, and-6 protein complex has now been shown to be required for the expression of many inducible genes in yeast (e.g. *PHO5*, *CUP1*, and *HIS3*), and also for the activity of the glucocorticoid receptor when it is expressed in yeast (72). A role for the activator complex in modulating chromatin structure is suggested by the isolation of suppressor mutations that allow inducible gene transcription in yeast strains deficient for *SWI* gene activities. These *sin* genes (suppressors that are *SWI* independent) have roles in chromatin structure. The *sin1* gene has some similarity to the gene encoding HMG1; this is a protein that recognizes and binds to linker DNA in chromatin (73). The *sin2* gene encodes a mutant of histone H3 in the globular C-terminal domain (8). Thus *sin1* mutants might influence the folding of DNA between nucleosomes, and *sin2* mutants might influence the integrity of the nucleosome core directly. An unfolding of chromatin could potentially remove the requirement for the function of the activator complex, and suggests that the activator complex might itself mediate such unfolding. The influence of *sin1* and *sin2*, histone stoichiometry, and the N-termini of histones H3 and H4 on transcriptional activation (see Table 1) clearly demonstrates that the molecular mechanisms regulating gene expression in eukaryotes must include specific roles for the histone proteins.

Table 1 Genes involved in the global control of transcription

Activation	Repression
H4	H2B, H3, H4
SWI 1,2, 3, snf 5, 6	sin1
Brahma	HP1, Polycomb
	SIR 2, 3, 4
	N-terminal acetyl transferase
	NAT1, ARD1

3.2 Genetic evidence for chromosomal domain function in transcription

The influence of chromosomal domains on gene expression has been explored in *Drosophila* through experiments that make use of the phenomenon of position effect variegation (74). When normally active genes are placed next to transcriptionally inactive, condensed heterochromatin domains, the heterochromatin structure is believed to spread into the normally active gene, leading to its repression. Thus the expression of a gene depends on its chromosomal position, hence the term 'position effect'. This phenomenon provides a useful screen for genes and their products that will suppress or enhance the repression of transcription due to the chromosomal position of a reporter gene (75).

The most useful and informative suppressors of chromosomal position effect are those that impact on chromatin structure directly. Deletion of a substantial fraction of the histone genes will suppress position effect (76) as will acetylation of the N-terminal tails of the core histones (77). Importantly, mutation of a class of specific *trans*-acting factor-encoding genes will also prevent the transcriptional repression of reporter genes placed within heterochromatin (78–81). Among these repressor proteins is heterochromatin protein 1, HP1, which was the first of the chromodomain (chromatin modification organizer) proteins to be characterized (78). The gene for HP1 has similarity to members of a gene family known as *Polycomb* (82). This family regulates the heritable repression of particular homeotic genes (83). Since neither Polycomb nor HP1 proteins will bind directly to DNA, they are proposed to be associated with chromatin via yet to be defined protein–protein interactions. Polycomb is distributed over at least 50 different domains in polytene chromosomes, suggesting that it may influence the expression of a wide variety of genes (84). Mammalian homologues of the chromodomain proteins have now been isolated, their activity might contribute to the highly variable expression of some transgenes in mice dependent upon the chromosomal position at which they are integrated (85, 86).

Evidence for position-effect variegation in *S. cerevisiae* also demonstrates the presence of functional domains within chromosomes. Genes integrated into yeast chromosomes near the silent mating-type loci or close to the telomeres are repressed in a way that reflects their proximity to these sites in the chromosome (87,

88). This silencing effect can spread over at least 5–10 kb, but not over as much as 20–30 kb (89, 90). Thus the level of chromatin organization contributing to position effect in yeast would appear to be the chromatin fibre, each turn of which will contain approximately 1 kb of DNA. Many of the proteins defined as influencing gene expression at the silent mating loci, have also been found to be involved in position-effect variegation in yeast. These proteins include histone H4, the N-terminal acetyl transferase components NAT1 and ARD1, and the proteins SIR2, -3, and -4, which have all been previously identified as necessary for the repression of the silent mating loci in yeast (see Table 1; 91). SIR3 has been previously implicated in the regulation of chromatin structure since mutations in this gene suppress the effect of certain mutations in histone H4 (65). A separate, but equally interesting link between the chromatin-mediated repression of transcription in yeast and *Drosophila* follows from the identification of supressors of mutations in *Polycomb* gene. One such suppressor is the *Drosophila* gene *brahma*, mutations of which result in derepression of homeotic genes that would normally be silenced through the action of Polycomb (92). The *brahma* gene is closely related to *SWI2*: both genes encode large proteins of 1600–1700 amino acids, which contain extended regions of homology over more than 600 amino acids. A *Drosophila* regulatory protein, fushi tarazu, has also been found to depend on the action of Swi1 (and presumably Swi2 and -3) to activate transcription in yeast (71). Thus it is possible that a similar activator complex exists in *Drosophila* to that found in yeast, and that this complex, including the brahma protein, is required to relieve chromatin-mediated transcriptional repression.

These genetic experiments provide conclusive evidence that the proteins that assemble DNA into chromatin have essential functions in regulating the transcription process *in vivo*. Together with the structural information reviewed earlier, we now have a firm foundation with which to interpret how the transcriptional regulation of specific genes employs chromatin structure not only for the purpose of gene repression, but also for gene activation.

4. Biochemical evidence of a role for chromatin structure in transcriptional control

4.1 *In vivo* evidence for chromatin-mediated effects on transcription

The first experiment to suggest that chromatin, rather than purified DNA, was required for accurate and efficient transcription of eukaryotic genes was done with templates isolated from *Xenopus* oocytes (93). RNA polymerase III initiated transcription accurately and efficiently at pre-initiation complexes assembled on the highly reiterated *Xenopus* 5S RNA genes. The bulk of nucleosome-associated DNA did not serve as a polymerase III substrate and was thus apparently invisible to the enzyme. These experiments were extended to demonstrate that chromatin isolated from different developmental stages during *Xenopus* embryogenesis retained *in vitro* the programme of gene activity seen *in vivo* (94). As development progresses a large family of 5S RNA genes is selectively repressed (94). The basis for the repression of these genes was found to be the presence of histone H1 in chromatin (95). However, folding of nucleosomal arrays into higher order structures such as the chromatin fibre was not necessary to maintain repression (96). More recently, it has been shown that the level of histone H1 increases dramatically from very small amounts in the early *Xenopus* embryo, to normal stoichiometries in chromatin at the time that repression of the 5S RNA genes is established (97). It is worth noting that *Xenopus* oocyte chromatin does not have histone H1 associated with its nucleosomes, instead a specialized linker histone is present. The addition of histone H1 to chromatin naturally deficient in the protein will lead to the selective repression of 5S RNA genes (98–100). This selective repression depends upon the specific organization of DNA around the core histones with respect to DNA sequence, i.e. nucleosome positioning (see below) (100). Genes may be silenced even though transcription factors remain associated with them, presumably because the formation of nucleosomes prevents the effective access of RNA polymerase to the pre-initiation complex. Similar transitions in linker histone content occur during early mammalian development (101) and have been correlated with the regulation of transcriptional activity (102).

The specific organization of DNA around the core histones with respect to DNA sequence is a common feature of chromatin structure around promoters and enhancers (103). This nucleosome positioning provides a local position effect that is critical for gene regulation. Both DNA sequence and *trans*-acting factors contribute to the positioning of nucleosomes (11, 12, 30, 31, 104). Manipulation of nucleosome positioning *in vivo* can selectively prevent *trans*-acting factor access to DNA (104).

Two excellent examples of a role for positioned nucleosomes in the regulation of inducible gene expression are the induction of transcription from the mouse mammary tumour virus (MMTV) promoter by glucocorticoids, and from the yeast *PHO5* (inorganic phosphatase) promoter in response to a reduction in inorganic phosphate concentration. The MMTV promoter is organized into an array of five positioned nucleosomes in the transcriptionally inactive state. Two of these nucleosomes incorporate essential promoter elements (see Fig. 6; 105, 106). Within the promoter are binding sites for the glucocorticoid receptor. Association of the receptor with the chromatin-bound promoter in the presence of glucocorticoids leads to the appearance of a segment of DNA that is preferentially cut by DNaseI, i.e. a DNaseI hypersensitive site that correlates with the activation of transcription (107). By the use of functional MMTV promoters on replicating episomes, it has been shown that the generation of a DNaseI hypersensitive site depends on the binding of the glucocorticoid receptor to DNA wrapped around the core histones, and the subsequent displacement of histone H1 from the nucleosome (108, 109). Disruption of the nucleosome in a replication-independent process is necessary to allow the binding of NF1 to DNA (109) and the subsequent assembly of a pre-initiation complex (see Chapter 1). Similar events appear to occur when the glucocorti-

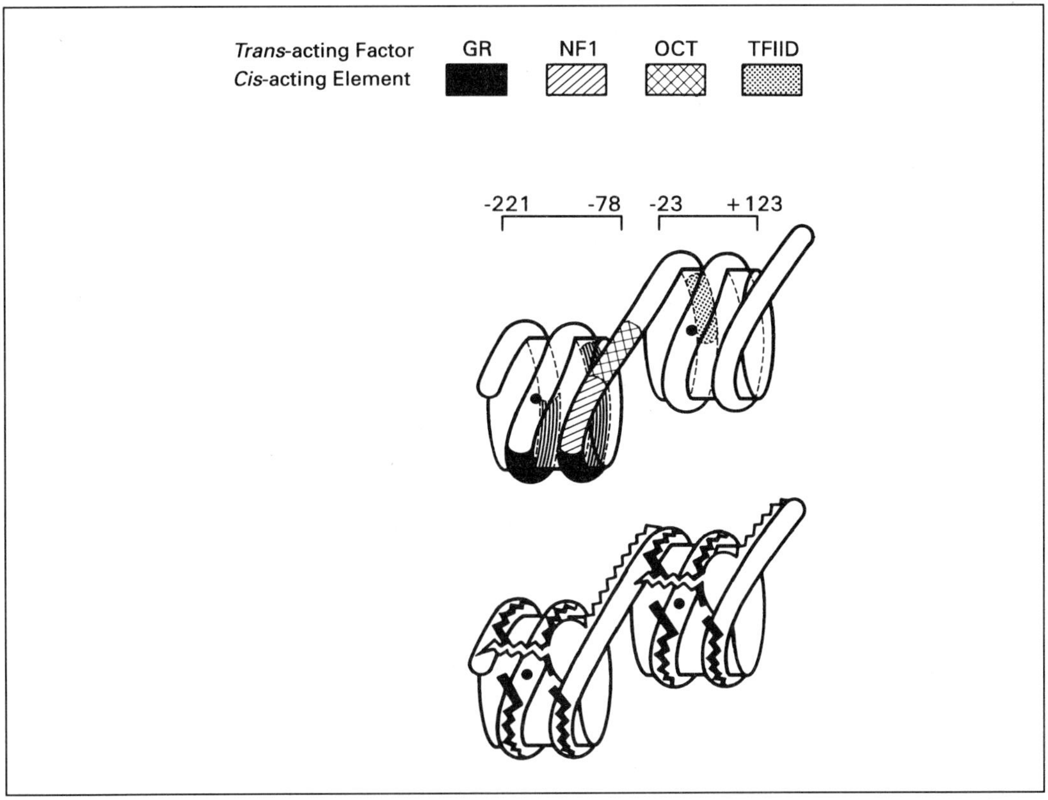

Fig. 6 Nucleosome structure of the MMTV promoter. The sites of transcription factor interaction are indicated on the surface of the core histones, together with their distance from the start site of transcription (+1). Transcription factors are: GR, glucocorticoid receptor; NF1, nuclear factor 1; OCT, octamer factor; and TFIID.

coid receptor binds to the rat tyrosine amino transferase promoter in chromatin (110).

Regulation of transcription at the *PHO5* promoter also makes use of defined chromatin structures (111–113). In this promoter, two nucleosomes are positioned to either side of an essential promoter element recognized by the transcription factor PHO4. A second binding site for PHO4 and a site for another transcription factor PHO2 are incorporated into one of the positioned nucleosomes. On induction all four of the positioned nucleosomes are disrupted. Analysis of mutants in the *trans*-acting factors reveals that both PHO2 and PHO4 are essential for both the transcriptional activation of the *PHO5* gene and the rearrangement of chromatin structure. These changes are believed to be initiated by the binding of PHO4 to regulatory elements between the two central nucleosomes. Changes in DNA sequence in the adjacent nucleosomes can influence transcriptional activation and chromatin rearrangement, suggesting that the chromosomal organization of the whole promoter region is essential for correct regulation (113, 114). Recent experiments have demonstrated that neither transcription nor replication is required for the disrup-

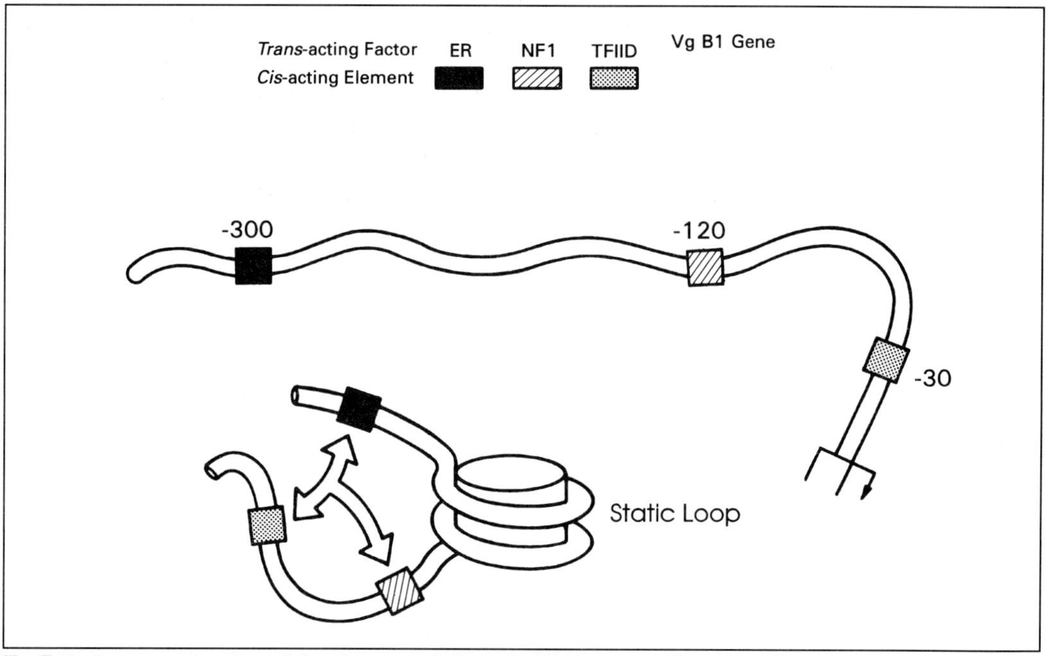

Fig. 7 Nucleosome structure of the *Xenopus* vitellogenin B1 promoter. The sites of transcription factor interactions are indicated on the surface of the core histones, together with their distance from the start site of transcription (+1, hooked arrow). The open arrow indicates possible interactions between transcription factors. Transcription factors are: ER, oestrogen receptor; NF1, nuclear factor 1; and TFIID.

tion of the positioned nucleosomes (115, 116). The exact mechanism may depend upon the interaction of activators with the N-terminus of histone H4 (67), histone acetylation (27), or depletion of histones from the nucleosome (117; see below).

Positioned nucleosomes have also been found in contexts where they might facilitate rather than repress the transcription process. In both the *Drosophila hsp26* and alcohol dehydrogenase gene promoters, positioned nucleosomes fold DNA into a three-dimensional structure in which widely separated regulatory elements are brought into juxtaposition (118, 119). Although a functional role for these nucleosomes is yet to be found, reconstruction of a similar positioned nucleosome *in vitro* between an enhancer and a promoter in the *Xenopus* vitellogenin gene facilitates enhancer activity approximately 10-fold. In addition, transcriptional repression is avoided, since inappropriate inhibitory histone–DNA contacts are prevented from occurring over the promoter when the nucleosome is positioned in this way (see Fig. 7; 26). *In vivo* data also support chromatin-mediated communication between distant regulatory elements (120). It is therefore established that the histones together with other chromatin structural proteins will have an important regulatory role in creating the correct scaffolding for transcriptional activation (121, 122).

In summary, *in vivo* experiments show that transcription factors and chromatin structure at the nucleosomal level work together to effect gene regulation. The type of histone, the modification of the histone, and the exact positioning of histone–DNA interactions can all have important regulatory consequences for the transcription process. These experiments that map the changes in the organization of chromatin at the nucleosomal level during the process of activating transcription provide the structural basis for those biochemists who seek to reconstruct gene regulation *in vitro* using defined components. The *in vitro* experiments should also be guided by the existing information concerning chromatin structure and the genetic information defining the major players involved in global gene activation.

4.2 *In vitro* reconstruction of chromatin-mediated effects on transcription

In vitro experiments on chromatin and transcription are severely limited at this time by the difficulty in correctly assembling chromatin (see 123, 124 for discussion). *In vivo* chromatin assembly is coupled to the replication process (25, 125). This coupling means that nucleosome assembly on a replicating template is much more repressive towards basal transcription than on a non-replicating template on which chromatin assembly is slow (25). These *in vivo* observations can be reproduced *in vitro* (24, 126–128). An important conclusion concerning the access of *trans*-acting factors to DNA during the replication process is that the nucleosome is assembled from the inside out (see Fig. 8): first, the $(H3/H4)_2$ tetramer (in which histone H4 is acetylated) is deposited, then histones H2A/H2B bind in a process taking several minutes, and finally histone H1 (or another linker histone) is incorporated. This latter process correlates with removal of acetyl groups from the core histone tails. The 5S RNA gene promoter is much more accessible to transcription factors during the early stages of chromatin assembly than in the later stages (27, 30, 126–128). Since replication disrupts pre-existing nucleoprotein complexes (129, 130), the staged assembly of chromatin following replication provides a window of opportunity for the reprogramming of genes. Nevertheless, if transcription factors do not bind rapidly and tightly to promoter elements a state of repression will be established. The class III gene basal transcriptional machinery has the capacity to programme certain genes at the replication fork in the face of chromatin assembly.

Most *in vitro* reconstruction experiments examining transcriptional regulation have not made use of replicating templates, or of histones that have the same modification state of newly assembled histones found *in vivo*, or of histones that generate correctly spaced nucleosomes, or of nucleosomes at all. However, it should be clearly stated that any form of histone–DNA complex, even that imperfectly assembled, is a closer approximation to the *in vivo* situation within the chromosome than no histones being present at all. Several useful conclusions can be drawn from existing work concerning both transcriptional activation and chromatin structure. Pre-existing histone–DNA interactions over a promoter can impede some transcription factors, such as TFIIIA and the heatshock transcription factor from access to DNA (117, 131), much more than others, such as the glucocorticoid receptor and GAL4 (131–135). The degree of inhibition depends on the exact position of the *trans*-acting factor binding site within the nucleosome for both the glucocorticoid receptor and

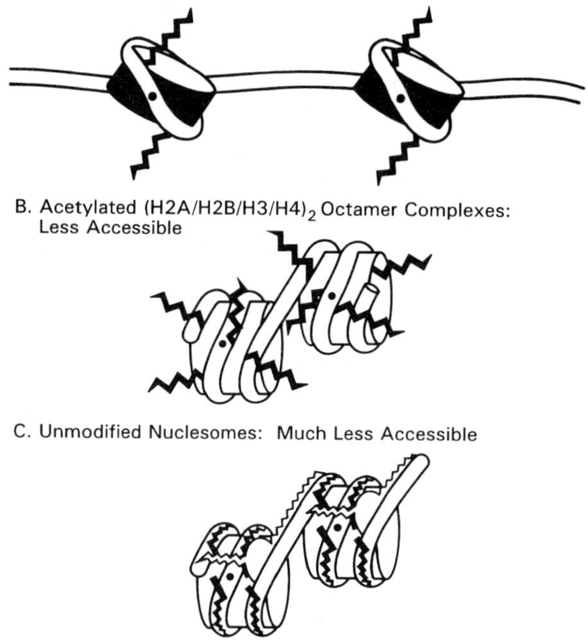

Fig. 8 Staged assembly of chromatin. A. Immediately following the synthesis of nascent DNA, acetylated histone $(H3/H4)_2$ tetramer complexes wrap DNA; B., in a time-dependent process (5–10 min) histones H2A/H2B are sequestered; and C., as histones are deacetylated the linker histones are bound. The accessibility of DNA to transcription factors is progressively reduced as chromatin matures.

GAL4. In most binding studies involving either of these proteins, multiple sites are usually present within a short DNA segment. Nevertheless, *in vitro*, the glucocorticoid receptor and GAL4 under certain circumstances can bind effectively to a nucleosomal template. This is the necessary first step in nucleosome disruption. The later steps have yet to be reconstructed, but presumably involve the activator complex (i.e. Swi1, -2, -3, snf5, and -6). Several experiments have documented the activation of transcription in crude extracts by GAL4-VP16 *in vitro* from chromatin templates: the activator works very well using templates in which histone H1 is bound to naked DNA (18), less well with templates in which incorrectly spaced nucleosomes are present (20), and poorly on templates in which chromatin assembly is coupled to replication and in which somatic histone H1 is presumably correctly bound (24). These results reflect the relative affinity with which histone H1 interacts with DNA and chromatin templates (see Fig. 9; 15). They also serve to emphasize the important role that both nucleosome and chromatin fibre structure might play in gene regulation *in vivo*. An important element of this regulation is the repression of basal transcription in the absence of gene-specific transcriptional activators such as GAL4-VP16. Under standard *in vitro* transcription conditions, cor-

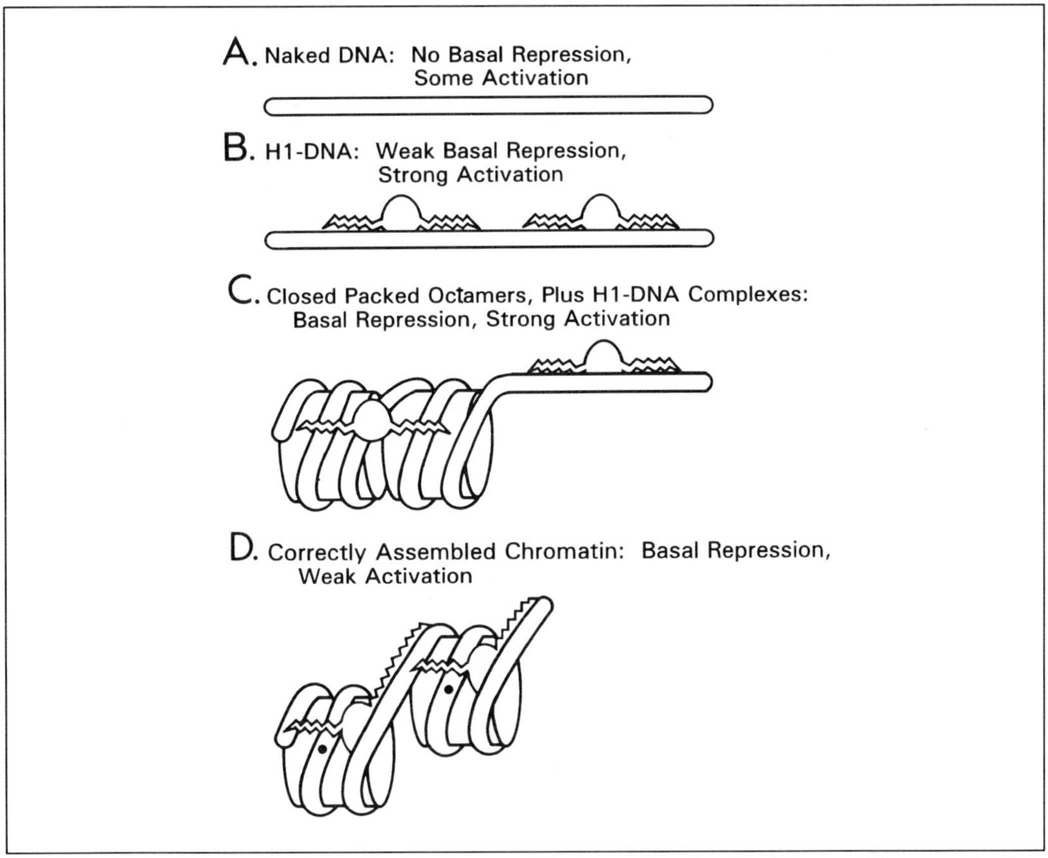

Fig. 9 The influence of chromatin structure reconstituted *in vitro* on basal transcription, and the capacity of an activator (GAL4-VP16) to stimulate the transcription process. A. Naked DNA is completely accessible to the basal transcription machinery and the only function of the activator is to stimulate recruitment and utilization of the basal transcriptional machinery by RNA polymerase. B. Histone H1–DNA complexes prevent the basal transcriptional machinery gaining access to DNA, the activator serves as an anti-repressor and easily overcomes the weak repression by H1–DNA to strongly activate transcription. C. Similar results to B are obtained with a mixture of closed-packed histone octamers (< 160 bp per histone octamer) and H1–DNA complexes. It is probable that the DNA associated with octamers cannot be transcribed. D. Correctly assembled chromatin containing H1 and linker DNA is very repressive to transcriptional activators. *In vivo* nucleosome positioning is used to maintain regulatory regions free from histone DNA contacts.

rectly reconstituted nucleosomal arrays will be folded into structures resembling the chromatin fibre, formation of such structures will also impair the transcription process (136). An important conclusion is that the final transcriptional activity will depend on the rate and stability with which the basal transcriptional machinery associates with chromatin (127, 137) and also on the rate of assembly and composition of chromatin (18–25, 27, 126-128).

It might be anticipated that future experiments will increasingly attempt to reconstruct the changes in nucleoprotein structure observed *in vivo*; in these studies precisely positioned nucleosomes will necessarily have a prominent role.

5. Conclusion

The regulation of transcription within the eukaryotic nucleus relies upon the histone proteins in several essential ways. The packaging of DNA into chromatin structures renders large segments of the genome invisible to *trans*-acting factors or polymerase, such that each regulatory complex that does form is clearly visible within the chromatin fibre to other regulatory molecules. At the promoter, incorporation of the binding sites for the basal transcriptional machinery into nucleosomes serves to repress transcription. Specific regulatory molecules, other than the basal transcriptional machinery, exist that can associate with nucleosomal DNA and initiate a chain of events to remove repressive nucleosomal structures. Many components in this pathway have been defined genetically and include the histones, the histone N-terminal tails, and a large activator complex. How they operate biochemically is, at this time, yet to be defined. The genetic data demonstrate that the disruption of histone–DNA complexes will play a major role in the induction of transcription from many genes. However, not all genes will be repressed by nucleosome assembly, certain promoters make use of the staged assembly of chromatin *in vivo* and a rapid and tight association of *trans*-acting factors with promoter elements to remain constitutively active. Moreover, nucleosome assembly is not necessarily repressive, and the folding of DNA by the histones can facilitate the activation of genes by bringing widely separated regulatory elements into juxtaposition.

The specific effects of chromatin structure rely on selective histone–DNA interactions. These are established by secondary structural features of DNA and by specific *trans*-acting factor histone interactions. In fact, it is remarkable to note the similar structural motifs adopted by *trans*-acting factors to those found in the histones. These similarities further suggest an intimate structural and functional interrelationship between chromatin structure and transcription factor function (138, 139). Although it is possible to describe mechanisms for the consequences of local nucleosome positioning with respect to DNA sequence, it is difficult at this time to attribute molecular mechanisms that explain chromosomal position effects. Genetic evidence clearly implicates chromosomal structure; however, how this impacts on gene function in a chromosomal context is not known. Possible mechanisms might include the presence of proteins (e.g. histone deacetylases) that prevent nucleosome disruption, the presence of proteins that stabilize the chromatin fibre, and the sequestration of a particular domain of chromatin in a transcriptionally inert or active region of the nucleus.

In summary, research on the influence of chromosomal structure on gene expression has undergone a major revolution. The long-standing dogma that histones merely package DNA away from any significant activity is seen to be as much of an over-simplification as that of gene regulation occurring on naked DNA. Understanding how both transcription factors and histones interact to regulate transcription presents molecular biologists with important questions, the answers to which will have general applicability for other processes such as DNA replication, recombination, and repair. The future offers considerable promise for reconstructing, and thus understanding, the correct regulation of genes in a natural chromosomal context.

Acknowledgements

I thank Ms Thuy Vo for the preparation of the manuscript.

References

1. Wolffe, A. P. (1992) *Chromatin structure and function*. Academic Press, London.
2. Wolffe, A. P. and Dimitrov, S. (1993) Histone modulated gene activity: developmental implications. *Crit. Rev. Euk. Gene Exp.*, **3**, 167.
3. Croston, G. E. and Kadonaga, J. T. (1993) Role of chromatin structure in the regulation of transcription by RNA polymerase II. *Curr. Opin. Cell Biol.*, **5**, 417.
4. Svaren, J. and Horz, W. (1993) Histones, nucleosomes and transcription. *Curr. Opin. Genet. Dev.*, **3**, 219.
5. Adams, C. C. and Workman, J. L. (1993) Nucleosome displacement in transcription. *Cell*, **72**, 1.
6. Workman, J. L. and Buchman, A. R. (1993) Multiple functions of nucleosomes and regulatory factors in transcription. *Trends Biochem. Sci.*, **18**, 90.
7. Grunstein, M. (1990) Histone function in transcription. *Ann Rev. Cell Biol.*, **6**, 643.
8. Herskowitz, I., Andrews, B., Kruger, W., Ogas, J., Sil, A., Coburn, C., and Peterson, C. (1992) Integration of multiple regulatory inputs in the control of HO expression in yeast. In *Transcriptional regulation* (eds S. McKnight and K. Yamamoto), Vol. 2, p. 949. Cold Spring Harbor Press, New York.
9. Winston, F. and Carlson, M. (1992). Yeast SNF/SWI transcriptional activators and the SPT/SIN chromatin connection. *Trends Genet.*, **8**, 387.
10. Hirschhorn, J. N., Brown, S. A., Clark, C. D., and Winston, F. (1992) Evidence that SNF2/SWI2 and SNF activate transcription in yeast by altering chromatin structure. *Genes Dev.*, **6**, 2288.
11. Shimizu, M., Roth, S. Y., Szent-Gyorgi, C., and Simpson, R. T. (1991) Nucleosomes are positioned with base pair precision adjacent to the alpha 2 operator in Saccharomyces cerevisiae. *EMBO J.*, **10**, 3033.
12. Roth, S. Y., Shimizu, M., Johnson, L., Grunstein, M., and Simpson, R. T. (1992) Stable nucleosome positioning and complete repression by the yeast α2 repressor are disrupted by amino-terminal mutation in histone H4. *Genes Dev.*, **6**, 411.
13. van Holde, K. E. (1988) Chromatin. In *Springer series in molecular biology* (ed. A. Rich). Springer Verlag, Berlin.
14. Hayes, J. J., Clark, D. J., and Wolffe, A. P. (1991) Histone contributions to the structure of DNA in the nucleosome. *Proc. Natl Acad. Sci. USA*, **88**, 6829.
15. Hayes, J. J. and Wolffe, A. P. (1993) Preferential and asymmetric interaction of linker histones with 5S DNA in the nucleosome. *Proc. Natl Acad. Sci. USA*, **90**, 6415.
16. Arents, G., Burlingame, R. W., Wang, B. W., Love, W. E., and Moudrianakis, E. N. (1991) The nucleosomal core histone octamer at 3.1Å resolution: a tripartite protein assembly and a left-handed superhelix. *Proc. Natl Acad. Sci. USA*, **88**, 10148.
17. Pruss, D. and Wolffe, A. P. (1993) Histone-DNA contacts in a nucleosome core containing a *Xenopus* 5S rRNA gene. *Biochemistry*, **32**, 6810.

18. Croston, G. E., Kenigan, K. A., Lira, L. M., Marshak, D. R., and Kadonaga, J. T. (1991) Sequence specific antirepression of histone H1-mediated inhibition of basal RNA polymerase II transcription. *Science*, **251**, 643.
19. Workman, J. L., Taylor, I. C. A., and Kingston, R. E. (1991) Activation domains of stably bound Gal4 derivatives alleviate repression of promoters by nucleosomes. *Cell*, **64**, 533.
20. Laybourn, P. J. and Kadonaga, J. T. (1991). Role of nucleosome cores and histone H1 in regulation of transcription by RNA polymerase II. *Science*, **254**, 238.
21. Lorch, Y., La Pointe, J. W., and Kornberg, R. D. (1992) Initiation on chromatin templates in a yeast RNA polymerase II transcription system. *Genes Dev.*, **6**, 2282.
22. Croston, G. E., Laybourn, P. J., Paranjape, S. M., and Kadonaga, J. T. (1992). Mechanism of transcriptional antirepression by Gal4-VP16. *Genes Dev.*, **6**, 2270.
23. Laybourn, P. J. and Kadonaga, J. T. (1992) Threshold phenomena and long distance activation of transcription by RNA polymerase II. *Science*, **257**, 1682.
24. Kamakaka, R. T., Bulger, M., and Kadonaga, J. T. (1993) Potentiation of RNA polymerase II transcription by Gal4-VP16 during but not after DNA replication and chromatin assembly. *Genes Dev.*, **7**, 1779.
25. Almouzni, G. and Wolffe, A. P. (1993) Replication coupled chromatin assembly is required for the repression of basal transcription *in vivo*. *Genes Dev.*, **7**, 2033.
26. Schild, C., Claret, F-X., Wahli, W., and Wolffe, A. P. (1993) A nucleosome-dependent static loop potentiates estrogen-regulated transcription from the *Xenopus* vitellogenin B1 promoter *in vitro*. *EMBO J.*, **12**, 423.
27. Lee, D. Y., Hayes, J. J., Pruss, D., and Wolffe, A. P. (1993) A positive role for histone acetylation in transcription factor binding to nucleosomal DNA. *Cell*, **72**, 73.
28. Richmond, T. J., Finch, J. T., Rushton, B., Rhodes, D., and Klug, A. (1984) Structure of the nucleosome core particle at 7Å resolution. *Nature*, **311**, 532.
29. Eickbusch, T. H., and Moudrianakis, E. N. (1978) The histone core complex: an octamer assembled by two sets of protein-protein interactions. *Biochemistry*, **17**, 4955.
30. Hayes, J. J., Tullius, T. D., and Wolffe, A. P. (1990) The structure of DNA in a nucleosome. *Proc. Natl Acad. Sci. USA*, **87**, 7405.
31. Hayes, J. J., Bashkin, J., Tullius, T. D., and Wolffe, A. P. (1991) The histone core exerts a dominant constraint on the structure of DNA in a nucleosome. *Biochemistry*, **30**, 8434.
32. Cary, P. D., Moss, T., and Bradbury, E. M. (1978) High resolution proton magnetic-resonance studies of chromatin core particles. *Eur. J. Biochem.*, **89**, 475.
33. Walker, I. O. (1984) Differential dissociation of histone tails from core chromatin. *Biochemistry*, 23, 5622.
34. Lambert, S. F. and Thomas, J. O. (1986) Lysine-containing DNA-binding regions on the surface of the histone octamer in the nucleosome core particle. *Eur. J. Biochem.*, **160**, 191.
35. Cary, P. D., Crane-Robinson, C., Bradbury, E. M., and Dixon, G. H. (1982) Effect of acetylation on the binding of N-terminal peptides of histone H4 to DNA. *Eur. J. Biochem.*, **127**, 137.
36. Singer, D. S. and Singer, M. F. (1976) Studies on the interaction of histone H1 with superhelical DNA: characterization of the recognition and binding regions of H1 histone. *Nucl. Acids Res.*, **3**, 2531.
37. Boulikas, T., Wiseman, J. M., and Garrard, W. T. (1980) Points of contact between histone H1 and the histone octamer. *Proc. Natl Acad. Sci. USA*, **77**, 127.
38. Widom, J. (1989) Toward a unified model of chromatin folding. *Ann. Rev. Biophys. Chem.*, **18**, 365.

39. Hansen, J. C. and Ausio, J. (1992) Chromatin dynamics and the modulation of genetic activity. *Trends Biochem. Sci.*, **17**, 187.
40. Clark, D. J. and Kimura, T. (1990) Electrostatic mechanism of chromatin folding. *J. Mol. Biol.*, **211**, 883.
41. Hill, C. S. and Thomas, J. O. (1990) Core histone-DNA interactions in sea urchin sperm chromatin: the N-terminal tail of H2B interacts with linker DNA. *Eur. J. Biochem.*, **187**, 145.
42. Bavykin, S. G., Usachenko, S. I., Zalensky, A. O., and Mirzabekov, A. D. (1990) Structure of nucleosomes and organization of internucleosomal DNA in chromatin. *J. Mol. Biol.*, **212**, 495.
43. Perry, C. A. and Annunziato, A. T. (1991) Histone acetylation reduces H1-mediated nucleosome interactions during chromatin assembly. *Exp. Cell. Res.*, **196**, 337.
44. Makavov, V. L., Lejnine, S., Bedoyan, J., and Langmore, J. P. (1993) Nucleosomal organization of telomere-specific chromatin in rat. *Cell*, **73**, 775.
45. Pluta, A. F., Cooke, C. A., and Earnshaw, W. C. (1990) Structure of the human centromere at metaphase. *Trends Biochem. Sci.*, **15**, 181.
46 Pardue, M. L. and Hennig, W. (1990) Heterochromatin: junk or collectors item? *Chromosoma*, **100**, 3.
47. Jeppesen, P. and Turner, B. M. (1993) The inactive X chromosome in female mammals is distinguished by a lack of histone H4 acetylation, a cytogenetic marker for gene expression. *Cell*, **74**, 281.
48. Morishima, A., Grumbach, M. M., and Taylor, J. H. (1962) Asynchronous duplication of human chromosomes and the origin of sex chromatin. *Proc. Natl Acad. Sci. USA*, **48**, 756.
49. Grant, S. G. and Chapman, V. M. (1988) Mechanisms of X-chromosome regulation. *Ann. Rev. Genet.*, 22, 199.
50. Turner, B. M., Birley, A. J., and Lavender, J. (1992) Histone H4 isoforms acetylated at specific lysine residues define individual chromosomes and chromatin domains in *Drosophila* polytene nuclei. *Cell*, **69**, 375.
51. Mohr, E., Trieschman, L., and Grossbach, U. (1989) Histone H1 in two subspecies of *Chironomus thummi* with different genome sizes: homologous sites differ largely in their content of a specific H1 variant. *Proc. Natl Acad. Sci. USA*, **86**, 9308.
52. Hebbes, T. R., Thome, A. W., and Crane-Robinson, C. (1988) A direct link between core histone acetylation and transcriptionally active chromatin. *EMBO J.*, **7**, 1395.
53. Kamakaka, R. T. and Thomas, J. O. (1990) Chromatin structure of transcriptionally competent and repressed genes. *EMBO J.*, **9**, 3997.
54. Jackson, D. A. (1991) Structure function relationships in eukaryotic nuclei. *BioEssays*, **13**, 1.
55. Scheer, U. and Benavente, R. (1990) Functional and dynamic aspects of the mammalian nucleolus. *BioEssays*, **12**, 14.
56. Jackson, D. A., Hassan, A. B., Errington, R. J., and Cook, P. R. (1993) Visualization of focal sites of transcription within human nuclei. *EMBO J.*, **12**, 1059.
57. Wansink, D. G., Schul, W., Kraan, I., Steensel, B., Driel, R., and de Jong, L. (1993) Fluorescent labeling of nascent RNA reveals transcription by RNA polymerase II in domains scattered throughout the nucleus. *J. Cell. Biol.*, **122**, 283.
58. Han, M., Chang, M., Kim, U.-J., and Grunstein, M. (1987). Histone H2B repression causes cell-cycle-specific arrest in yeast: effects on chromosomal segregation, replication and transcription. *Cell*, **48**, 589.
59. Han, M. and Grunstein, M. (1988) Nucleosome loss activates yeast downstream promoters *in vivo*. *Cell*, **55**, 1137.

60. Kim, U., Han, M., Kayne, P., and Grunstein, M. (1988) Effects of histone H4 depletion on the cell cycle and transcription of *Sacchromyces cerevisiae. EMBO J.*, **7**, 2211.
61. Durrin, L. K., Mann, R. K., and Grunstein, M. (1992) Nucleosome loss activates CUP1 and HIS3 promoters to fully induced levels in the yeast *Sacchromyces cerevisiae. Mol. Cell Biol.*, **12**, 1621.
62. Kayne, P. S., Kim, U. J., Han, M., Mullen, J. R., Yoshizaki, F., and Grunstein, M. (1988) Extremely conserved histone H4 N-terminus is dispensible for growth but essential for repressing the silent mating loci in yeast. *Cell*, **55**, 27.
63. Park, E. C. and Szostak, J. W. (1990) Point mutations in the yeast histone H4 gene prevent silencing of the silent mating locus HML. *Mol. Cell Biol.*, **10**, 4932.
64. Megee, P. C., Morgan, B. A., Mittman, B. A., and Smith, M. M. (1990) Genetic analysis of histone H4: essential role of lysines subject to acetylation. *Science*, **247**, 4932.
65. Johnson, L. M., Kayne, P. S., Kahn, E. S., and Grunstein, M. (1990) Genetic evidence for an interaction between SIR3 and histone H4 in the repression of silent mating loci in Saccharomyces cerevisiae. *Proc. Natl Acad. Sci. USA*, **87**, 6286.
66. Johnson, L. M., Fisher-Adams, G., and Grunstein, M. (1992) Identification of a non-basic domain in the histone H4 N-terminus required for repression of the yeast silent mating loci. *EMBO J.*, **11**, 2201.
67. Durrin, L. K., Mann, R. K., Kayne, P. S., and Grunstein, M. (1991) Yeast histone H4 N-terminal sequence is required for promoter activation *in vivo. Cell*, **65**, 1023.
68. Mann, R. K. and Grunstein, M. (1992) Histone H3 N-terminal mutations allow hyperactivation of the yeast GAL1 gene *in vivo. EMBO J.*, **11**, 3297.
69. Abrams, E., Neigeborn, L., and Carlson, M. (1986) Molecular analysis of SNF2 and SNF5, genes required for expression of glucose repressible genes in *Saccharomyces cerevisiae. Mol. Cell Biol.*, **6**, 3643.
70. Laurent, B. C., Treitel, M. A., Carlson, M. (1991) Functional interdependence of the yeast SNF2, SNF5, and SNF6 proteins in transcriptional activation. *Proc. Natl Acad. Sci. USA*, **88**, 2687.
71. Peterson, C. L. and Herskowitz, I. (1992) Characterization of the yeast SWI1, SWI2 and SWI3 genes which encode a global activator of transcription. *Cell*, **68**, 573.
72. Yoshinaga, S. K., Peterson, S. L., Herskowitz, I., and Yamamoto, K. R. (1992) Roles of SWI1, SWI2 and SWI3 proteins for transcriptional enhancement by steroid receptors. *Science*, **258**, 1598.
73. Kruger, W. and Herskowitz, I. (1991) A negative regulator of HO transcription, SIN1 (SPT2), is a non specific DNA binding protein related to HMG1. *Mol. Cell. Biol.*, **11**, 4135.
74. Eissenberg, J. C. (1989) Position effect variegation in *Drosophila*: towards a genetics of chromatin assembly. *BioEssays*, **11**, 14.
75. Schaffer, C. D., Wallrath, L. L., and Elgin, S. C. R. (1993) Regulating genes by packaging domains: bits of heterochromatin in euchromatin. *Trends Genet.*, **9**, 35.
76. Moore, G. D., Sinclair, D. A., and Grigliatti, T. A. (1983) Histone gene multiplicity and position-effect variegation in *Drosophila melanogaster. Genetics*, **101**, 327.
77. Mottus, R., Reeves, R., and Grigliatti, T. A. (1980). Butyrate suppression of position effect variegation in *Drosophila melanogaster. Mol. Gen. Genet.*, **178**, 465.
78. James, T. C. and Elgin, S. C. (1986) Identification of a non histone chromosomal protein associated with heterochromatin in *Drosophila melanogaster* in its gene. *Mol. Cell. Biol.*, **6**, 3862.
79. Garzino, V., Pereira, A., Laurenti, P., Graba, Y., Levis, R.W., Le Parco, Y., and Pradel, J.

(1992) Cell lineage-specific expression of *modulo*, a dose dependent modifier of variegation in *Drosophila*. *EMBO J.*, **11**, 4471.

80. Reuter, G., Giarre, M., Farah, J., Gausz, J., Spieror, A., and Spieror, P. (1990) Dependence of position effect variegation in *Drosophila* on dose of a gene encoding an unusual zinc finger protein. *Nature*, **334**, 219.
81. Eissenberg, J. C., James, T. C., Foster, H. D. M., Hartnett, T., Ngan, V., and Elgin, S. C. (1990) Mutation in a heterochromatin-specific chromosomal protein is associated with suppression of position effect variegation in *Drosophila melanogaster*. *Proc. Natl Acad. Sci. USA*, **87**, 9923.
82. Paro, R. and Hogness, D. S. (1991) The Polycomb protein shares a homologous domain with a heterochomatin associated protein of *Drosophila*. *Proc. Natl Acad. Sci. USA*, **88**, 263.
83. Jurgens, G. (1985) A group of genes controlling the spatial expression of the bithorax complex in *Drosophila*. *Nature*, **316**, 153.
84. Zink, B. and Paro, R. (1989) *In vivo* binding pattern of a *trans* regulator of homeotic genes in *Drosophila melanogaster*. *Nature*, **337**, 468.
85. Singh, P. B., Miller, J. R., Pearce, J., Rothany, R., Burton, R. D., Paro, R., James, T. C., and Gaunt, S. J. (1991) A sequence motif found in a *Drosophila* heterochromatin protein is conserved in animals and plants. *Nucl. Acids Res.*, **19**, 789.
86 Grosveld, F., Blom van Assendelft, G., Greaves, D., and Kollias, G. (1987) Position-independent, high level expression of the human β-globin gene in transgenic mice. *Cell*, **51**, 975.
87. Laurensen, P. and Rine, J. (1992) Silencers, silencing and heritable transcriptional states. *Microb. Rev.*, **56**, 543.
88. Sandell, L. E. and Zakian, V. A. (1992) Telomeric position effect in yeast. *Trends Cell Biol.*, **2**, 10.
89. Gottschling, D. E., Aparicio, O. M., Billington, B. L., and Zakian, V. A. (1990) Position effect at *S. cerevisiae* telomeres: reversible repression of pol II transcription. *Cell*, **63**, 751.
90. Mastrangelo, M. F., Weinstock, K. G., Shafer, B. K., Hedge, A. M., Garfinkel, D. J., and Strathern, J. N. (1992) Disruption of a silencer domain by a retrotransposon. *Genetics*, **131**, 519.
91. Aparicio, O. M., Billington, B. L., and Gottschling, D. E. (1991) Modifiers of position effect are shared between telomeric and silent mating-type loci in *S. cerevisae*. *Cell*, **66**, 1279.
92. Tamkun, J. W., Deuring, R., Scott, M. P., Kissinger, M., Patlatucci, A. M., Kaufman, T. C., and Kennison, J. A. (1992) Brahma: a regulator of *Drosophila* homeotic genes structurally related to the yeast transcriptional activator SNF2/SWI2. *Cell*, **68**, 561.
93. Parker, C. S. and Roeder, R. G. (1977) Selective and accurate transcription of the *Xenopus* 5S RNA genes in isolated chromatin by purified RNA polymerase III. *Proc. Natl Acad. Sci. USA*, **74**, 44.
94. Wormington, W. M. and Brown, D. D. (1983) On set of 5S RNA gene regulation during *Xenopus* embryogenesis. *Develop. Biol.*, **99**, 248.
95. Schlissel, M. S. and Brown, D. D. (1984) The transcriptional regulation of *Xenopus* 5S RNA genes in chromatin: the roles of active stable transcription complexes and histone H1. *Cell*, **37**, 903.
96. Gurdon, J. B., Dingwall, C., Laskey, R. A., and Korn, L. J. (1982) Developmental inactivity of 5S RNA genes persists when chromosomes are cut between genes. *Nature*, **299**, 652.

97. Dimitrov, S., Almouzni, G., Dasso, M., and Wolffe, A. P. (1993) Chromatin transitions during early *Xenopus* embryogenesis: changes in histone H4 acetylation and in linker histone type. *Develop. Biol.*, **160**, 214.
98. Wolffe, A. P. (1989) Transcriptional activation of *Xenopus* class III genes in chromatin isolated from sperm and somatic nuclei. *Nucl. Acids Res.*, **17**, 767.
99. Wolffe, A. P. (1989) Dominant and specific repression of *Xenopus* oocyte 5S RNA genes and satellite I DNA by histone H1. *EMBO J.*, **8**, 527.
100. Chipev, C. C. and Wolffe, A. P. (1992) Chromosomal organization of *Xenopus laevis* oocyte and somatic 5S rRNA genes *in vivo*. *Mol. Cell. Biol.*, **12**, 45.
101. Clark, H. J., Oblin, C., and Bustin, M. (1992) Developmental regulation of chromatin composition during mouse embryogenesis: somatic histone H1 is first detectable at the 4-cell stage. *Development*, **115**, 791.
102. Majumder, D., Miranda, M., and De Pamphilis, M. (1993) Analysis of gene expression in mouse preimplantation embryos demonstrates that the primary role of enhancers is to relieve repression of promoters. *EMBO J.*, **12**, 1131.
103. Simpson, R. T. (1991) Nucleosome positioning: occurrence, mechanisms and functional consequences. *Prog.Nucl. Acids Res. and Mol. Biol.*, **40**, 143.
104. Simpson, R. T. (1990) Nucleosome positioning can affect the function of a *cis*-acting DNA element *in vivo*. *Nature*, **343**, 387.
105. Richard-Foy, H. and Hager, G. L. (1987) Sequence-specific positioning of nucleosomes over the steroid-inducible MMTV promoter. *EMBO J.*, **6**, 2321.
106. Cordingley, M. G., Riegel, A. T., and Hager, G. L. (1987) Steroid dependent interaction of transcription factors with the inducible promoter of the mouse mammary tumor virus *in vivo*. *Cell*, **48**, 271.
107. Zaret, K. S. and Yamamoto, K. R. (1984). Reversible and persistent changes in chromatin structure accompany activation of a glucocorticoid dependent enhancer element. *Cell*, **38**, 29.
108. Bresnick, E. H., Bustin, M., Marsaud, V., Richard-Foy, H., and Hager, G. L. (1992) The transcriptionally-active MMTV promoter is depleted of histone H1. *Nucl. Acids Res.*, **20**, 273.
109. Archer, T. K., Lefebvre, P., Wolford, R. G., and Hager, G. L. (1992) Transcription factor loading on the MMTV promoter: a bimodal mechanism for promoter activation. *Science*, **255**, 1573.
110. Reik, A., Schutz, G., and Stewart, A. F. (1991) Glucocorticoids are required for establishment and maintenance of an alteration in chromatin structure: induction leads to a reversible disruption of nucleosomes over an enhancer. *EMBO J.*, **10**, 2569.
111. Almer, A. and Horz, W. (1986) Nuclease hypersensitive regions with adjacent positioned nucleosomes mark the gene boundaries of the PHO5/PHO3 locus in yeast. *EMBO J.*, **5**, 2681.
112. Almer, A., Rudolph, H., Hinnen, A., and Horz, W. (1986) Removal of positioned nucleosomes from the yeast PHO5 promoter upon PHO5 induction releases additional activating DNA elements. *EMBO J.*, **5**, 2689.
113. Fascher, K. D., Schmitz, J., and Horz, W. (1990). Role of *trans*-activating proteins in the generation of active chromatin at the PHO5 promoter in *S. cerevisiae*. *EMBO J.*, **9**, 2523.
114. Straka, C. and Horz, W. (1991) A functional role for nucleosomes in the repression of a yeast promoter. *EMBO J.*, **10**, 361.
115. Schmid, A., Fascher, K. D., and Horz, W. (1992) Nucleosome disruption at the PHO5 promoter upon PHO5 induction occurs in the absence of DNA replication. *Cell*, **71**, 853.

116. Fascher, K. D., Schmitz, J., and Horz, W. (1993) Structural and functional requirements for the chromatin transition at the PHO5 promoter in *Saccharomyces cerevisiae* upon PHO5 activation. *J. Mol. Biol.*, **231**, 658.
117. Hayes, J. J. and Wolffe, A. P. (1992) Histones H2A/H2B inhibit the interactions of transcription factor IIIA with the *Xenopus borealis* somatic 5S RNA gene in a nucleosome. *Proc. Natl Acad. Sci. USA*, **89**, 1229.
118. Thomas, G. H. and Elgin, S. C. R. (1988) Protein/DNA architecture of the DNase I hypersensitive region of the *Drosophila* hsp26 promoter. *EMBO J.*, **7**, 7291.
119. Jackson, J. R. and Benyajati, C. (1992) *In vivo* stage- and tissue-specific DNA-protein interactions at the *D.melanogaster* alcohol dehydrogenase distal promoter and adult enhancer. *Nucl. Acids Res.*, **20**, 5413.
120. Cullen, K. E., Kladde, M. P., and Seyfred, M. A. (1993) Interaction between transcription regulatory regions of prolactin chromatin. *Science*, **261**, 203.
121. Thanos, D. and Maniatis, T. (1992) The high mobility group protein HMGI (Y) is required for NF-κB-dependent virus induction of the human IFN-β gene. *Cell*, **71**, 777.
122. Paull, T. T., Haykinson, M. J., and Johnson, R. C. (1993) The non specific DNA-binding and -bending proteins HMG1 and HMG2 promote the assembly of complex nucleoprotein structures. *Genes Dev.*, **7**, 1521.
123. Wolffe, A. P. and Schild, C. (1991) Chromatin assembly. *Methods Cell Biol.*, **3**, 461.
124. Wolffe, A. P. (1991) Activating chromatin. *Curr. Biol.*, **1**, 364.
125. Worcel, A., Han, S., and Wong, M. L. (1978) Assembly of newly replicated chromatin. *Cell*, **15**, 969.
126. Almouzni, G., Clark, D. J., Mechali, M., and Wolffe, A. P. (1990) Chromatin assembly on replicating DNA *in vitro*. *Nucl. Acids. Res.*, **18**, 5767.
127. Almouzni, G., Méchali, M., and Wolffe, A. P. (1990) Competition between transcription complex assembly and chromatin assembly on replicating DNA. *EMBO J.*, **9**, 573.
128. Almouzni, G., Mechali, M., and Wolffe, A. P. (1991) Transcription complex disruption caused by a transition in chromatin structure. *Mol. Cell. Biol.*, **11**, 655.
129. Sogo, J. M., Stahl, H., Koller, T., and Knippers, R. (1986) Structure of the replicating SV40 minichromosomes: the replication fork, core histone segregation and terminal structures. *J. Mol. Biol.*, **189**, 189.
130. Wolffe, A. P. and Brown, D. D. (1986) DNA replication *in vitro* erases a *Xenopus* 5S RNA gene transcription complex. *Cell*, **47**, 217.
131. Taylor, I. C. A., Workman, J. L., Schuetz, T. J., and Kingston, R. E. (1991) Facilitated binding of GAL4 and heat shock factor to nucleosomal templates: differential function of DNA-binding domains. *Genes Dev.*, **5**, 1285.
132. Perlmann, T. and Wrange, O. (1988) Specific glucocorticoid receptor binding to DNA reconstituted in a nucleosome. *EMBO J.*, **7**, 3073.
133. Archer, T. K., Cordingley, M. G., Wolford, R. G., and Hager, G. L. (1991) Transcription factor access is mediated by accurately positioned nucleosomes on the mouse mammary tumor virus promoter. *Mol. Cell. Biol.*, **11**, 688.
134. Pina, B., Bruggemeier, U., and Beato, M. (1990) Nucleosome positioning modulates accessibility of regulatory proteins to the mouse mammary tumor virus promoter. *Cell*, **60**, 719.
135. Workman, J. L. and Kingston, R. E. (1992) Nucleosome core displacement *in vitro* via a metastable transcription factor/nucleosome complex. *Science*, **258**, 1780.
136. Hansen, J. and Wolffe, A. P. (1992) Influence of chromatin folding on transcription initiation and elongation by RNA polymerase III. *Biochemistry*, **31**, 7977.

137. Workman, J. L., Abmayr, S., Cromlish, W. A., and Roeder, R. G. (1988) Transcriptional regulation by the immediate early protein of pseudorabies virus during *in vitro* nucleosome assembly. *Cell*, **55**, 211.
138. Clark, K. L., Halay, E. D., Lai, E. and Burley, S. K. (1993) Co-crystal structure of the HNF-3/fork head DNA recognition motif resembles histone H5. *Nature*, **364**, 412.
139. Ramakrishnan, V., Finch, J. T., Graziano, V., Lee, P. L., and Sweet, R. M. (1993) Crystal structure of globular domain of histone H5 and its implications for nucleosome binding. *Nature*, **362**, 219.

3 | CREM: a bZip transcription factor central to the cAMP nuclear response

NICHOLAS S. FOULKES and PAOLO SASSONE-CORSI

1. Introduction

Co-ordinated programmes of gene expression are central to the complex processes of cell growth and differentiation. From extensive studies it has emerged that the basic features of gene regulation are highly conserved and apply to both prokaryotes and eukaryotes. The rate of transcription of a gene by RNA polymerase is controlled by regulatory sequences that surround and extend 5′ of the transcriptional start site (the promoter) and can also lie thousands of base pairs 5′ or 3′ from the gene (enhancers). Transcription factors bind specifically to short sequence elements within promoters and enhancers, and are able to either activate or repress transcription initiation by RNA polymerase (see Chapter 1). The function of these factors can in turn be modulated and, indeed, many constitute targets of signal transduction pathways. Thus the complement of factors bound to promoter and enhancer elements integrate information from DNA sequence and signal transduction to control the rate of transcription initiation. An impressive number of these transcription factors have now been characterized at both the structural and functional level and several clear features have emerged.

1.1 Modularity

Transcription activators are intrinsically modular. At the simplest level they each include a DNA-binding domain and an activation domain. These domains can be interchanged between different factors and still retain their functional properties. Furthermore, within both the DNA-binding and activation domains there exist co-operating subdomains that in many cases are also interchangeable. This modularity suggests that during evolution, increasing complexity of transcriptional control may have resulted not only by duplication and divergence of existing genes, but also by a domain shuffling process to generate factors with novel properties. Modularity has also been a great asset for transcriptionologists and has enabled rapid progress in our dissection of transcription control mechanisms.

1.2 Dimerization

Many factors bind to DNA as dimers. Heterodimeric complexes with related factors as well as homodimer formation provides for more diversity of DNA-binding specificity as well as activation function.

1.3 Activation domains

Activation domains fall into various categories and include regions rich in glutamine, proline, or acidic amino acids. Correspondingly, there is considerable variation in the ability of domains to activate transcription and they can also be the basis for repressor function. Furthermore, post-translational modifications, such as phosphorylation, frequently modulate the function of these domains. They are thought to function by making contacts either directly or indirectly with the basal transcription machinery (see Chapter 1).

1.4 DNA-binding domains

DNA binding is mediated by a domain with a flexible extending structure that can protrude into the major or minor groove of the DNA double helix and thereby make hydrophobic or electrostatic contacts with the purine and pyrimidine bases. The geometry of these contacts is defined by the specific nucleotide sequence and thus defines the sequence-binding specificity. Comparisons of the structure of transcription factors have permitted a classification based upon the type of DNA-binding domain (1):

- The helix–turn–helix (HTH) motif. This is a component of the homeobox domain of eukaryotes and various prokaryotic regulatory proteins and mediates dimerization as well as DNA binding (2, 3; discussed in more detail in Chapter 6).
- The zinc-finger domain. First defined in the factor TFIIIA, which regulates transcription of 5S ribosomal RNA. Subsequently, at least three additional subtypes have been described, which include those found in steroid receptor, the yeast GAL4 factor, and GATA1 (4–6).
- The helix–loop–helix (HLH). This functions principly as a dimerization domain in combination with a region rich in basic amino acids, which acts as the DNA-binding domain. bHLH factors include MyoD1 (7; discussed in Chapter 7).
- The basic–leucine zipper (bZip). The DNA-binding/dimerization domain that is the subject of this review.
- The bHLH–Zip domain. In some factors, C-terminal and adjacent to a bHLH motif, there lies a leucine zipper region. Here the leucine zipper and HLH domain co-operate to direct dimerization while the basic region directs DNA binding. These so-called bHLH–Zip factors include the factors Myc and Max (8).

In this chapter, we will concentrate on the biology of bZip transcription factors. These factors have been the focus of much attention partly because the oncoproteins cFos and cJun belong to this class of factor and thus they represented one of the first demonstrations that transcription factors are crucial players in tumourigenesis and transformation. However, bZip factors also represent perfect examples of the biology of transcription factors in general.

There have been several comprehensive reviews on the precise structure and function of the bZip domain itself (9–11). In this review we will only briefly outline the current state of knowledge of this structure. In addition, we will focus our attention on the function of one bZip factor in particular, CREM (CRE modulator). The CREM gene generates a family of closely related bZip factors that are nuclear targets of the cAMP signal transduction pathway and that have been implicated in many aspects of the physiology of the neuroendocrine system.

2. The bZip motif: the hallmark of a family of transcription factors

2.1 Historical perspectives

The leucine zipper was originally described by Landschulz *et al.* for the mammalian transcription factor CCAAT/enhancer-binding protein (C/EBP) (12). Alignment of the primary structure with that of Fos, Jun, and the yeast regulatory protein GCN4 revealed a common region of homology consisting of a heptad repeat of leucine residues: the leucine zipper (see Fig. 1; 12). Central to the model that was subsequently developed to explain the structure of this region was the notion that all of these factors bind to DNA as dimeric complexes.

The heptad leucine repeat region was predicted to form an amphipathic α-helix with the leucine residues aligned along one ridge (12). Furthermore, by modelling, it was demonstrated that two of these helices could associate in a coiled-coil conformation (12). This central feature has been confirmed by a series of studies investigating the structural properties of GCN4 leucine zipper peptides (13–15). The overall structure of the bZip domain has been predicted by two similar models; termed the 'scissors-grip' and 'induced helical fork' models (16, 17). In both, the repeating leucines lie at the interface between two parallel helices and thus are reminisent of the teeth of a zipper. Immediately N-terminal to the leucine zipper lies a 30 amino acid conserved region, rich in basic amino acids. The predicted α-helical conformation of the leucine zipper extends through the basic region, and thus the coiled-coil region of the dimer juxtaposes the paired basic regions to form a Y-shaped structure, where the arms are the basic regions and the stem is formed by the coiled-coil (16, 17). In this structure the α-helical basic region is able to wrap around the DNA helix, contacting bases in the major groove. The 'scissors-grip' and 'induced helical fork' models differ in the details of the basic domain/DNA interactions. Consistent with these models is that the spacing between the basic region and the leucine zipper is a highly conserved feature, and thus the orientation of basic region helices relative to the leucine zipper helices appears critical.

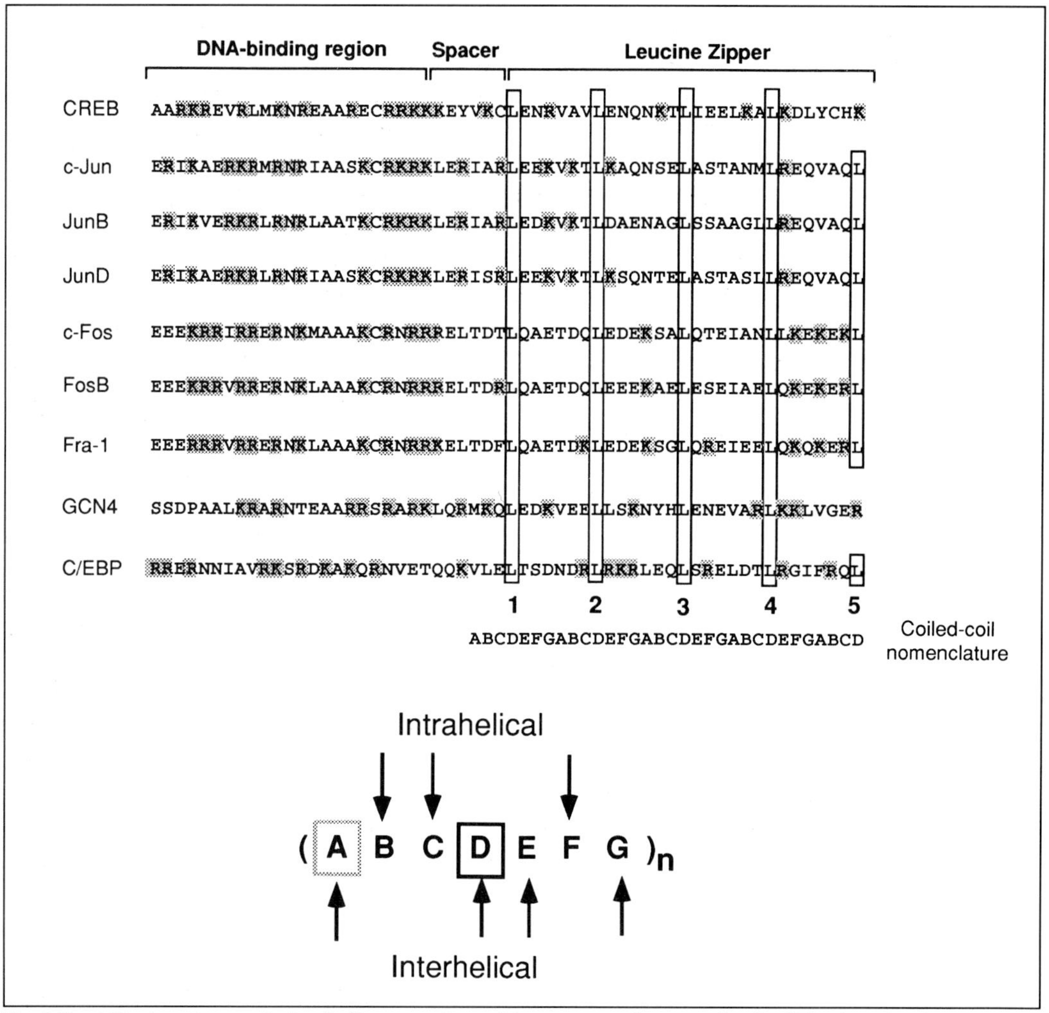

Fig. 1 The bZip domain: a coiled-coil. Above: amino acid sequence alignment for the bZip domain of nine regulatory proteins (151). The extent of the basic DNA-binding region, the leucine zipper, and the intervening six amino acid spacer region are indicated. The component leucines of the leucine zipper are indicated by boxing and are numbered. Basic amino acids are shaded. Immediately beneath, the amino acids of the leucine zipper are labelled with respect to their position in the coiled-coil structural motif by the $(ABCDEFG)_n$ nomenclature. Below: one unit of the coiled-coil heptad repeat with the amino acids involved in interhelical and intrahelical interactions indicated by labelled arrowheads. The black box indicates the D position, which is occupied by leucine, while the shaded box indicates amino acids implicated in defining the specificity of heterodimerization.

2.2 The leucine zipper: an α-helical coiled-coil

The bZip motif has now been encountered in the context of both tissue-specific and more ubiquitously expressed factors of animals, plants, and fungi. In numerous cases it has been functionally delimited by deletions as the region required for both DNA binding and dimerization. As described in the previous section, the leucine zipper consists of a two-stranded, parallel, α-helical, coiled-coil (18). Coiled-coils are central components of fibrous proteins, where two or four helices can form this structure and can be in parallel or anti-parallel orientations (19, 20). The heptad repeat of leucines underlies a general heptad repeat, which within the α-helix is the structural unit of the coiled-coil. By convention the residues of this repeat are designated as $(A, B, C, D, E, F, G)_n$ (see Fig. 1). Leucines occupy position 'D' and invariably a β-branched hydrophobic residue (such as valine, threonine, or isoleucine) occupies position 'A'. Thus, each leucine within the zipper lies three residues N-terminal and four residues C-terminal to another hydrophobic amino acid (21). This so-called 4,3 hydrophobic repeat creates a continuous hydrophobic interface between the coiled helices and favours dimer rather than tetramer formation (22). From the crystal structure of the GCN4 leucine zipper it has been shown that intrahelical contacts are made by residues 'B', 'C', and 'F', while interhelical interactions are formed by residues 'A', 'D', 'E', and 'G' (14). Interestingly, in the central heptad repeat of many leucine zippers, the 'A' hydrophobic position is substituted by an asparagine residue, or another polar amino acid. In the case of GCN4 this asparagine appears to constrain the dimerized helices to lie in parallel and in register since in the antiparallel conformation, or if the helices are not in register, it will be buried in a hydrophobic environment (14). This residue also restricts GCN4 to form dimers with leucine zippers that include a polar residue at the equivalent position. Furthermore, thermodynamic calculations show that the presence of the asparagine residue in the GCN4 zipper significantly destabilizes the structure relative to an uninterrupted coiled-coil. This feature possibly enables a more dynamic reassortment of dimer combinations (14).

The structure of the bZip domain has also been analysed in order to explain the ability of the leucine zipper to mediate heterodimerization between non-identical helices. An important feature of heterodimerization is specificity; namely, only certain heterodimer combinations are possible where the two α-helices can align in register to generate a symmetrical coiled-coil. The most extensively studied case has been that of Fos and Jun (23–27). While Fos–Jun heterodimers and Jun–Jun homodimers are possible, Fos–Fos homodimers do not form. Mutagenesis studies have implicated the glutamic acid residues at three consecutive 'E' and 'G' positions of the Fos zipper in blocking putative interhelical electrostatic interactions and so preventing homodimerization (28). These findings, and others, have implicated the electrostatic interactions of amino acids at positions 'A', 'E', and 'G' as crucial in defining the specificity of dimerization (29, 30).

3. bZip factors and the cAMP nuclear response

An important step forward in the understanding of the biology of bZip transcription factors has been the discovery that many such factors constitute the final targets of specific signal transduction pathways activated intracellularly by various signals at the cell surface. The two major signal transduction systems are those that utilize cAMP and diacylglycerol (DAG) as secondary messengers (31). Each path-

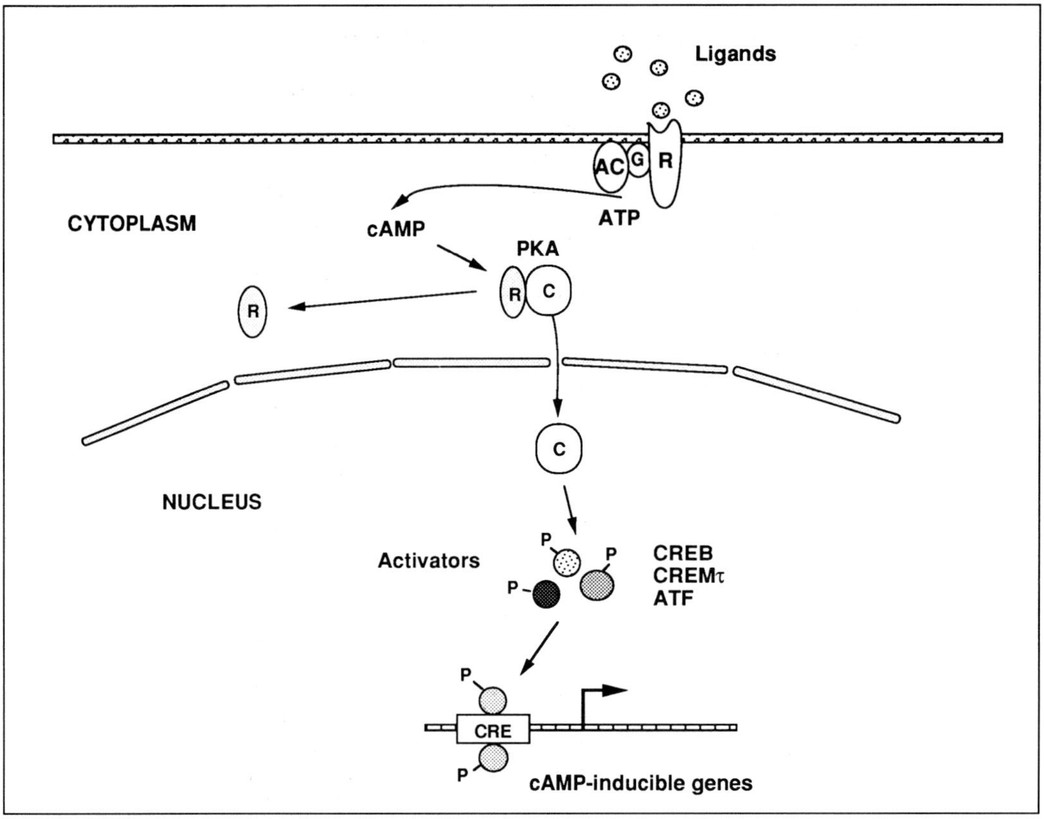

Fig. 2 The cAMP signal transduction pathway. Schematic representation of the route whereby ligands at the cell surface interact with membrane receptors (R) and result in altered gene expression. Ligand binding activates coupled G-proteins (G), which in turn stimulates the activity of the membrane-associated adenylate cyclase (AC). This converts ATP to cAMP, which causes the dissociation of the inactive tetrameric protein kinase A (PKA) complex into the active catalytic subunits (C) and the regulatory subunits (R). Catalytic subunits migrate into the nucleus where they phosphorylate (P) and thereby activate transcriptional activators such as CREB, CREMτ, and ATF. These then interact with the cAMP-responsive enhancer element (CRE) found in the promoters of cAMP-responsive genes to activate transcription.

way is also characterized by specific protein kinases (protein kinase A and protein kinase C respectively) and specific target DNA control elements [cAMP-responsive element (CRE) and TPA-responsive element (TRE) respectively]. Although initially characterized as distinct systems, accumulating evidence points towards extensive cross-talk between these two pathways (32–34).

Intracellular levels of cAMP are regulated primarily by adenylate cyclase (see Fig. 2). This enzyme is in turn modulated by various extracellular stimuli mediated by receptors and their interaction with G proteins (35). The binding of a specific ligand to a receptor results in the activation or inhibition of the cAMP-dependent pathway, ultimately influencing the transcriptional regulation of various genes through distinct promoter-responsive sites (36). Increased cAMP levels directly influence the function of the tetrameric protein kinase A (PKA) complex (37).

Table 1 Sequence comparison of CRE and ATF sites

Promoter	Sequence	Location
Somatostatin	TGACGTCA	–48/–41
c-fos	TGACGTTT	–66/–59
	TGACGCAG	–297/–290
Proenkephalin	TGACGCAG	–86/–93
Tyrosine hydroxylase	TGACGTCA	–39/–46
Fibronectin	TGACGTCA	–173/–166
PEPCK	TTACGTCA	–83/–90
VIP	TGACGTCT	–70/–77
Parathyroid hormone	TGACGTCA	–68
α-Chorionic gonadotropin	TGACGTCA	–117/–124; –135/–142
c-Jun	TGACATCA	–71/–64
Jun-D	TGACCTCA	–53/–46
GHF-1	TGACGTCA	–200/–193
	GTACGTCA	–157/–150
Krox-20	CGACGTCA	+248/+255
Tα-1	TGACGTCA	TCR α 3' enhancer
LDH-A	TGACGTCA	–48/–41
Aα (MHC II)	TGACGTCA	'X-box' element
IL-6	TGACGTCC	–158/–165
Tyrosine aminotransferase	CTGCGTCA	–3651/–3644
TPA	TGACATCA	–114/–107
β_2-Adrenergic receptor	GTACGTCA	–57/–50
HTLV-I LTR	TGACGACA	–97/–90
	TGACGTGT	–196/–189
	TGACGTCT	–245/–238
HTLV-II LTR	TGACGATT	–83/–76
	TGACGTCC	–176/–169
	TGACGTCT	–216/–209
BLV LTR	TGACGGCA	–125/–118
	TGACGTCT	–153/–160
E2A	TGACGTAG	–74/–67
E3	TGACGAAA	–55/–62
E4	TGACGTAA	–44/–51
	TGACGATT	–144/–137
	TGACGTAA	–168/–161
	TGACGTAG	–234/–227
	TGACGTGG	–266/–259
CARE1 (ICER)	TGACCTGCA	–91/–83
CARE2	TGATGGCA	–79/–72
CARE3	TGACGTCA	–59/–52
CARE4	TGATGTCA	–48/–41

Comparison of cAMP-responsive elements in various promoter regions. For most of these sites, cAMP inducibility has been demonstrated. There is a higher conservation of the 5' half of the palindrome (TGACG) with respect to the 3' sequence (TCA). The CRE element is extremely similar to the consensus TRE (TGACTCA), the site of binding for the oncoproteins Fos–Jun. The position of the CARE elements within the P2 CREM promoter is depicted in Fig. 12.

Binding of cAMP to two PKA regulatory subunits releases the catalytic subunits enabling them to phosphorylate target proteins. In the nucleus, the phosphorylation state of bZip transcription factors and related proteins directly modulates their function and thus the expression of cAMP-inducible genes.

The analysis of promoter sequences of several genes allowed the identification of promoter elements that could mediate the transcriptional response to changes in the levels of intracellular cAMP. A number of sequences have been identified of which the best characterized is the CRE. A consensus CRE site is constituted by an 8-bp palindromic sequence (TGACGTCA, see Table 1). Several genes that are regulated by a variety of endocrinological stimuli contain similar sequences in their promoter regions, although at different positions. The binding-site specificity appears to require 18–20 bp, since the five or so bases flanking the core consensus have been shown to dictate, in some cases, the permissivity of transcriptional activation (38). The CRE consensus sequence has also been found in the context of other distinct promoter elements where they apparently confer different transcriptional properties. For example, CRE-like sites are present in the X-box motif associated with the major histocompatibility complex (MHC) class II genes (39); this element acts as a non-cAMP-inducible pre-B-cell enhancer. The 21-bp enhancer in the long terminal repeat (LTR) of HTLV-I contains consensus CRE sites. However, in this context they are implicated in mediating transcriptional activation by the viral tax protein (40, 41). Another extensively studied example is the ATF (activating transcription factor) element that is present in the early promoters of adenovirus and mediates transcriptional activation by the viral oncogenic protein E1A (42–44). The study of the factors that bind to these CRE elements has provided an important demonstration of the versatility of bZip factors and their control of gene expression.

3.1 A multi-gene family

The first CRE-binding factor to be characterized was CREB (CRE binding protein) but subsequently at least ten additional CRE-binding factor cDNAs have been cloned (see Fig. 3). They were obtained by screening a variety of cDNA expression libraries with CRE or ATF sites, with the HTLV-I LTR 21-bp repeat, or with the MHC class II X-box sequences. In many cases the same cDNAs were isolated by independent laboratories using different strategies, thus creating some confusion over nomenclature (see legend Fig. 3). However, certain clear features have emerged:

(1) They all belong to the bZip transcription factor class. The number of leucines in the zipper heptad repeat varies from four to six; in many of the factors there is the possibility for histidine to substitute for a leucine at the C-terminus of the zipper structure (see Fig. 3). Noteworthy is ATF6 where, at two positions in the zipper, leucine is substituted for valine and alanine. Consistent with the disruption of the leucine repeat, the binding of this factor to a single ATF site was demonstrated to be weak (45).

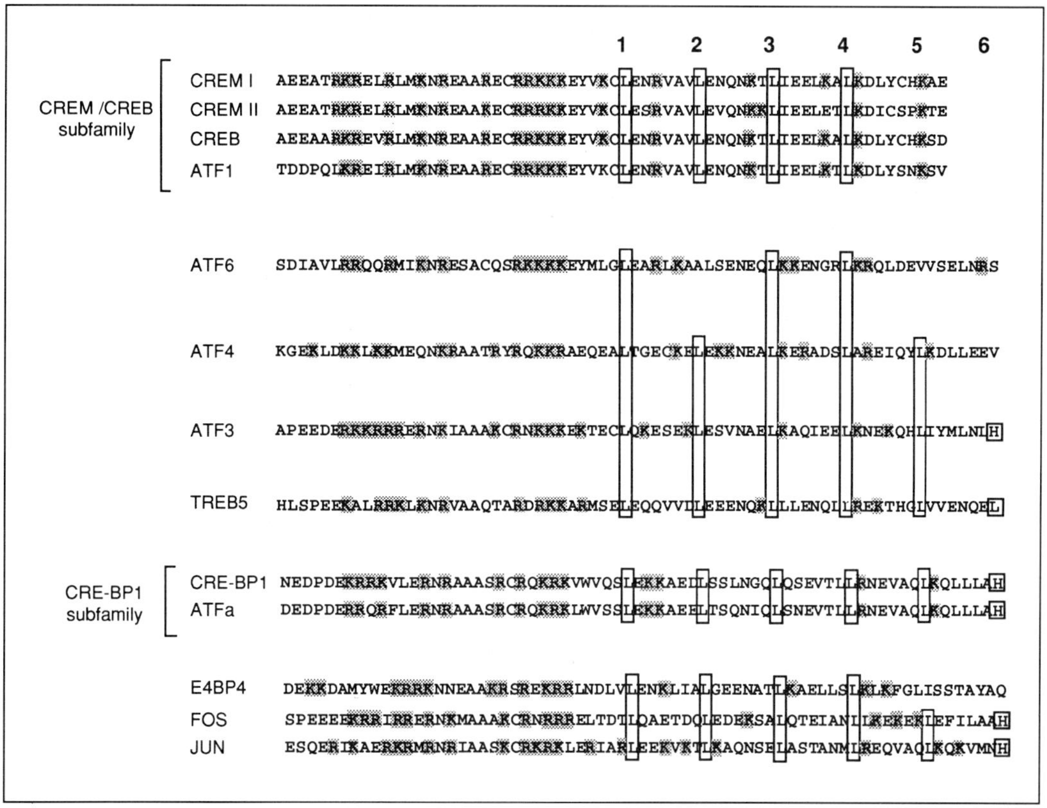

Fig. 3 Sequence comparison of the bZip domains of all CRE-binding proteins described to date. The sequences of c-Fos and c-Jun are included for comparison. The aligned sequences are presented as in Fig. 1. The boxed histidine residues at the C-terminus of some of the sequences may substitute for leucine to extend the zipper structure (151). At position 2 in the leucine zipper of ATF6, an alanine substitutes for a leucine. Related sequences constituting the CREM/CREB and CREBP1 subfamilies are indicated. Clones apparently derived from the same genes have been isolated by independent groups. ATF1 (45) is identical to *tax*-responsive element-binding protein-36 (TREB36; 40): TREB5 (40) has the same sequence as hXBP1 (39), TREB7 (40), HB16 (152), mXBP/CRE-BP2 (47), and ATF2 (45) seem to be derived from the CREBP1 gene (153). The recently described LRF1 (133) is possibly a splicing variant of ATF3 (45). Of these factors, only ATF4, ATF6 (45), and CREM (46) have been originally described once.

(2) Outside the bZip region, homology between these factors is very poor. Based upon regions of sequence similarity, however, they can be divided into subfamilies. For example, CREB, ATF1, and CREM (CRE modulator; see Section 4) share extensive homology (45, 46). Notably, the bZip domains for all three factors are at an equivalent position at the C-terminus of the protein. This CREM/CREB subfamily is, however, clearly distinct from ATFa and CRE-BP1, which constitute another subfamily (47, 48).

(3) The different factors are able to heterodimerize with each other but only in certain combinations. A 'dimerization code' exists that seems to be a property of the leucine zipper structure of each factor; for example: ATF1 can dimerize with

CREB but not with CREBP1 or ATF3 while CREBP1 can heterodimerize with ATF3 but not with CREB (45, 49, 50). It has also been demonstrated that some ATF/CREB factors are able to heterodimerize with Fos and Jun, again with a dimerization code. Heterodimerization seems to affect the stability of specific heterodimer–DNA complexes. Fos–Jun heterodimers bind with greater affinity to a TRE than a CRE (51); however, heterodimers between Jun and ATF3, ATF4, and CREBP1 bind more strongly to a CRE than a TRE (52). This property is likely to reside in the similarity between the CRE (TGACGTCA) and TRE (TGACTCA) sequences (34). Theoretically, this possibility for pairwise combinations of different factors dramatically increases the complexity and thereby the versatility of the transcriptional response to signal transduction. In addition, the possibility for targets of the cAMP/PKA pathway to interact with Fos and Jun (targets for the DAG/PKC pathway) provides an important route for different intracellular signals to be integrated at the level of transcriptional control.

(4) Alternative splicing of CRE-binding bZip factors seems common. For example, mXBP/CREBP2 is likely to be a splicing variant of CREBP1: this mouse clone carries a deletion relative to the human CREBP1, within a proline-, serine-, threonine-rich domain; outside this region, the clones are >99% identical at the amino acid level. Also, there are reports of additional splicing variants of CREBP1 (47). Alternative splicing of the α-region in the CREB cDNA was the first to be identified in this group of transcription factors (53). This 14 amino acid segment lies within the activation domain of CREB and its inclusion or exclusion defines the two CREB isoforms CREB341 and CREB327, respectively (see Section 3.2.2). Subsequently, a large family of alternatively spliced CREB mRNA isoforms have been described, although the functional significance of the majority of these isoforms remains a mystery (54, 55; see Section 3.3). The CREM gene also generates a large family of alternatively spliced transcripts although, unlike CREB, the function and physiological role of many of these isoforms has been determined (46, 56–58; see Section 4). Alternative splicing reflects the multi-exonic, modular structure of these genes (55, 59, 60).

(5) These bZip factors act as both activators and repressors of transcription. Thus, while CREB, CREMτ, and ATF1 mediate transcriptional activation (54, 60), repressors of cAMP-induced transcription include the CREM isoforms CREMα, β, and γ (46, 59) and the recently described inducible cAMP early repressor (ICER) proteins (see Section 5; 61). In addition, the bZip protein E4BP4 is able to bind to CRE sites *in vitro* and contains a C-terminal domain that actively represses transcription (62). Also, CREB-2 (most probably equivalent to ATF4; 45) is a ubiquitously expressed CRE-binding bZip protein that counteracts the PKA-induced activation of the human enkephalin promoter (63). Interestingly, deletion studies show that the bZip domain is sufficient for full antagonism (63), as previously described for the transcriptional antagonists CREM (59) and the ICER proteins (61).

3.2 Phosphorylation-directed transcriptional activation

Important steps in the understanding of the control of bZip transcription factor function by phosphorylation have been made with the characterization of the transcriptional activators CREB and the CREMτ isoform (56, 64, 65). Figure 4 shows the general structure of the activator proteins. Interestingly, the relative position of the different functional domains is well conserved among various other CRE-binding proteins such as ATF1 and ATF2 (45). The transcriptional activation domain contains two independent regions. The first, known as the phosphorylation-box (P-box) or kinase-inducible domain (KID), contains several consensus phosphorylation sites for various kinases, such as PKA, PKC, glycogen synthase kinase-3, and casein kinases (CK) I and II (Fig. 4). The second region, divided into two parts, flanks the P-box, and is constituted by domains rich in glutamine residues (66).

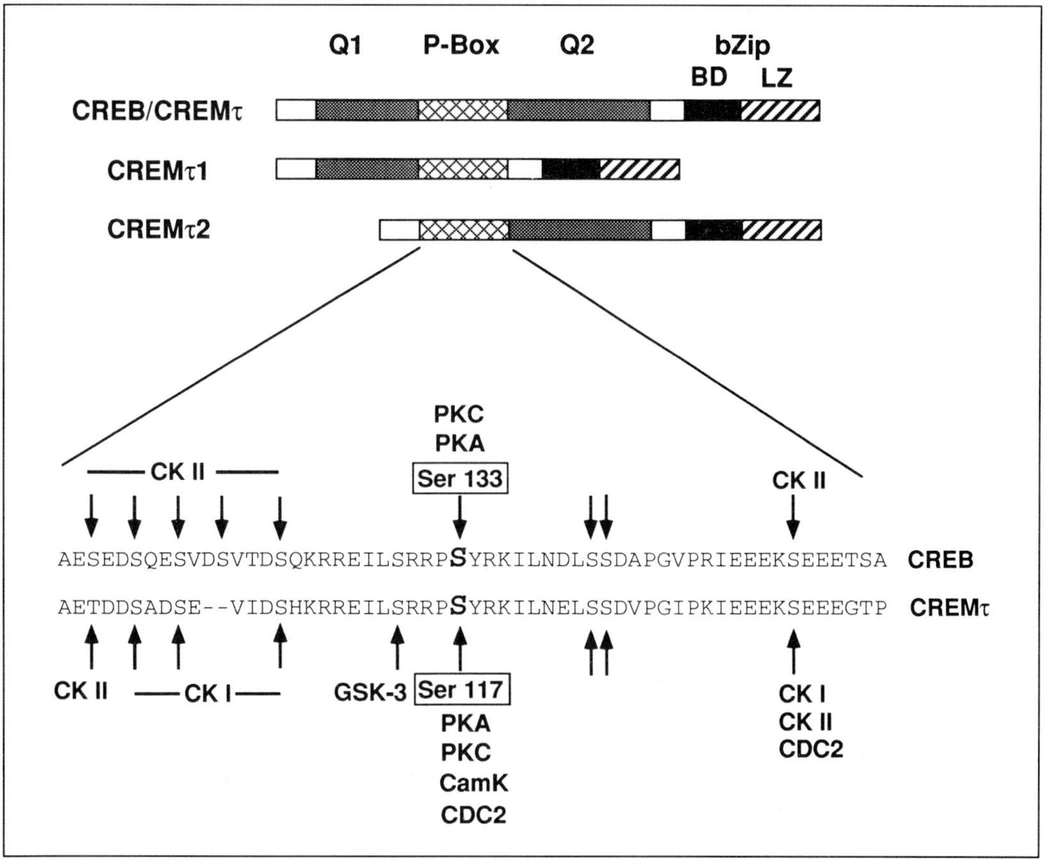

Fig. 4 The P-box domain. Above: schematic representation of the transcriptional activators CREB and CREM (CREMτ, -τ1, and -τ2). The two glutamine-rich domains (Q1 and Q2) and the bZip region (BD and LZ) are indicated, in addition to the P-box. In all these factors, the P-box is flanked by glutamine-rich domains. Below: the detailed amino acid sequence of the CREM and CREB P-box domains. Arrows indicate the serine and threonine residues in CREB and CREM that have been demonstrated *in vivo* or *in vitro* to be phosphorylated by the indicated kinases (65).

3.2.1 The P-box: modulation by phosphorylation

Upon activation of the adenylate cyclase pathway, a serine residue at position 133 of CREB and at position 117 of CREMτ is phosphorylated by PKA (see Fig. 4) (64, 65). The phosphorylation appears indispensable for activation and it is demonstrated that the phosphoserine cannot be substituted by other negatively charged residues (64, 67). In addition, ATF1 activates transcription upon phosphorylation by the catalytic subunit of PKA (68, 69); however, although ATF1 can be phosphorylated by PKA *in vitro, in vivo* phosphorylation has yet to be demonstrated. The major effect of phosphorylation is to convert CREB and CREMτ into powerful transcriptional activators. Within the P-box, serine 133/117 is located in a region of about 50 amino acids that contains an abundance of phosphorylatable serines and acidic residues and that was shown to be essential for transactivation by CREB and CREMτ (65, 67, 70). It is apparent that this domain represents the convergence point for phosphorylation events stimulated by several signal transduction pathways. Indeed, it has been shown that the residue Ser 117 in CREM is the target for phosphorylation by PKC, CamK, $p34^{cdc2}$ as well as PKA *in vitro* (see Fig. 4; 65). These phosphorylation events are relevant *in vivo*, since treatment with forskolin, TPA, or the Ca^{2+} ionophore A23187 all lead to the enhanced phosphorylation of Ser 117 (65). In PC12 cells, increases in the levels of intracellular Ca^{2+} by membrane depolarization have been shown to increase the phosphorylation of serine 133 in CREB and a concomitant induction of c-*fos* gene expression mediated by a CRE in the c-*fos* promoter (71–73). Although Ca^{2+}-dependent CamK was shown to be able to phosphorylate serine 133 *in vitro* (73, 74), the *in vivo* significance remains unclear, since PKA also seems to be necessary for c-*fos* induction mediated by Ca^{2+} influx in PC12 cells (75). Significantly, co-operativity for phosphorylation by casein kinases I and II of Thr and Ser residues within the P-box of CREM has been documented (65): phosphorylation of Thr 94 by CKII makes Ser 97 a much more efficient phosphorylation target for CKI. Interestingly, phosphorylation of P-box domain sites by CKII enhances CREMτ DNA binding *in vitro* (65). Whether phosphorylation by PKA influences the DNA-binding function of CREB and CREMτ remains a controversial point. In a study by Nichols *et al.* (76), a two- to three-fold increase in DNA binding by CREB was observed upon phosphorylation by PKA. In contrast, others failed to detect any effect of the phosphorylation upon DNA binding or dimerization for both CREM and CREB (65, 77). Similarly, PKC-mediated phosphorylation was reported to increase binding through enhancement of dimer formation (76); other studies, however, failed to demonstrate any such effect (65).

Although phosphorylation of serine 133 appears indispensable for activation by CREB, it is not sufficient for full activity. An acidic region just downstream of serine 133 (140-DLSSD) has been shown to be important for CREB function (67, 70). In addition, deletion of a region called α2, containing several sites that can be phosphorylated by CKII *in vitro*, causes a decrease in CREB activity, although differences in the magnitude of this decrease have been reported (67, 70).

Dephosphorylation appears to represent a key mechanism in the negative regulation of CREB function. It has been proposed that CREB activity can be attenuated following induction by specific phosphatases (78). After the initial burst of phosphorylation in response to cAMP, CREB is dephosphorylated *in vivo* by protein phosphatase-1 (PP-1). However, the situation is more complex since Nichols *et al.* (76) show that both PP-1 and PP-2A can dephosphorylate CREB *in vitro*, resulting in an apparent decreased binding to low-affinity CRE sites. Therefore, the precise role of PP-1 and PP-2A in the dephosphorylation of CREB remains to be determined.

3.2.2 Modularity of the activation domain

The two domains flanking the P-box contain about three times more glutamine residues than in the remainder of the protein in both CREB and CREM. Glutamine-rich domains have been characterized in other factors, such as AP-2 and Sp1 (79, 80) as transcriptional activation domains. The current notion is that they constitute surfaces of the protein that can interact with other components of the transcriptional machinery, such as RNA polymerase II (discussed in Chapter 1). The Q2 domain appears to make a more significant contribution to the transactivation function than Q1. This is demonstrated by the properties of the two naturally occuring CREM isoforms CREMτ1 and CREMτ2 (59; also named CREMε, see 81) and artificially generated deletion mutants of CREB (81). CREMτ1 and -τ2 incorporate singly, the Q1 and Q2 domains, respectively (Fig. 4), and CREMτ2 appears to be a stronger transcriptional activator upon PKA induction than CREMτ1 (59). In agreement with these results on CREM, deletion of the Q2 region in CREB dramatically abolishes activation function (81). Furthermore, ATF1 lacks a counterpart of the Q1 domain, and still functions as an efficient transcription activator (45). Thus, the P-box region and the Q2 domain are sufficient to mediate cAMP-induced transcription. The role of the α-domain, which lies between the P-box and the Q1 domain, remains controversial. Some groups claim that CREB341 (including the α-domain) has enhanced transcriptional activation relative to CREB327 (lacking the α-domain) while others have detected no such effect (53, 55, 60).

It is apparent that the activation domain is inherently a modular structure. Thus, the Q2 domain, when fused to a heterologous DNA-binding domain, still retains a non-inducible activation function (81). In addition, the P-box is able to confer PKA inducibility on a heterologous acidic activation domain (e.g. GAL4; 81). Furthermore, the P-box is able to confer this inducibility *in trans* as well as *in cis* (81). Thus, theoretically, the P-box could be involved not only in the regulation of the adjacent Q domains, but also in controlling the activation function of other factors bound to separate promoter elements (81). Further steps towards an understanding of the mechanism of action of the P-box has come with the identification of a 265-kDa, 2441-amino acid protein, CBP (CREB-binding protein) that is able to interact specifically with the phosphorylated CREB P-box domain (82). The CBP sequence reveals two zinc-finger domains, a glutamine-rich domain at its C-terminus, and a single consensus PKA recognition site. The role of CBP in mediating CREB-enhanced transcription is discussed in detail in Chapter 1.

3.3 Functional redundancy in the cAMP nuclear response

The picture that has emerged from studies of CRE-binding proteins is the existence of a plethora of bZip factors with different activation properties, which execute transcriptional control from CRE sites. Further complexity arises from the various heterodimeric combinations of these factors. Thus, the final transcriptional response to cAMP is the outcome of a complex interplay of nuclear targets activated by signal transduction events. Several recent findings indicate that there is significant functional redundancy amongst CRE-binding factors.

The CREB gene encodes a family of alternatively spliced transcripts (see Fig. 5; 83). All these isoforms share certain exons, such as those that encode the P-box and the Q1 domain, and they differ by the complement of the remaining exons (for a review see 54). However, several of these isoforms include novel exons that feature stop codons in all three reading frames and thereby encode C-terminally truncated CREB proteins that lack the bZip domain (see Fig. 5; 55, 84, 85). Thus, for example, CREB transcripts including exon γ (55; equivalent to Y, see 85) and exon ψ (55) encode truncated proteins that consist of the Q1 domain or only the N-terminal segment, respectively. The transcripts that incorporate exon ω (84) encode larger proteins that contain Q1, Q2, and the P-box (see Fig. 5). In most cell lines and tissues tested, the CREB341 and CREB327 isoforms constitute the major products of the gene. During spermatogenesis, however, the truncated forms are more abundant and there is evidence that the splicing that generates these transcripts is developmentally regulated (55, 84, 85). The function of these truncated CREB proteins is an enigma since they do not appear to be localized in the nucleus and their expression has no detectable effect upon cAMP-stimulated transcription (55). It is conceivable that they represent non-functional by-products of the CREB gene and as such might be a demonstration of functional redundancy amongst CRE-binding factors. Functional redundancy is further supported by the phenotype of mice in which the CREB gene has been altered by homologous recombination (G. Schütz, personal communication). The presence of a large family of CRE-binding factors suggests that there is considerable overlap of cAMP-regulated activation function and thus it could be predicted that only homologous recombination experiments inactivating more than one factor might produce a mutant phenotype.

The advantage for a signal transduction pathway of having a family of transcriptional regulators with overlapping functions is not only that it confers functional versatility but also that it protects the pathway during evolution; namely, mutations affecting individual factors, which might radically alter their function, do not automatically disable the whole pathway. Furthermore, in this way such mutations can persist in the gene pool and subsequently might be able to confer a selective advantage. In such a system, where negative selection upon individual factors is not too stringent, it might not be surprising to encounter the expression of products such as the C-terminal truncated, non-functional CREB isoforms. Furthermore, in a gene that employs extensive alternative splicing to diversify its function, non-functional splicing products such as these might be unavoidable. In this light, the strik-

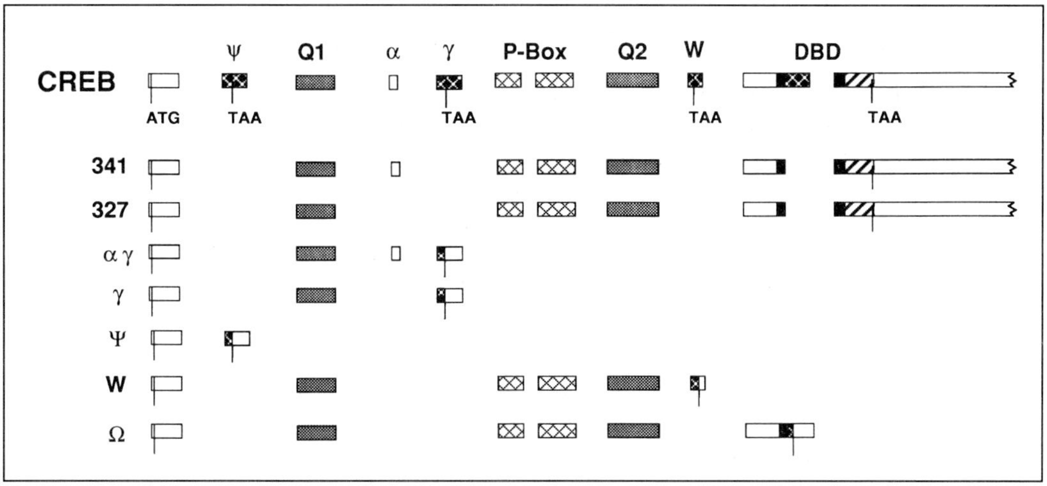

Fig. 5 The non-DNA-binding isoforms of CREB. Above: schematic representation of the CREB gene exon structure and the location of the exons ψ, γ, Ω, and ω, which all bear in-frame stop codons. Below: the cDNA structures of the various truncated CREB isoforms in addition to those of the CREB341 and -327 activators.

ing conservation of the sequence of exons Ψ and W between rat and human (84, 85) could reflect a positive selection against possible frame-shift mutations, which might incorporate detrimental C-terminal sequences.

4. CREM: key player in neuroendocrine physiology

The discovery of the CREM gene opened a new dimension in the study of the transcriptional response to cAMP (46). The dynamic and versatile pattern of CREM expression, combined with its tissue- and developmental-specific pattern, contrasts with that of the remaining members of the CRE-binding factor family, which seem to be constant and ubiquitous (45, 72). This offered the first clue that CREM occupied a privileged position amongst this group of factors and pointed further to it fulfilling a pivotal role in the nuclear response to cAMP.

4.1 The modularity of the CREM gene: differential RNA processing generates a family of CREM proteins

Three different CREM cDNAs were initially characterized that represent alternatively spliced transcripts (46). Subsequent studies have now firmly established that differential transcript processing is central to the regulation of CREM expression. Control seems to be exerted at three different levels: alternative splicing, alternative polyadenylation, and alternative translational initiation. The importance of these mechanisms is reinforced by the fact that all of the CREM isoforms that incorporate the P-box exons (CREMα, -β, -γ, -ατ, -τ, -τ1, and -τ2, see Fig. 6) are generated from a GC-rich promoter (P1), which has been shown to behave as a housekeeping

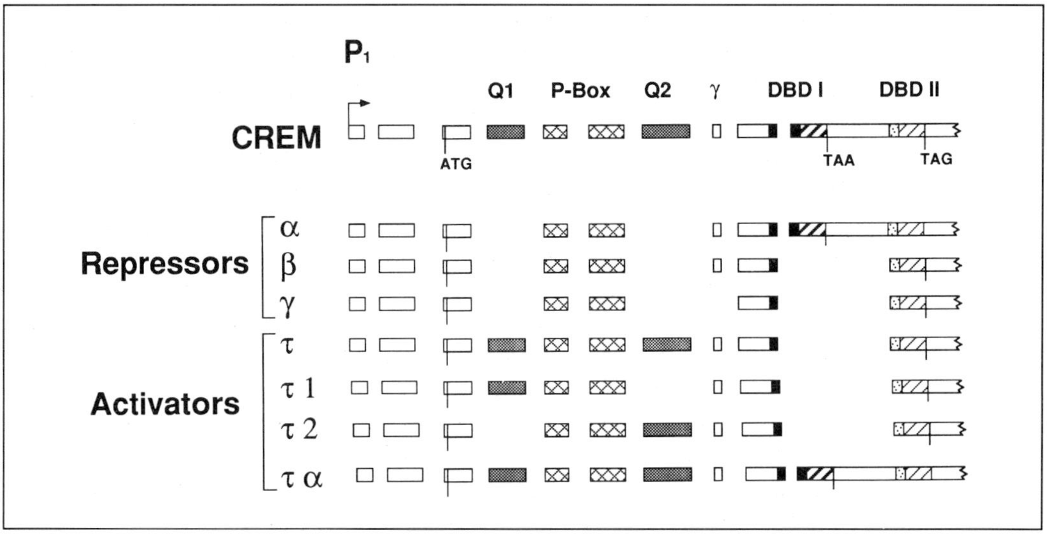

Fig. 6 Activators and repressors from the same gene. Above: schematic representation of the CREM gene. Below: the various activator and repressor isoforms that have been described to date. All transcripts represented in this figure are generated from the P1 promoter, which is GC-rich and directs a non-inducible pattern of expression.

promoter directing a non-inducible pattern of expression (58, 61 and see Section 5.1).

4.1.1 Alternative splicing

The first CREM cDNA, termed CREMα, was isolated from a mouse pituitary cDNA library (46). The most striking feature of the CREMα cDNA is the presence of two bZip DNA-binding domains (DBDI and DBDII, Fig. 6). DBDI is complete and contains a leucine zipper and basic region very similar to CREB; DBDII is located in the 3′ untranslated region of the gene, out of phase with the main coding region, and contains a half of the basic region and a leucine zipper more divergent from CREB. By reverse transcriptase-polymerase chain reaction (RT-PCR) analysis, two additional mRNA isoforms were identified, CREMβ and CREMγ (46). In these isoforms the DBDII is incorporated into the main open reading frame by a process of alternative splicing; specifically, an alternative splicing acceptor site lies at the 5′ boundary of DBDII. The cDNA CREMγ differs from CREMβ by a short in-frame deletion of 12 amino acids just upstream of the bZip domain. The three encoded CREM proteins feature extensive homology with the activator CREB; they are, however, significantly smaller (46, 55). Notably, while they share the highly conserved bZip domain and the P-box region, they lack the glutamine-rich domains. CREMα, -β, and -γ function as antagonists of cAMP-induced transcription either by binding to CRE sites as non-activating homodimers or by heterodimerizing with, and thereby blocking, CRE-binding activators (46). In addition, phosphorylation of sites within the P-box has been demonstrated to modulate the degree of antagonist function (59).

The CREM gene also encodes a transcriptional activator that differs from the antagonists by the insertion of the two glutamine-rich domains flanking the P-box (56). cDNAs have been characterized that incorportate singly either the N-terminal or C-terminal glutamine-rich domains and that still function as activators (59). The activators, in common with the antagonists, exhibit differential splicing of the DNA-binding domains and the γ-exon (see also Section 4.2.1).

Characterization of the genomic organization of the CREM gene has revealed the molecular basis for this extensive family of isoforms. The gene is multi-exonic with the coding region divided into nine exons that are distributed over more than 80 kb of DNA (59, 86, Foulkes and Sassone-Corsi, in preparation). The most striking feature of the structure is that the exons accurately define functional domains (Fig. 6; 59, 86). For example, the two glutamine-rich domains are encoded by two distinct exons and the differentially spliced bZip domain is composed of a common 5′ exon that encodes the 5′ half of the basic region, while the alternative 3′ exon sequences encode the remainder of the basic region and the leucine zipper. This modular structure combined with extensive differential splicing permits the CREM gene to encode a family of transcription factors with different activation properties. Regulation of the processing of the CREM primary transcript thus appears to be a key step in defining the contribution of CREM products to the cAMP nuclear response.

4.1.2 Alternative polyadenylation

Additional versatility is obtained by the possibility to generate transcripts with different 3′ ends (57). Ten copies of the sequence AUUUA are distributed throughout the 3′ untranslated region; this element has been demonstrated to confer mRNA instability in other genes (87, 88). By the use of alternative polyadenylation sites, the CREM gene can generate transcripts bearing different numbers of these elements and thus having different stability (see Section 4.3). During spermatogenesis the use of the most 5′ polyadenylation site is hormonally regulated (Section 4.3); transcripts polyadenylated at this site are much more stable because nine of the AUUUA elements are absent (57). Furthermore, the choice of polyadenylation site in some cases appears to be linked to the pattern of exon splicing (57, Foulkes and Sassone-Corsi, in preparation). Thus, the relative abundance of different isoforms can also be controlled by RNA processing.

4.1.3 Alternative translational initiation

By alternative usage of translation initiation sites a single CREM mRNA can generate both an activator and a repressor (89). The use of an internal AUG in the CREMτ transcript generates S-CREM, a protein that acts as a powerful repressor of cAMP-induced transcription. It is puzzling that CREM, which already makes extensive use of differential splicing to generate both activators and antagonists, utilizes an additional mechanism to generate factors with opposite functions. It is important to note, however, that S-CREM is distinct from the antagonist CREM forms generated by alternative splicing. It does not contain the P-box, but instead contains a single glutamine-rich domain. This is in contrast to the CREMα/β/γ antagonists, which contain the P-box but no glutamine-rich domains (46). It is reasonable to hypothesize that the structural differences between these various down-regulators have functional significance, and that they may interact differently with other components of the transcriptional machinery. In addition, the differential presence of the P-box among these various CREM down-regulators could suggest that the function of some of them might be modulated by phosphorylation.

4.2 Expression and function of CREM in the testis

4.2.1 A striking developmental switch

Expression in the testis provided the first clues that CREM might be distinct from the other bZip factors mediating the cAMP response. CREM was shown to constitute a highly abundant transcript in adult testis, while in pre-pubertal animals CREM is expressed at very low levels (Fig. 7). Thus, in testis CREM is the subject of a developmental switch in expression (56). Furthermore, characterization of the isoforms that were expressed revealed that the abundant CREM transcript exclusively encodes the CREMτ activator, while in pre-pubertal testis only the repressor forms were detected. In order to address the precise role played by CREM in the testis, the expression pattern of the RNA and protein have been defined in relation to spermatogenesis (90).

Spermatogenesis corresponds to the sequence of cytological events that results in the formation of haploid spermatogonia from precursor cells (91). This process begins by mitotic division of germ-cell spermatogonia to give rise to diploid spermatocytes, which themselves replicate their DNA content before undergoing the two successive meiotic divisions, which results in the production of haploid, round spermatids. The latter germ cells are sculptured into the shape of the mature spermatozoa in the process of spermiogenesis, which involves an extensive biochemical and morphological restructuring (92). In mice, the entire development from stem cell to spermatozoan takes 35 days (93). Spermatogenesis is a process occurring in a precise and co-ordinated manner within the seminiferous tubules (94). During this entire developmental process the germ cells are maintained in intimate contact with the somatic Sertoli cells (95). Indeed, the germ cells are encapsulated within the Sertoli cells and in this way they are supplied with growth factors and nutrients. In addition, cellular debris generated during spermiogenesis is processed by the Sertoli cells. As the spermatogonia mature they move from the periphery towards the lumen of the tubule until the mature spermatozoa are conducted from the lumen to the collecting ducts.

In the seminiferous tubules of rodents, spermatogenesis proceeds in a highly co-ordinated fashion. Precursor cells in a defined segment of the tubule differentiate in synchrony and these zones of differentiation progress along the length of the tubules as a wave (92). A cross-section through a given point of the tubule thus reveals that the proportion of cells at the various stages will change according to its

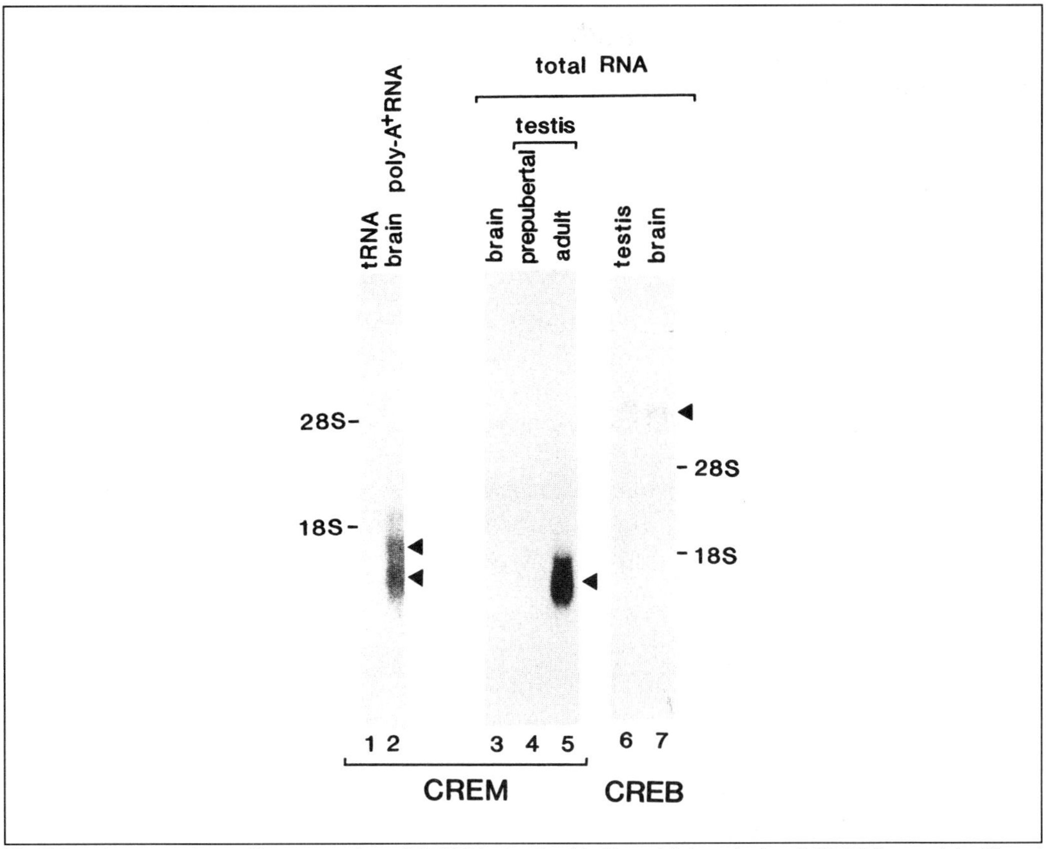

Fig. 7 Developmental-specific expression of CREM in testis. Northern blot analysis of CREM (lanes 1 to 5) and CREB (lanes 6 and 7), expression in poly-A^+ (lane 2), and total RNA samples (lanes 3 to 7). Filters containing 5 μg of brain poly-A^+RNA (lane 2) or 10 μg of total RNA (lanes 3, 4, and 5, respectively) were hybridized with a CREMα cDNA clone. A full length CREB cDNA fragment was hybridized to a filter containing 10 μg each of testis and brain total RNA (lanes 6 and 7). Arrowheads indicate CREM-specific bands in brain, the single 2-kb CREM band visible in testis RNA, and the position of the 7-kb CREB transcript.

position relative to the waves of differentiation. The cycle of the seminiferous epithelium has been defined as the series of changes occurring in a given area of the seminiferous epithelium between two successive appearances of the same cellular association. Using the rat testis, LeBlond and Clermont (96) and Perey *et al.* (97) described 14 (stages I–XIV) different cellular associations within the seminiferous epithelium and, subsequently, this nomenclature has been used widely. Furthermore, during the pre-pubertal development of the testis, the germ cells involved in the first round of spermatogenesis are synchronized in their development. Thus, 3–5 days after birth in the mouse, the spermatagonial progenitor cells start to proliferate mitotically; by 9–10 days the first wave of cells has differentiated into preleptotene spermatocytes; at day 13–14 these cells have differentiated into pachytene spermatocytes; by day 18 meiosis is complete and the cells constitute spermatids; finally, by 30–32 days the cells are forming condensing spermatids. The other testis cells, namely the Sertoli and Leydig cells, are present at all these stages of testicular development; however, the proportion of the total cell population that they constitute decreases as the germ-cell population proliferates. Thus, the day of appearance of a gene product during this early stage of development can be used to estimate in which cell type it is expressed (98).

The finding that the switch in CREM expression did not occur either in the testis of mice irradiated *in utero* or in the the mice mutant *olt*, but was still intact in the mutant *Tfm*, assigned CREMτ as a germ-cell-specific transcript (56). Irradiation *in utero* is known to selectively kill all germ cells and leave only the somatic cells intact, while the genetic mutants *olt* and *Tfm* have defects in spermatogenesis that arrest germ-cell development at the spermatocyte and spermatid stages, respectively (99–101). Subsequently, it was defined that the abundant CREMτ transcript appears by day 14. This, combined with analysis of CREMτ transcript expression in fractionated germ cells and *in situ* hybridization analysis, determined that CREMτ constitutes an abundant transcript from the pachytene spermatocyte stage onwards (56).

4.2.2 Function of the CREM activator during spermatogenesis

By immunohistochemistry and Western blot analysis with a CREM-specific antibody, CREMτ protein is not detected in pachytene spermatocytes but in more mature germ cells that have undergone meiosis (Fig. 8). Specifically, CREMτ protein is restricted to round spermatids, mainly at stages VII–VIII of seminiferous tubule differentiation (90). In mouse, overall transcription ceases at about stage IX, when the transition proteins and protamines replace the histones in order to compact and condense the chromatin (102). Thus, since there is no CREM protein in the spermatozoa, CREM transactivator function must be restricted to the late phase of transcription before the compaction of the DNA. The absence of the CREM protein in the spermatocytes could be ascribed to a translational delay of CREM mRNA. Translational control is an important regulatory mechanism of gene expression during spermatogenesis (for review see 98, 103). In particular, it has been reported that protamine gene mRNAs are accumulated at high levels, to be translated only at a later time, during the differentiation process (104, 105; for review see 106, 107).

The abundance of CREMτ protein begs the question as to the role that this abundant transactivator plays in haploid germ cells. Several genes have been identified that are transcribed at the time of appearance of the CREM protein and that include CRE-like sequences in their promoter regions (90; see Table 2). Furthermore, in the case of the testis-specific RT7 gene promoter, there are several lines of evidence directly implicating CREMτ as a tissue- and temporal-specific regulator (90, 108). CREM binds to the CRE-related sequence in the RT7 gene and is able to activate its expression in transient transfection assays. In addition, in an *in vitro* transcription system using a nuclear extract of seminiferous tubules, the addition of a CREM-specific antibody as well as an excess of CRE competitor decreases RT7 transcription (90). These results suggest that, by recognizing various CRE sequences, CREM

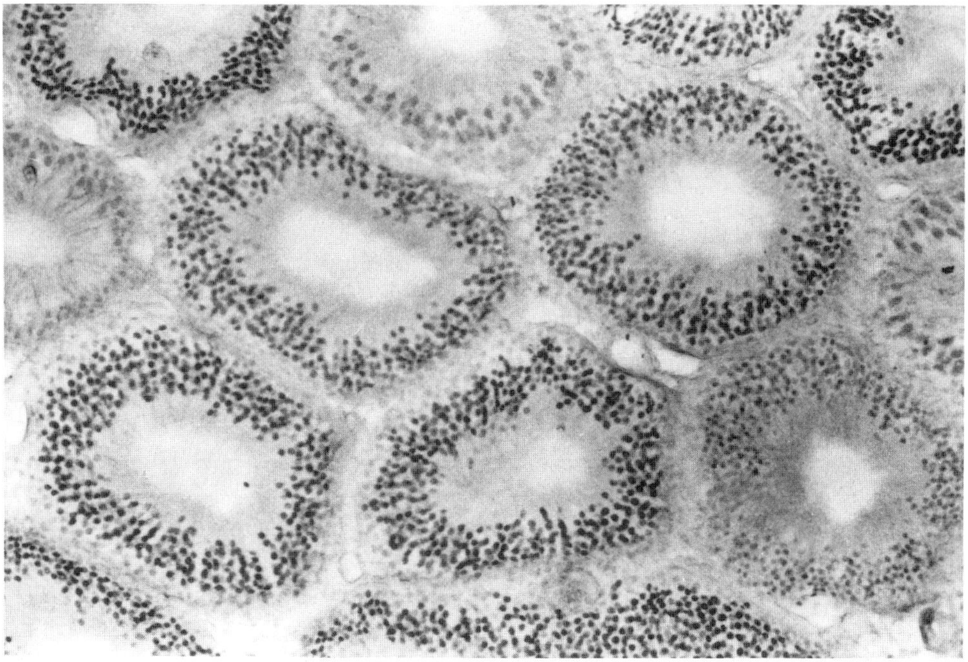

Fig. 8 CREMτ is expressed stage-specifically in developing spermatids. Peroxidase staining of rat seminiferous tubules showing expression of the CREM protein in spermatids. The CREM antibody used for this experiment was prepared against a bacterially produced CREMτ protein. Note the differential intensity of staining in the various tubules, indicating that CREM expression is developmentally regulated.

might participate in the testis-specific promoter activation of numerous haploid-expressed genes. Interestingly, most of the genes activated at the same time as the appearance of CREM encode structural proteins (90). For example, transition protein and protamine are detectable around day 22 during mouse spermatogenesis (103, 106, 107), exactly when CREM proteins are synthesized. In addition, the transition protein and protamine transcripts are visualized at the same stage of spermiogenesis (stage VII) in round spermatids (107, 109).

Recently CREB has also been implicated in playing specific roles during spermatogenesis. Namely, in addition to the expression of the novel, C-terminally truncated CREB proteins (see Section 3.3; 55, 84, 85) the levels of the CREB327 and CREB341 mRNAs have been documented to rise in Sertoli cells during certain parts of the 12-day cycle of the seminiferous tubule in rat (84). However, it should be noted that the levels of CREB transcript in testis are dramatically less than that of the abundant CREMτ transcript (56) and thus the impact of the reported CREB induction is questionable. Furthermore, the reported immunohistochemical analysis with CREB antibodies suggests that the distribution and levels of the CREB protein were similar to those of CREMτ (84, 90), with levels extremely high in the nuclei of spermatids during stages II–VIII. Unfortunately, in these experiments there was no

Table 2 Target CRE sites for CREMτ in testis

CRE sequence	Corresponding gene
TTGGC **TGACGTCA**GAGAG	Somatostatin
AATTGGG **TGAGGTCA**CTTAA	RT7
AGCTTCCTCTT **TGACTTCA**TAATTCCT	Protamine 1
TGGGCCG **ACAGGTCA**CAGTGGGG	Protamine 2
TATGTAG **TGACGTCA**CAAGAGAG	Transition protein 1
AGGACGT **GCTGGTCA**CCCCCAAA	Thymosin
TTGATCT **AGACGTCA**AAATTCCT	Tpx-1a
CTTCCTA **ATACGTCA**CAGCATCT	Tpx-1b

CRE-like sites that bind to CREMτ and are located in promoters activated in round spermatids. The eight base pair CRE-like elements are highlighted in boldface. The somatostatin CRE sequence is shown for comparison (154). In the Tpx-1 promoter (155), two motifs have been identified (designated 1a and 1b). The complete promoter sequences for RT7, protamine 1 and 2, transition protein 1, and thymosin are reported elsewhere (156–158).

control for the specificity of the CREB antibodies used (84), and thus it cannot be excluded that they were detecting the CREM protein. Indeed, the antibodies were raised against a synthetic peptide correponding to the P-box domain, which is highly conserved between CREB and CREMτ (56, 84). This casts further doubt on the functional significance of CREB expression in testis.

4.3 Hormonal control of CREM in testis: the key role of FSH

The spermatogenic developmental programme is under the tight control of the hypothalamus, which regulates the level of pituitary gonadotropins, luteinizing hormone (LH), and follicle-stimulating hormone (FSH) (110, 111). Hormonal stimulation, rather than acting directly on germ cells, is mediated by the Sertoli and Leydig cells (112, 113). In response to LH stimulation, the Leydig cells, located outside the seminiferous tubules in the interstitial area, produce steroid hormones. FSH and steroid hormones regulate Sertoli cell function depending on the differentiation stage of the tubule. The effects exerted on Sertoli cells by FSH and on Leydig cells by LH are mediated by the cAMP-dependent protein kinase pathway (113). Stimulation of adenylate cyclase-coupled receptors induces intracellular cAMP accumulation, which leads to the activation of the cAMP-dependent protein kinase A (PKA). Although no adenylate cyclase-coupled receptors have been characterized on germ cells, PKA activity changes during germ-cell differentiation to reach a maximum level in spermatozoa (114). Interestingly, testis-specific isoforms of the catalytic and regulatory subunits of PKA have been described, suggesting a difference in the substrate specificity of the PKA expressed in germ cells compared with the one found in somatic cells (115–117). The regulation of CREM function in testis seems to be intricately linked to FSH, both at the level of the control of transcript processing and at the level of protein activity.

CREMτ protein is phosphorylated by a cAMP-stimulated PKA activity in round spermatid extracts (90). Phosphorylation occurs at serine 117, the previously described phospho-acceptor site required for transactivation function (Fig. 4; 65). Thus, CREMτ could represent a developmentally regulated target of the cAMP-dependent PKA, since the increase in cAMP concentration elicited by the action of FSH must activate the testis-specific kinase (118). In this respect, it is noteworthy that FSH-stimulated cAMP production is stage-dependent in the seminiferous epithelium (119). At stage VII–VIII a significant FSH-stimulated cAMP production has been reported (119). Furthermore, it should also be noted that both the PKA regulatory subunit transcripts (RIIα and RIIβ) accumulate at high levels in seminiferous tubules at the same stages as the CREM protein (120).

Surgical removal of the pituitary gland leads to loss of CREMτ expression in the rat adult testis (57). Furthermore, in pre-pubertal animals, hypophysectomization prevents the switch in CREM expression at the pachytene spermatocyte stage, thus implicating the pituitary directly in the maintenance, as well as the switch to high levels, of CREMτ expression. The hormonal stimulus originating from the pituitary has been defined by physiological studies using the golden hamster (57). In the hamster, seasonal variation of gonadal activity is characterized by cessation of spermatogenesis during winter. This modulation is dependent on the photoperiod and can be experimentally reproduced by artificial lighting (121). Constant short-photoperiod (SP) causes complete gonadal atrophy within eight weeks (Fig. 9A; 122, 123). Atrophy is accompanied by a reduction in testosterone and pituitary hormone production (124, 125). Restoration of long-photoperiod (LP) causes a sharp rise in FSH, LH, and prolactin levels and then a progressive increase in testis size and recovery of spermatogenesis (Fig. 9A; 124, 126). In hamsters maintained under a constant short-photoperiod, the testes remain in the atrophied state for up to 12 weeks; after this period, however, they spontaneously begin to regenerate. CREMτ expression ceases during the SP-induced gonadal atrophy and then increases upon restoration of long-photoperiod. The absence of CREMτ expression in the atrophied testis coincides with low levels of circulating gonadotrophic hormones. Atrophied testes still contain spermatocytes and early spermatids (121), thus CREM expression is not solely under developmental control.

Injections of FSH into hamsters maintained in short-photoperiod with fully atrophied testes led to a rapid and significant induction of the CREMτ transcript (Fig. 9B; 57). Indeed, CREMτ expression in the FSH-induced animals is equivalent to sexually active adult animals. Nuclear run-on assay confirmed that the induction of CREMτ expression in testis was not associated with an increase in the rate of transcription initiation. Instead, the abundant CREMτ transcript is characterized by the use of an alternative polyadenylation site, which lies only 50 bp downstream of the stop codon (Fig. 10; 57). The use of this site generates a more stable transcript because nine out of the ten AUUUA destabilizer elements are excluded (see also Section 4.1.2). CREM is also the first example of a gene whose expression is modulated by a pituitary hormone during spermatogenesis (57). The effect on CREM expression described here could be mediated by yet another hormonal messenger

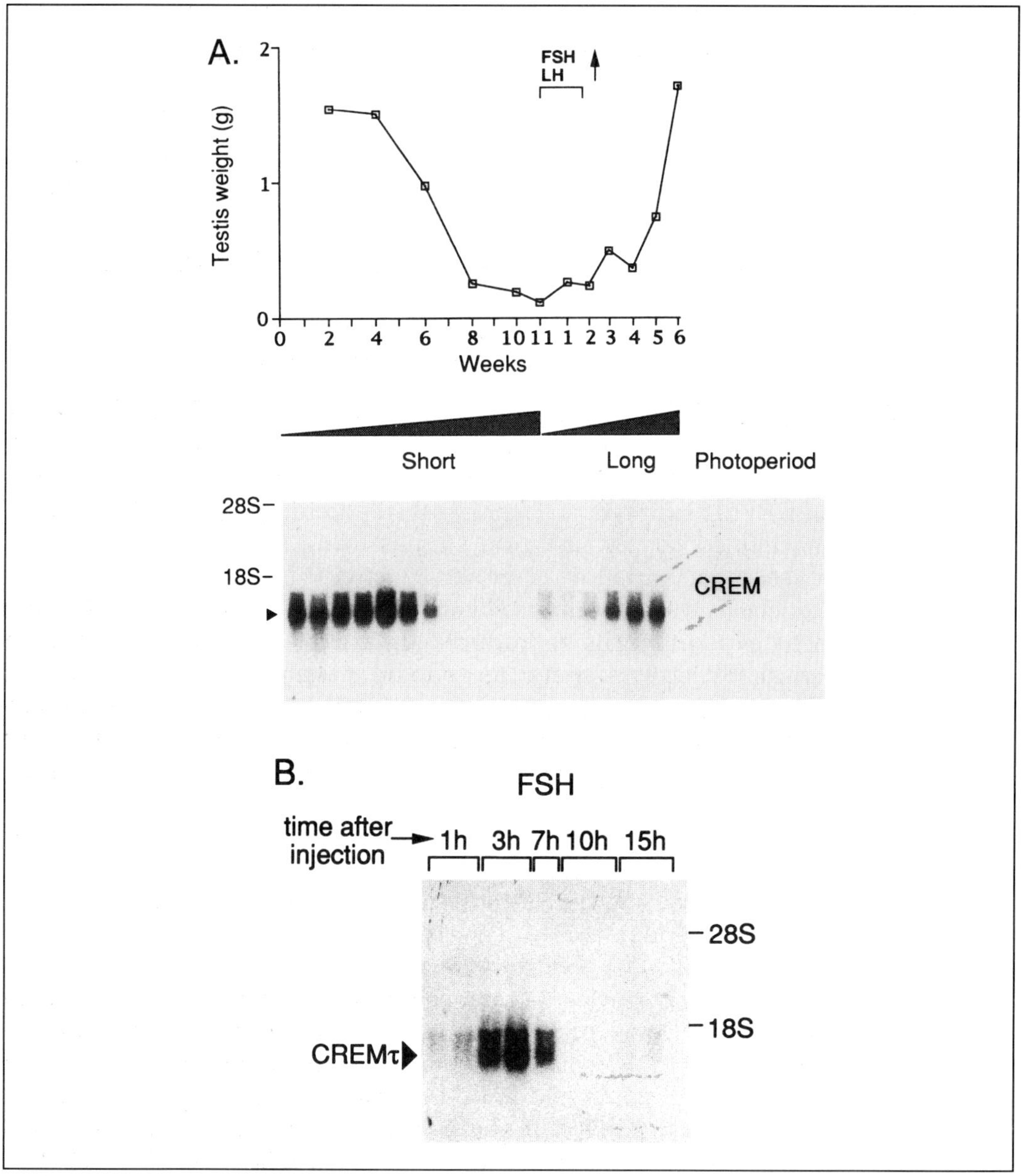

Fig. 9 Hormonal control of CREM expression in testis. A. Photoperiod control of CREM expression in hamster testis. Golden hamsters were exposed initially to short photoperiods for up to 12 weeks and then switched to long-photoperiod conditions for up to 6 weeks. The weight of the dissected testis was then determined (g) and plotted against time (weeks) [top panel]. The time period during which FSH and LH are induced following the change to long-photoperiod conditions is indicated by a bracket. Total RNA prepared from each of the animals was analysed by Northern blotting using a CREM probe. CREM expression is dramatically reduced in the atrophied testis. There is a recovery of expression upon restoration of the long photoperiod. B. Animals mantained in short photoperiod were injected with FSH and then sacrificed 1, 3, 7, 10, or 15 hours later. Testis total RNA was prepared and analysed for CREMτ expression by Northern blotting. CREMτ bands are indicated. There is a significant rise in CREMτ expression 3 hours after FSH injection

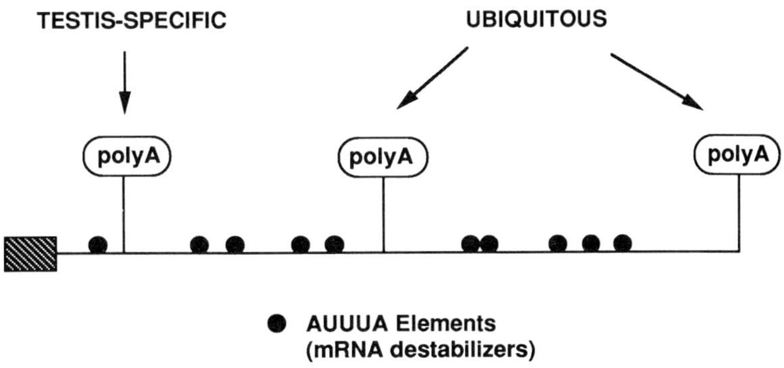

Fig. 10 In germ cells the abundant CREMτ transcript is polyadenylated at an alternative site, which confers increased stability. Schematic representation of the 3' untranslated region of CREM (46). The 3' end of the coding sequence is represented by a shaded box. The three polyadenylation signals (polyA) are non-canonical and are indicated; each AUUUA destabilizer (87) is represented by a circle. Use of the testis-specific site generates a transcript with a truncated 3' untranslated region and only one instability element. This transcript is intrinsically more stable than those containing multiple copies of AUUUA.

whose function would be to rapidly convey the FSH-induced signal from Sertoli to germ cells. Indeed, intimate interactions between these cell types have been described (127). One important implication of the FSH function illustrated here is that hormones can regulate gene expression at the level of RNA processing and stability. The wider consequence of these observations is that post-transcriptional events can constitute important targets for intracellular signal transduction pathways.

5. ICER: a new dimension to the neuroendocrine cAMP pathway

During studies of CREM expression within the neuroendocrine system, an unexpected new facet emerged; namely, the transcription of the CREM gene is inducible by cAMP (58, 61). Furthermore, the kinetics of this induction are those of an early response gene (128, 129). This important finding further reinforces the notion that CREM products play a fulcral role in the nuclear response to cAMP since the expression of no other CRE-binding factor has been shown to be inducible to date. For example, the recently characterized CREB promoter is GC-rich and reminiscent of the promoters of constitutively expressed housekeeping genes (60, 130). Upon detailed analysis of the induced CREM products, more surprises were in store. The promoter that directs expression of the previously characterized CREM isoforms (P1) is not cAMP inducible. Instead, an alternative promoter, lying within an intron near the 3′ end of the gene, directs the cAMP-induced transcription of a novel truncated CREM product, termed ICER (58, 61). Importantly, ICER is the smallest bZip factor yet described, functions as a powerful repressor of cAMP-induced transcription, and, furthermore, negatively autoregulates the ICER promoter (61). The expression of ICER was first described in the pineal gland where it is the subject of a dramatic circadian pattern of expression (58). However, recent data implicate dynamic ICER expression as a general feature of neuroendocrine systems (58).

5.1 Inducibility of the CREM gene: use of an alternative intronic promoter

Clues that the CREM gene was cAMP-inducible first came from the demonstration that adrenergic signals direct CREM transcription in the pineal gland (58). The inducibility phenomenon was then characterized in detail in the pituitary corticotroph cell line AtT20 (61). In unstimulated cells the level of CREM transcript is below the threshold of detectability. However, upon treatment with forskolin (or cAMP analogues), CREM transcript levels are detectable within 30 minutes, peak after 2 hours, and then progressively decline to basal levels by 5 hours. This characteristic kinetic classifies CREM as an 'early response gene' and thus directly implicates the cAMP pathway in the early response for the first time. CREM inducibility is specific for the cAMP pathway since it is not inducible by TPA or dexamethasone treatment (61).

In order to further characterize the cAMP-induced CREM transcript, a battery of exon-specific probes was used in systematic Northern blotting, RNAase protection, and RT-PCR analyses. In this way it was demonstrated that all of the previously characterized exons located 5′ to the γ-exon were absent from the induced CREM transcript. The 5′ boundary of the γ-exon is defined by a consensus splicing acceptor site. Thus, in order to identify the remaining exons constituting the 5′ end, a rapid amplification of cDNA ends (RACE)-PCR cloning strategy was used. By this approach a series of short, overlapping cDNA clones defined a novel cDNA, which is termed ICER [induced cAMP early repressor (58, 61)]. An 82-base pair sequence lying 5′ of the γ-exon boundary extends the CREM open reading frame upstream by eight amino acids to a consensus Kozak ATG codon and includes a short 5′ untranslated region. Full length cDNA clones isolated from a pineal cDNA library, together with RNAase protection assays, confirmed the results of RACE-PCR while primer extension analysis demonstrated that the 5′ end of the ICER clones correspond to a transcription start site (58, 61). The ICER cDNA clones also revealed differential splicing of the two alternative DNA-binding domains and of the γ-exon, as previously described for the CREMα and -β isoforms (Fig. 11). In addition, by Northern blot and RNase protection analyses it is apparent that the ICER transcripts employ alternative polyadenylation sites in a cell-specific fashion; thus, cAMP induction of the CREM gene generates a family of transcripts (61).

In order to locate the start point of transcription of the ICER transcripts relative to the promoter of the previously described isoforms, an overlapping series of phage clones encompassing the entire CREM gene were screened with a probe for the 5′ ICER-specific exon. Hybridization located the start of transcription (P2) within the

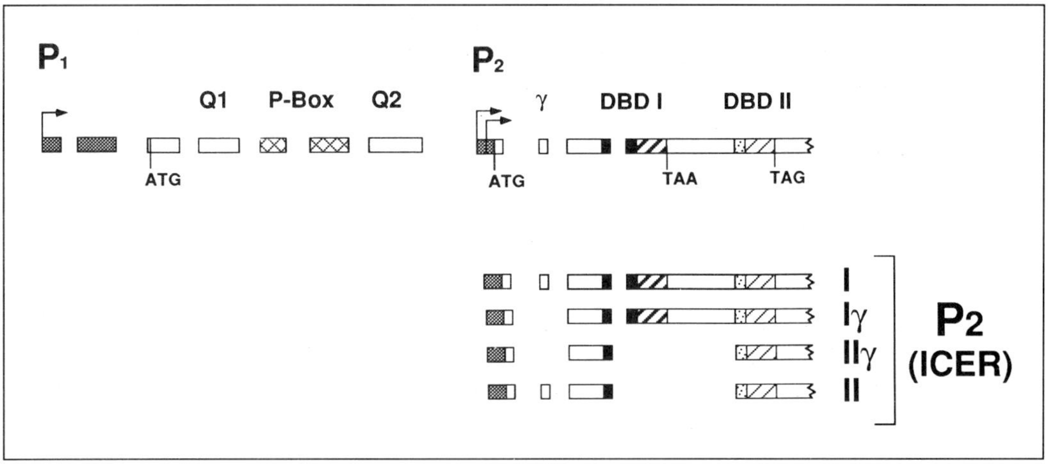

Fig. 11 The ICER family. Schematic representation of the ICER transcripts in relation to the CREM gene structure. Exons encoding the glutamine-rich domains (Q1 and Q2), the P-box, the γ domain (γ), and the two alternative DNA-binding domains (DBDI and DBDII) are shown. All the ICER transcripts are derived from an internal start site of transcription (P2) located between the Q2 and γ-exon. A family of four types of ICER transcript is generated by alternative splicing of the DBDs and γ-domain exons; ICER-I, ICER-Iγ, ICER-II, ICER-IIγ.

10-kb intron that lies between the Q2 glutamine-rich domain exon and the γ-exon (Fig. 12; 58, 61).

In contrast to the promoter that generates all of the previously characterized CREM isoforms (P1), which is GC-rich and not inducible by cAMP (N. Foulkes, in preparation), the P2 promoter has a normal A-T and G-C content and is strongly inducible by cAMP. It contains two pairs of closely spaced CRE elements (cAMP autoregulatory responsive elements, CAREs; Table 1) organized in tandem. Furthermore, the separation between the CAREs in each pair is only three nucleotides. These features make P2 unique amongst cAMP-regulated promoters and are suggestive of co-operative interactions among the factors binding to these sites (61). Tandemly repeated pairs of CRE elements have been previously described in the promoters of the Pit-1 and α-CG genes (131, 132), but the individual elements are more widely spaced. The P2 promoter directs transcription from two alternative start points (S1 and S2), and contains two TATA-like elements upstream from each start site.

5.2 ICER, the smallest bZip factor

The ICER open reading frame is constituted by the C-terminal segment of CREM. The predicted open reading frame encodes a small protein of 120 amino acids with an expected molecular weight of 13.4 kDa. This protein, compared with the previously described CREM isoforms, essentially consists of only the DNA-binding domain, which is constituted by the leucine zipper and basic region. The provocative structure of ICER is suggestive of its function and makes it one of the smallest transcription factors ever described (61). The intact DNA-binding domain directs

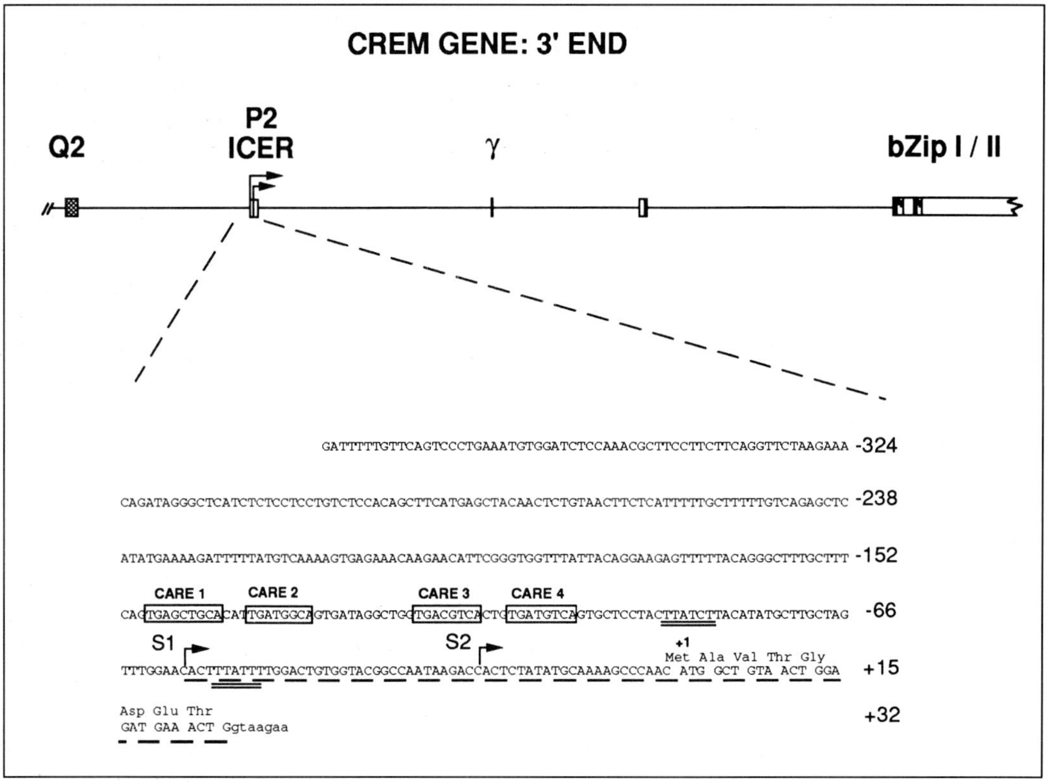

Fig. 12 The ICER P2 promoter. Above: schematic representation of the ICER exon and P2 promoter in the context of the genomic structure of the 3' end of the CREM gene. Flanking exons encoding the Q2 glutamine-rich domain, the γ-domain, and the 3' terminal exon incorporating the alternative bZip I/II domains are indicated. Below: 400 bp genomic sequence including the ICER 5' exon. Dashed underlining delineates the ICER 5' exon. Lower case sequence represents the beginning of the first intron of the ICER transcript. The two alternative start sites of transcription, S1 and S2, are shown. Putative TATA elements upstream of S1 and S2 are indicated by double underlining while the four CRE-like elements (CARE) are boxed and labelled. Position +1 corresponds to the A of the Kozak ATG initiation codon.

specific ICER binding to a consensus CRE element. Consistent with previous analysis of the other CREM isoforms, ICER fails to bind to AP-1 or Sp1 sites. Importantly, ICER is able to heterodimerize with CREMτ, as well as with the other CREM proteins and with CREB. ICER functions as a powerful repressor of cAMP-induced transcription in transfection assays using an extensive range of reporter plasmids carrying individual CRE elements or cAMP-inducible promoter fragments (58). Interestingly, ICER-mediated repression is obtained at sub-stoichiometric concentrations, similarly to the previously described CREM antagonists (46). However, it should be noted that ICER is significantly more potent when compared with CREMα or CREMβ (58).

The small size of the ICER products is striking and has possibly masked their presence in previously performed analyses. The smallest CRE-binding factor described previously was the liver-regenerating factor (LRF; 133). Based on homology, it is thought that it represents a splicing variant of ATF3 (45), although it is also conceivable that it could be generated by the use of an alternative promoter as in the case of ICER. This 21-kDa bZip protein functions as an activator of transcription in a heterodimeric complex with c-Jun, and its expression is also stimulated by mitogens. The major feature of ICER proteins is that they lack the N-terminal domain shared by the other CREM isoforms (Fig. 11). Specifically, the P-box domain, which is retained in all the P1 promoter-generated CREM isoforms, even those acting as repressors (CREMα, -β, and -γ), is absent from ICER (see Fig. 6 and Section 4.1). In CREMα, -β, and -γ, phosphorylation by PKA was shown to modulate the degree of repression activity (59). In contrast, ICER escapes from PKA-dependent phosphorylation and thus constitutes a new category of CRE-binding factor, for which the principle determinant of the activity is intracellular concentration and not the degree of phosphorylation.

5.3 Negative autoregulation

Upon co-treatment with cycloheximide, the kinetics of CREM gene induction by forskolin are altered in that there is a significant delay in the post-induction decrease in the transcript; elevated levels persist for as long as 12 hours. This implicates a *de novo* synthesized factor that might down-regulate CREM transcription (61). This observation, combined with the presence of CRE elements in the P2 promoter (see Section 5.1), suggested that the transient nature of the inducibility could be due to ICER. Consistently, the CARE elements in the P2 promoter have been shown to bind to the ICER proteins. Detailed studies have demonstrated that the ICER promoter is indeed a target for ICER negative regulation (61). Thus, there exists a negative autoregulatory mechanism controlling ICER expression (see Fig. 13).

6. Rhythmic expression

6.1 Day–night switch in ICER expression in the pineal gland

A crucial step in the understanding of the physiological significance of ICER came with the observation that it is expressed in a rhythmic fashion in the pineal gland (58). Crucial elements for the synchronization of biological rhythms in mammals are the pineal gland (134, 135) and the suprachiasmatic nucleus (SCN) (136, 137). Environmental lighting conditions are transduced by the pineal gland from a neuronal to an endocrine message, the rhythmic secretion of melatonin (Fig. 14; 134, 135, 138). This hormone synthesis is controlled by the SCN, being elevated at night and low during the day (135, 138). The cAMP-dependent signal transduction pathway serves as a relay to stimulate melatonin synthesis (138–140). Thus, from neuronal pathways that include the retina and the SCN, the pineal gland acts as a temporal regulator for the function of the hypothalamic–pituitary–gonadal axis (135). During the night, ICER constitutes an abundant transcript while during the

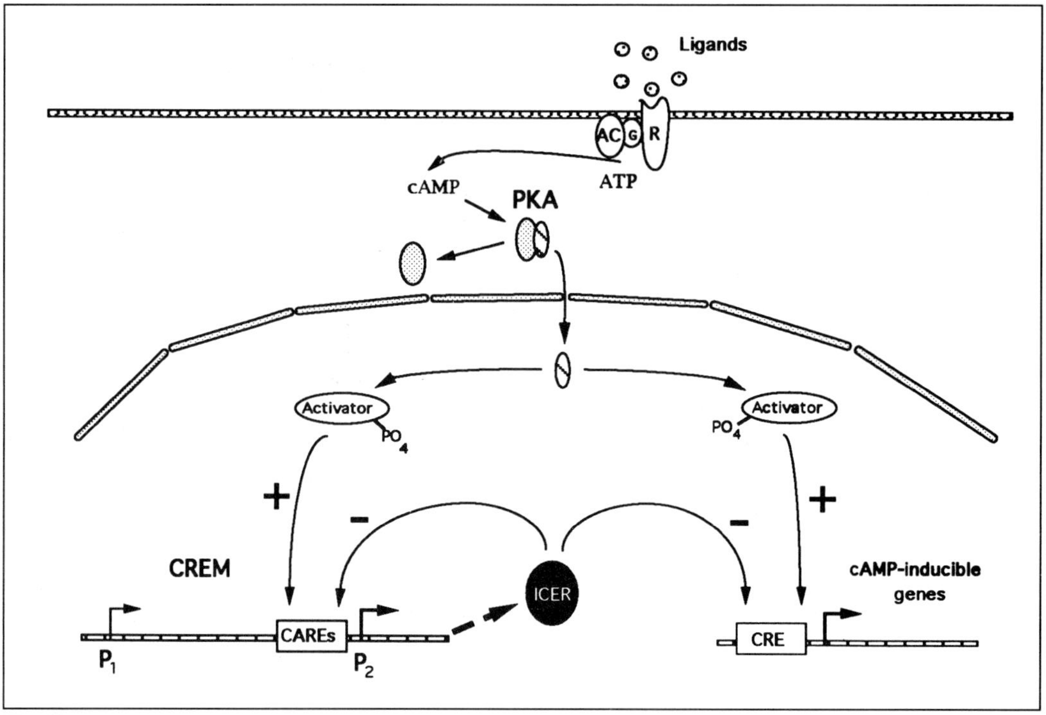

Fig. 13 The role of ICER in the regulation of gene expression by cAMP. Schematic representation of the cAMP signal transduction pathway operating from the cell membrane, through the cytoplasm, and into the nucleus. Ligands interacting with transmembrane receptors (R) stimulate the enzyme adenylate cyclase (AC) via interactions with G-proteins (G). The subsequent rise in intracellular cAMP concentration results in the dissociation of the regulatory and catalytic subunits of PKA and the translocation of active catalytic subunits into the nucleus. PKA phosphorylates and thereby stimulates transcriptional activators binding to CREs (e.g. CREB, CREMτ, and ATF1), which induce transcription from the promoters of cAMP-responsive genes. These factors activate transcription from the CREM P_2 promoter via the CARE elements and ultimately lead to a rapid increase in ICER protein levels. ICER represses cAMP-induced transcription, including that from its own promoter. The consequent fall in ICER protein levels eventually leads to a release of repression and permits a new cycle of transcriptional activation.

day it is present at low levels (Fig. 15; 58). An observation initially made by *in situ* hybridization was subsequently elaborated by RNase protection assay. For this analysis rats were sacrificed at hourly time points during a 24-hour cycle of light (12 hours) and dark (12 hours). The transcript shows a very characteristic and reproducible kinetic of expression (Fig. 15). By a series of physiological experiments, the mechanism controlling this pattern of ICER expression was determined (58). In animals maintained in constant darkness (DD) the temporal switch in ICER expression is conserved, which demonstrates that the ICER rhythm is driven by an endogenous clock mechanism (Fig. 16). In contrast, under constant light treatment (LL) no ICER switch is observed. LL conditions are known to dissociate the co-ordination of the clock output (Fig. 14; 141) and to suppress melatonin synthesis (142, 143). In addition, a light pulse at night, known to decrease melatonin production (144), also dramatically blocks ICER induction (Fig. 16; 58).

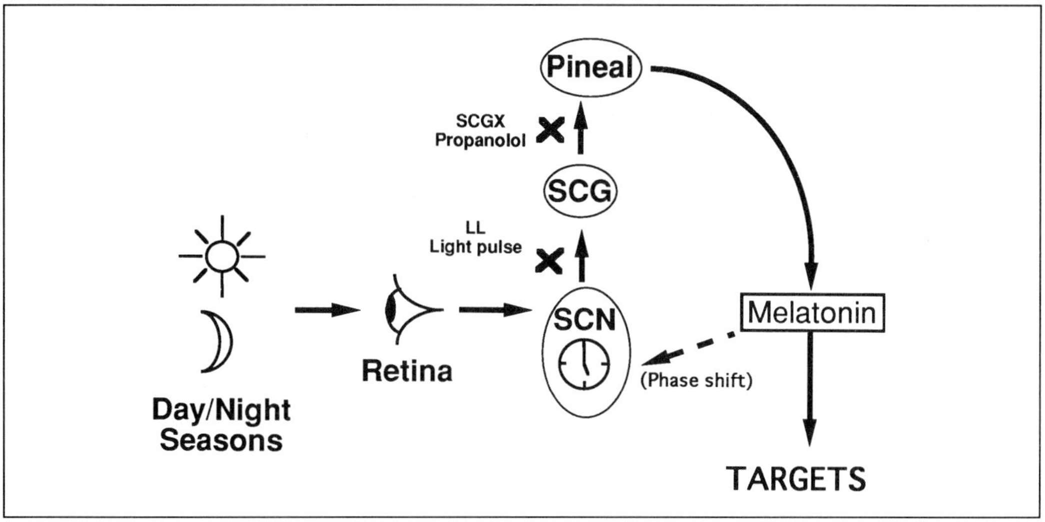

Fig. 14 Regulation of pineal gland function. Schematic representation of the pathway whereby day/night and seasonal changes in lighting conditions are detected by the retina and interpreted by the suprachiasmatic nucleus (SCN—the location of the biological clock; 118, 122). Rhythmic adrenergic signals originating from the SCN stimulate the pineal to produce the hormone melatonin. The nerve fibres linking the SCN and the pineal, pass through the superior cervical ganglion (SCG). While the specific targets for melatonin are still unclear, it is known to regulate the function of the SCN by phase-shifting the biological clock. In this way melatonin production synchronizes the biological clock with external time cues. Constant light (LL) or a light pulse at night blocks production of melatonin by inhibiting rhythmic stimulation by the SCN. Surgical lesioning of the SCG (SCGX) and the β-adrenergic antagonist propanolol block the normal β-adrenergic signals from stimulating the pineal at the level of the SCG and the pineal gland, respectively.

The next step was to define the signal whereby ICER circadian expression is driven by clock-distal elements. At night, post-ganglionic fibres originating from the superior cervical ganglia (SCG) release noradrenaline (145, 146), which in turn regulates melatonin synthesis via β-adrenergic receptors (Fig. 14; 138). Injection of the β-adrenergic antagonist propranolol prior to the onset of darkness blocks accumulation of ICER mRNA and melatonin synthesis (Fig. 16; 58, 147). The same effect was obtained by chronic denervation of pineal glands with bilateral removal of the SCG (SCGX). Furthermore, the β-adrenergic agonist isoproterenol (Iso) induces ICER expression in ganglionectomized animals, concomitantly with melatonin synthesis (Fig. 16; 58, 147).

Isoproterenol injections in intact rats are known to stimulate melatonin synthesis independently of the time of day (147). Unexpectedly, while induction during the night is observed, ICER is not inducible upon isoproterenol administration during the day (58). This is in contrast with SCGX animals, where isoproterenol induces ICER expression during both day and night. ICER represents the first case of day/night disparity in cAMP inducibility. In conclusion, signals from the SCN dictate the competence of CREM inducibility (58).

The question of possible targets for down-regulation by ICER in the gland can only be a matter for speculation at the moment. However, it has been proposed that

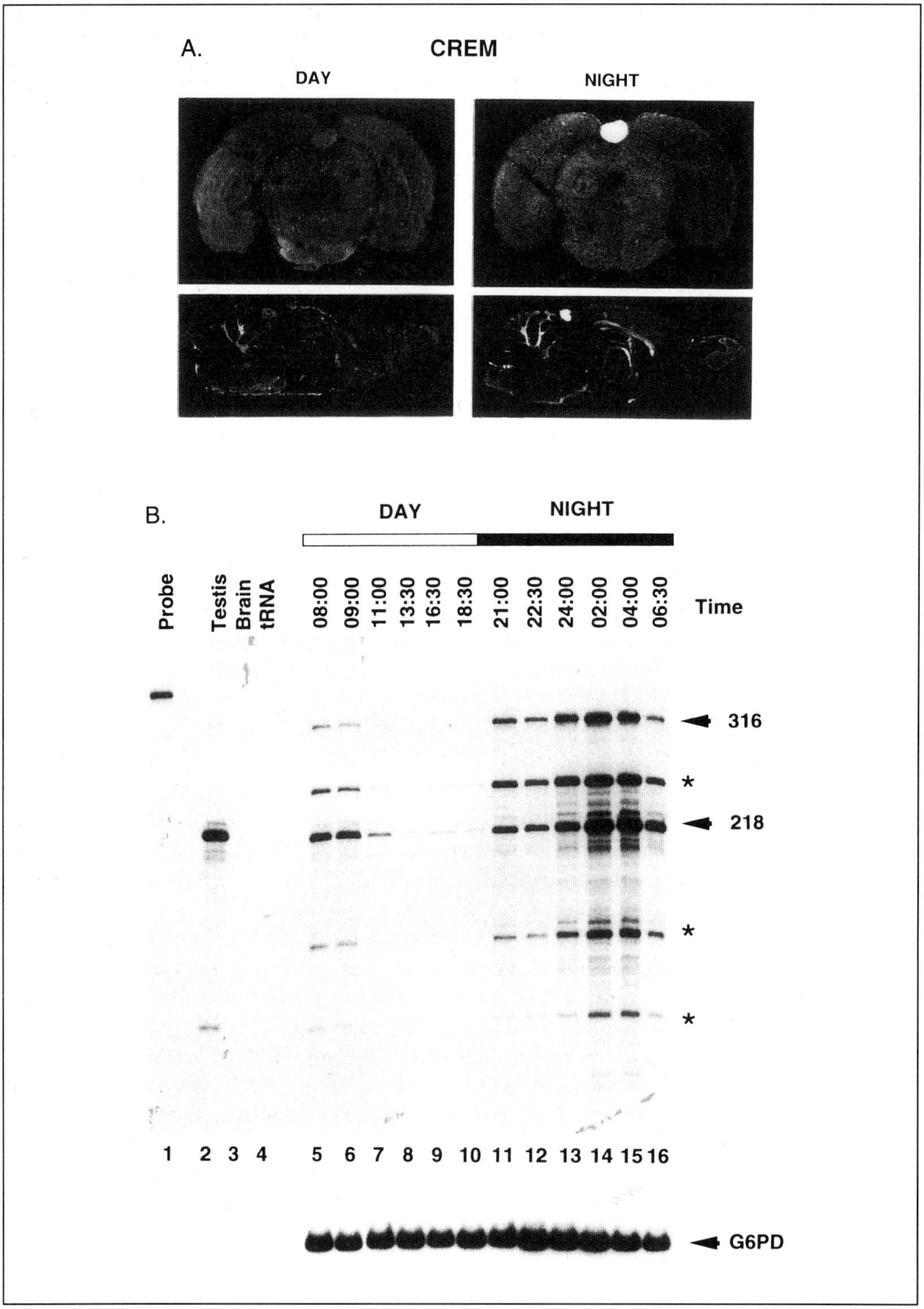
A.
CREM
DAY
NIGHT
B.
DAY
NIGHT
Probe
Testis
Brain
tRNA
08:00
09:00
11:00
13:30
16:30
18:30
21:00
22:30
24:00
02:00
04:00
06:30
Time
316
*
218
*
*
1 2 3 4 5 6 7 8 9 10 11 12 13 14 15 16
G6PD

Fig. 15 CREM expression is elevated at night in the pineal gland. A. *In situ* hybridization of brain sections with an antisense CREM-specific riboprobe. Rats maintained in 12h light/12h dark conditions (light on 07:00) were sacrificed at consecutive time points. Representative sections from 12:00 (DAY) and 02:00 (NIGHT) are shown. An intense hybridization signal is present in the pineal gland at night. B. RNase protection analysis using pineal gland total RNA at times indicated (lanes 5–16), or total RNA from mouse brain and testis (lanes 2–3). The sizes of bands corresponding to the specific CREM-protected fragments are shown; asterisks indicate bands generated by mismatches between rat RNA and the mouse CREM probe. The bands of 316nt and 218nt correspond to CREM RNA incorporating the first or second bZip, respectively.

a reasonable target could be the enzyme that catalyses the rate-limiting step of melatonin synthesis (148); namely, *N*-acetyl transferase (for which the gene has yet to be cloned) or factors that regulate its activity.

ICER expression was also analysed in a range of rat tissues by RNase protection assay (58). Dramatically, although the overall amount of CREM message varies considerably between different tissues, in each sample the major protected fragment corresponds to ICER, with the exception of adult testis (where the vast majority of the CREM mRNA corresponds to the P1-derived CREMτ transcript; see Section 4.2 and 58). An additional striking observation from this RNase protection analysis is a tissue-specific pattern of ICER expression. ICER is expressed at high levels predominantly in tissues of neuroendocrine origin, namely the pineal, pituitary, and adrenal glands. In all other neuronal and non-endocrine tissues tested, ICER is expressed in uniform low amounts (58).

6.2 CREM: transcriptional clock?

In the light of the known properties of ICER, it is tempting to speculate on the molecular basis of the non-inducibility in the pineal gland during the day. One possible explanation is that during the night the ICER protein accumulates and blocks cAMP-inducible transcription from the ICER promoter. With no ICER *de novo* production, the levels of ICER protein progressively decline until, by the beginning of the next night, the levels are sufficiently low to permit cAMP-directed activation. Activation is mediated by PKA, which phosphorylates pre-existing CRE-binding activators, which in turn re-stimulate the expression of ICER. In this scenario, the cyclic repression of ICER can be extended beyond the ICER promoter to include all other cAMP-controlled genes (Fig. 13). The efficiency of repression of a particular gene by ICER would dictate at what time, relative to ICER protein induction, the gene can be activated once again. In this way it is possible to envisage a cycle of transcriptional inducibility associated with rhythmic activation of the cAMP pathway. Thus, ICER could be thought of as a transcriptional clock. Given the specific expression of ICER in neuroendocrine systems, the phenomenon first observed in the pineal gland may well apply to other tissues. In other cases, the refractory period of non-inducibility that follows stimulation by cAMP may play

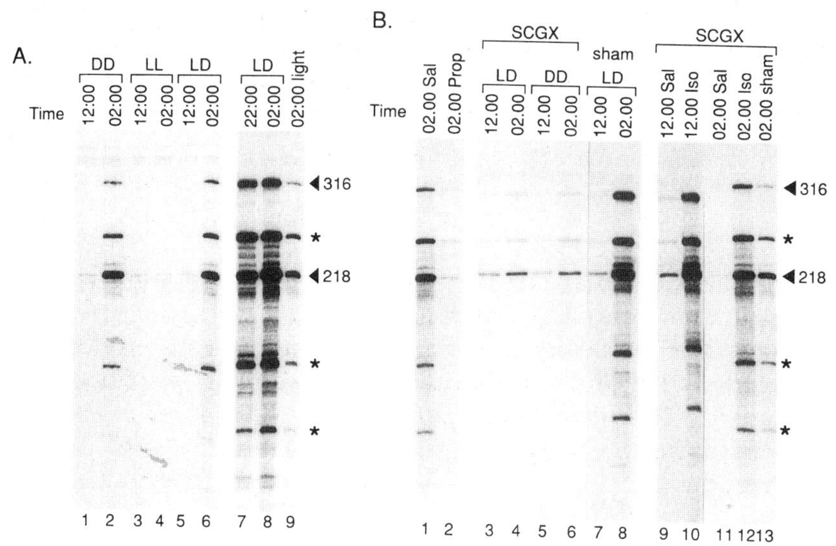

Fig. 16 CREM expression is under the control of the endogenous clock. A. RNase protection of pineal RNA from animals maintained in different lighting conditions (DD: constant darkness; LL: constant light; LD: 12h light/12h dark; lanes 1–6). Animals were sacrificed during the day (12:00) or night (02:00). In addition, animals kept under LD (lanes 7–8) show a dramatic reduction in CREM expression when exposed to 4 h bright white light (lane 9), starting at 22:00. B. Injection of rats with the β-adrenergic antagonist propranolol prior to the onset of darkness (18:45) and sacrificed at 02:00 caused a dramatic reduction in CREM expression (lane 2); injection of a saline (Sal) solution is shown as control (lane 1). Lanes 3–6: CREM expression at day and night in animals in which, 10 days prior to experimentation, the superior cervical ganglion was removed bilaterally (SCGX). One group of SCGX animals was transferred one day prior to experimentation into constant darkness (DD, lanes 5–6). Rats were sacrificed at times indicated. One group of animals was sham operated (lanes 7–8). Lanes 9–12: effect of injection of the β-adrenergic agonist isoproterenol (Iso) on CREM expression. SCGX animals received the injection 2 h before indicated time of sacrifice. Lane 13: pineal RNA from a control group of intact rats sacrificed at 02:00. The higher level of CREM expression in the injected SCGX animals (compare lanes 12 and 13) is likely to be due to supersensitivity of the pineal β-adrenergic receptors.

a central role in interpreting and responding to repetitive signals impinging on neuroendocrine cells.

7. Conclusions and perspectives

To date, much of the research in transcription factor biology has been devoted to understanding the structure and function relationship of these factors. Progress has been extremely rapid and now the basic principles of transcription factor function are close to being elucidated. As a result of this work new questions have been raised and so it is clear we still have a long way to go before we understand completely how the promoters and enhancers of genes execute transcriptional control. However, a much greater challenge lies ahead, and that is to relate transcriptional control mechanisms to the physiology and biology of the organism. The use of homologous recombination to inactivate specific gene products offers a powerful tool to address such questions. Paradoxically, however, in some cases it has complicated our understanding since it is clear that many important factors operate in the context of networks where there is considerable overlap of function. As in the case of the CRE-binding factors, it is likely that it will be necessary to knock out several genes in order to obtain a phenotype. Furthermore, it is possible that the phenotype obtained by loss of a single factor could reflect more the compensatory adjustments made by other factors in its network rather than the function of the target factor itself. In this context, it is important to stress that the inactivation by homologous recombination of a non-inducible constitutive CRE activator, the CREB gene, demonstrates redundancy within this family of transcription factors (159). In contrast, CREM appears to play a central role in the physiology of the neuroendocrine system (149, 150). The generation of a mouse with an inactivated CREM gene should therefore represent another major step forward in our understanding since it is likely to provide an informative phenotype.

Acknowledgments

We wish to thank all the members of the CREM group: E. Lalli, J. S. Lee, L. Monaco, E. Zazopoulos, K. Tamai, M. Lamas, F. Nantel, and also P. Pévet, J. Stehle, E. Borrelli, P. Garcia-Villalba, B. Boulay, J. P. Montmayeur, R. P. deGroot, B. M. Laoide, and J. R. Naranjo. The work in the P. Sassone-Corsi laboratory is supported by CNRS, INSERM, ULP, ARC, CHUR, FRM, and Rhône-Poulenc Rorer.

References

1. Harrison, S. C. (1991) A structural taxonomy of DNA-binding domains. *Nature*, **353**, 715.
2. Pabo, C. O. and Sauer, R. T. (1984) Protein-DNA recognition *Ann. Rev. Biochem.*, **53**, 293.
3. Gehring, W. J., Müller, M., Affolter, M., Percival-Smith, A., Billeter, M., Qian, Y. Q., Otting, G., and Wüthrich, K. (1990) The structure of the homeodomain and its functional implications. *Trends Genet.*, **6**, 323.
4. Miller, J., McLachlan, A. D., and Klug, A. (1985) Repetitive zinc-binding domains in the protein transcription factor IIIA from *Xenopus* oocytes. *EMBO J.*, **4**, 1609.
5. Evans, R. M. and Hollenberg, S. M. (1988) Zinc fingers: gilt by association. *Cell*, **52**, 1.
6. Orkin, S. H. (1990) Globin gene regulation and switching: circa 1990. *Cell*, **63**, 665.
7. Murre, C., Schonleber-McCaw, P., and Baltimore, D. (1989) A new DNA binding and dimerization motif in immunoglobulin enhancer binding, *daughterless*, MyoD, and myc proteins. *Cell*, **56**, 777.
8. Blackwood, E. M. and Eisenman, R. N. (1991) Max, a helix-loop-helix protein that forms a sequence-specific DNA-binding complex with Myc. *Science*, **251**, 1211.
9. Alber, T. (1992) Structure of the leucine zipper. *Curr. Opin. Genet. Dev.*, **2**, 205.
10. Kerppola, T. K. and Curran, T. (1991) Transcription factor interactions: basic on zippers. *Curr. Opin. Struct. Biol.*, **1**, 71.

11. Pathak, D. and Sigler, P. B. (1992) Updating structure-function relationships in the bZIP family of transcription factors. *Curr. Opin. Struct. Biol.*, **2**, 116.
12. Landschulz, W. H., Johnson, P. F., and McKnight, S. L. (1988) The leucine zipper: a hypothetical structure common a new class of DNA binding protein. *Science*, **240**, 1759.
13. O'Shea, E. K., Rutkowski, R., and Kim, P. S. (1989) Evidence that the leucine zipper is a coiled-coil. *Science*, **243**, 538.
14. O'Shea, E. K., Klemm, J. D., Kim, P. S., and Alber, T. (1991) X-ray structure of the GCN4 leucine zipper, a two-stranded, parallel coiled-coil. *Science*, **254**, 539.
15. Saudek, V., Pastore, A., Castiglione-Morelli, M. A. Frank, R., Gausepohl, H., and Gibson, T. (1991) The solution structure of a leucine-zipper motif peptide. *Protein Engr.*, **4**, 519.
16. Vinson, C. R., Sigler, P. B., and McKnight, S. L. (1989) Scissor-grip model for DNA recognition by a family of leucine zipper proteins. *Science*, **246**, 911.
17. O'Neil, K. T., Hoess, R. H., and Delgado, W. F. (1990) Design of DNA-binding peptides based on the leucine zipper motif. *Science*, **249**, 774.
18. Baxevanis, A. D. and Vinson, C. R. (1993) Interactions of coiled-coils in transcription factors: where is the specificity? *Curr. Opin. Genet. Dev.*, **3**, 278.
19. Cohen, C. and Parry, D. A. D. (1990) α-helical coiled-coils and bundles how to design an α-helical protein. *Proteins*, **7**, 1.
20. Crick, F. H. C. (1953) The packing of α-helices: simple coiled-coils. *Acta Cristallogr.*, **6**, 689.
21. Hu, J. C. and Sauer, R. T. (1992) The basic-region leucine-zipper family of DNA binding proteins. In *Nucleic acids and molecular biology* (ed. F. Eckstein and D. Lilley). Springer-Verlag, New York.
22. Conway, J. F. and Parry, D. A. D. (1990) Structural features in the heptad substructure and longer range repeats of two-stranded α-fibrous proteins. *Int. J. Biol. Macromol.*, **12**, 328.
23. Kouzarides, T. and Ziff, E. B. (1988) The role of the leucine zipper in the fos-jun interaction. *Nature*, **336**, 646.
24. Sassone-Corsi, P., Ransone, L. J., Lamph, W. W., and Verma, I. M. (1988) Direct interaction between fos and jun oncoproteins: role of the leucine zipper domain. *Nature*, **336**, 692.
25. Schuermann, M., Neuberg, M., Hunter, J. B., Jenuwein, T., Ryseck, R. P., Bravo, R., and Muller, R. (1989) The leucine zipper motif in Fos protein mediates complex formation with Jun/AP-1 and is required for transformation. *Cell*, **56**, 507.
26. Gentz, R., Rauscher, F. J., Josephs, S. F., and Curran, T. (1989) Parallel association of Fos and Jun leucine zippers juxtaposes DNA binding domains. *Science*, **243**, 1695.
27. Turner, R. and Tjian, R. (1989) Leucine repeats and an adjacent DNA binding domain mediate the formation of functional c-fos-c-jun heterodimers. *Science*, **243**, 1689.
28. Nicklin, M. J. H. and Casari, G. (1991) A single site mutation in a truncated Fos protein allows it to interact with the TRE *in vitro*. *Oncogene*, **6**, 173.
29. O'Shea, E. K., Rutkowski, R., and Kim, P. S. (1992) Mechanism of specificity in the Fos-Jun oncoprotein heterodimer. *Cell*, **68**, 699.
30. Schuermann, M., Hunter, J. B., Hennig, G., and Muller, R. (1991) Non-leucine residues in the leucine zipper region of Fos and Jun contribute to the stability and determine the specificity of dimerization. *Nucl. Acids Res.*, **19**, 739.
31. Nishizuka, Y. (1986) Studies and perspectives of protein kinase C. *Science*, **233**, 305.
32. Cambier, J. C., Newell, N. K., Justement, L. B., McGuire, J. C., Leach, K. L., and Chen, Z.

Z. (1987) Ia binding ligands and cAMP stimulate nuclear translocation of PKC in B lymphocytes. *Nature*, **327**, 629.
33. Yoshimasa, T., Sibley, D. R., Bouvier, M., Lefkowitz, R. J., and Caron, M. G. (1987) Cross-talk between cellular signalling pathways suggested by phorbol ester adenylate cyclase phosphorylation. *Nature*, **327**, 67.
34. Masquilier, D. and Sassone-Corsi, P. (1992) Transcriptional cross-talk: nuclear factors CREM and CREB bind to AP-1 sites and inhibit activation by Jun. *J. Biol. Chem.*, **267**, 22460.
35. McKnight, S. G., Clegg, C. H., Uhler, M. D., Chrivia, J. C., Cadd, G. G., Correll, L. A., and Otten, A. D. (1988) Analysis of the cAMP-dependent protein kinase system using molecular genetic approaches. *Rec. Progr. Horm. Res.*, **44**, 307.
36. Montmayeur, J. P. and Borrelli, E. (1991) Transcription mediated by a cAMP-responsive promoter element is reduced upon activation of dopamine D2 receptors. *Proc. Natl Acad. Sci. USA.*, **88**, 3135.
37. Krebs, E. G. and Beavo, J. A. (1979) Phosphorylation-dephosphorylation of enzymes. *Ann. Rev. Biochem.*, **48**, 923.
38. Deutsch, P. J., Hoeffler, J. P., Jameson, J. L., and Habener, J. F. (1988) Cyclic AMP and phorbol ester-stimulated transcription mediated by similar DNA elements that bind distinct proteins. *Proc. Natl Acad. Sci. USA*, **85**, 7922.
39. Liou, H.-C., Boothby, M. R., and Glimcher, L. H. (1988) Distinct cloned class II MHC DNA-binding proteins recognize the X-box transcription element. *Science*, **242**, 69.
40. Yoshimura,T., Fujisawa, J.-I., and Yoshida, M. (1990) Multiple cDNA clones encoding nuclear proteins that bind to the *tax*-dependent enhancer of HTLV-I: all contain a leucine zipper structure and basic amino acid domain. *EMBO J.*, **9**, 2537.
41. Wagner, S. and Green, M. R. (1993) HTLV-I *tax* protein stimulation of DNA-binding of bZip proteins by enhancing dimerization. *Science*, **262**, 395.
42. Lin, Y. and Green, M. R. (1988) Interaction of a common cellular transcription factor, ATF, with regulatory units in both E1A and cyclic AMP inducible promoters. *Proc. Natl Acad. Sci. USA*, **85**, 3396 .
43. Sassone-Corsi, P. (1988) Cyclic AMP induction of early adenovirus promoters involves sequences required for E1A-transactivation. *Proc. Natl Acad. Sci. USA.*, **85**, 7192.
44. Lee, K. A. W., Fink, S. J., Goodman, R. H., and Green, M. R. (1989) Distinguishable promoter elements are involved in transcriptional activation by E1A and cyclic AMP. *Mol. Cell. Biol.*, **9**, 4390 .
45. Hai, T.-Y., Liu, F., Coukos, W. J., and Green M. R. (1989) Transcription factor ATF cDNA clones: an extensive family of leucine zipper proteins able to selectively form DNA binding heterodimers. *Genes Dev.*, **3**, 2083.
46. Foulkes, N. S., Borrelli, E., and Sassone-Corsi, P. (1991) CREM gene: use of alternative DNA binding domains generates multiple antagonists of cAMP-induced transcription. *Cell*, **64**, 739.
47. Ivashkiv, L. B., Liou, H. C.,Kara, C. J., Lamph, W. W., Verma, I. M., and Glimcher, L. H. (1990) mXBP/CRE-BP2 and c-*jun* form a complex which binds to the cAMP, but not to the 12-O-tetradecanoyl phorbol-13-acetate response element. *Mol. Cell. Biol.*, **10**, 1609.
48. Gaire, M., Chatton, B., and Kedinger, C. (1990) Isolation and characterisation of two novel, closely related ATF cDNA clones from HeLa cells. *Nucl. Acids Res.*, **18**, 3467.
49. Hurst, H. C., Totty, N. F., and Jones, N. C. (1991) Identification and functional characterization of the cellular activating transcription factor 43 (ATF-43) protein. *Nucl. Acids Res.*, **19**, 4601.

50. Hoeffler, J. P., Lustbader, J. W., and Chen, C.-Y. (1991) Identification of multiple nuclear factors that interact with cyclic adenosine 3′, 5′-monophosphate response element-binding protein and activating transcription factor-2 by protein-protein interactions. *Mol. Endocrinol.*, **5**, 256.
51. Sassone-Corsi, P., Ransone, L. J., and Verma, I. M. (1990) Cross-talk in signal transduction: TPA-inducible factor Jun/AP-1 activates cAMP responsive enhancer elements. *Oncogene*, **5**, 427.
52. Hai, T.-Y. and Curran, T. (1991) Cross-family dimerization of transcription factors Fos:Jun and ATF/CREB alters DNA binding specificity. *Proc. Natl Acad. Sci. USA*, **88**, 3720.
53. Yamamoto, K. K., Gonzales, G. A., Menzel, P., Rivier, J., and Montminy, M. R. (1990) Characterisation of a bipartite activator domain in transcription factor CREB. *Cell*, **60**, 611.
54. Meyer, T. E. and Habener, J. F. (1993) Cyclic adenosine 3′5′-monophosphate response element binding protein (CREB) and related transcription-activating deoxyribonucleic acid binding proteins. *Endocrine Reviews*, **14**, 269.
55. Ruppert, S., Cole, T. J., Boshart, M., Schmid, M., and Schütz, G. (1992) Multiple mRNA isoforms of the transcription activator protein CREB: generation by alternative splicing and specific expression in primary spermatocytes. *EMBO J.*, **11**, 1503.
56. Foulkes, N. S., Mellström, B., Benusiglio, E., and Sassone-Corsi, P. (1992) Developmental switch of CREM function during spermatogenesis: from antagonist to transcriptional activator. *Nature*, **355**, 80.
57. Foulkes, N. S., Schlotter, F., Pévet, P., and Sassone-Corsi, P. (1993) Pituitary hormone FSH directs the CREM functional switch during spermatogenesis. *Nature*, **362**, 264.
58. Stehle, J. H., Foulkes, N. S., Molina, C. A., Simonneaux, V., Pévet, P., and Sassone-Corsi, P. (1993) Adrenergic signals direct rhythmic expression of transcriptional repressor CREM in the pineal gland. *Nature*, **365**, 314.
59. Laoide, B. M., Foulkes, N. S., Schlotter, F., and Sassone-Corsi, P. (1993) The functional versatility of CREM is determined by its modular structure. *EMBO J.*, **12**, 1179.
60. Cole, T. J., Copeland, N. Cr., Gilbert, D. J., Jenkins, N. A., Schütz, G., and Ruppert, S. (1992) The mouse CREB (cAMP responsive element binding protein) gene: structure, promoter analysis and chromosomal localization. *Genomics*, **13**, 974.
61. Molina, C. A., Foulkes, N. S., Lalli, E., and Sassone-Corsi P. (1993) Inducibility and negative autoregulation of CREM: an alternative promoter directs the expression of ICER, an early response repressor. *Cell*, **75**, 875.
62. Cowell, I. G., Skinner, A., and Hurst, H. C. (1992) Transcriptional repression by a novel member of the bZIP of transcription factors. *Mol. Cell. Biol.*, **12**, 3070.
63. Karponski, B. A., Morle, G. D., Huggenvik, J., Uhler, M. D., and Leiden, J. M. (1992) Molecular cloning of human CREB-2: an ATF/CREB transcription factor that can negatively regulate transcription from the cAMP response. *Proc. Natl Acad. Sci. USA*, **89**, 4820.
64. Gonzalez, G. A. and Montminy, M. R. (1989) Cyclic AMP stimulates somatostatin gene transcription by phosphorylation of CREB at ser 133. *Cell*, **59**, 675.
65. de Groot, R. P., den Hertog, J., Vandenheede, J. R., Goris, J., and Sassone-Corsi, P. (1993) Multiple and cooperative phosphorylation events regulate the CREM activator function. *EMBO J.*, **12**, 3903.
66. Habener, J. (1990) Cyclic AMP response element binding proteins: a cornucopia of transcription factors. *Mol. Endocrionol.*, **4**, 1087.

67. Lee, C. Q., Yun, Y., Hoeffler, J. P., and Habener, J. F. (1990) Cyclic-AMP-responsive transcriptional activation involves interdependent phosphorylated subdomains. *EMBO J.*, **9**, 4455.
68. Rehfuss, R. P., Walton, K. M., Loriaux, M. M., and Goodman, R. H. (1991) The cAMP-regulated enhancer-binding protein ATF-1 activates transcription in response to cAMP-dependent protein kinase A. *J. Biol. Chem.*, **266**, 18431.
69. Flint, K. J. and Jones, N. C. (1991) Differential regulation of three members of the ATF/CREB family of DNA-binding proteins. *Oncogene*, **6**, 2019.
70. Gonzalez, G. A., Menzel, P., Leonard, J., Fischer, W. H., and Montminy, M. R. (1991) Characterization of motifs which are critical for activity of the cyclic AMP-responsive transcription factor CREB. *Mol. Cell. Biol.*, **11**, 1306.
71. Sassone-Corsi, P., Visvader, J., Ferland, L., Mellon, P. L., and Verma, I. M. (1988) Induction of proto-oncogene *fos* transcription through the adenylate cyclase pathway: characterization of a cAMP-responsive element. *Genes Dev.*, **2**, 1529.
72. Sheng, M., McFadden, G., and Greenberg, M. E. (1990) Membrane depolarization and calcium induce c-*fos* transcription via phosphorylation of transcription factor CREB. *Neuron*, **4**, 571.
73. Sheng, M., Thompson, M. A., and Greenberg, M. E. (1991) CREB: a Ca^{2+}-regulated transcription factor phosphorylated by calmodulin-dependent kinases. *Science*, **252**, 1427.
74. Dash, P. K., Karl, K. A., Colicos, M. A., Prywes, R., and Kandel, E. R. (1991) cAMP response element-binding protein is activated by Ca^{2+}/calmodulin- as well as cAMP-dependent protein kinase. *Proc. Natl Acad. Sci. USA*, **88**, 5061.
75. Ginty, D. D., Glowacka, D., Bader, D. S., Hidaka, H., and Wagner, J. A. (1991) Induction of immediate early genes by Ca^{2+} influx requires cAMP-dependent protein kinase in PC12 cells. *J. Biol. Chem.*, **266**, 17454.
76. Nichols, M., Weih, F., Schmid, W., DeVack, C., Kowenz-Leutz, E., Luckow, B., Boshart, M., and Schütz, G. (1992) Phosphorylation of CREB affects its binding to high and low affinity sites: implications for cAMP-induced gene transcription. *EMBO J.*, **11**, 3337.
77. Yamamoto, K. K., Gonzalez, G. A., Briggs, III, W. H., and Montminy, M. R. (1988) Phosphorylation-induced binding and transcriptional efficiency of nuclear factor CREB. *Nature*, **334**, 494.
78. Hagiwara, M., Alberts, A., Brindle, P., Meinkoth, J., Feramisco, J., Deng, T., Karin, M., Shenolikar, S., and Montminy, M. (1992) Transcriptional attenuation following cAMP induction requires PP-1-mediated dephosphorylation of CREB. *Cell*, **70**, 105.
79. Williams, T., Admon, A., Luscher B., and Tjian, R. (1988) Cloning and expression of AP-2, a cell-type-specific transcription factor that activates inducible enhancer elements. *Genes Dev.*, **2**, 1557.
80. Courey, A. J. and Tjian, R. (1989) Analysis of Sp1 *in vivo* reveals multiple transcriptional domains, including a novel glutamine activation motif. *Cell*, **55**, 887.
81. Brindle, P., Linke, S., and Montminy, M. R. (1993) Protein-kinase-A-dependent activator in transcription factor CREB reveals new role for CREM repressors. *Nature*, **364**, 821.
82. Chrivia, J. C., Kwok, R. P. S., Lamb, N., Haniwawa, M., Montminy, M. R., and Goodman, R. H. (1993) Phosphorylated CREB binds specifically to the nuclear protein CBP. *Nature*, **365**, 855.
83. Hoeffler, J. P., Meyer, T. E., Waeber, G., and Habener, J. F. (1990) Multiple adenosine 3′-5′-monophosphate response element DNA binding proteins generated by gene diversification and alternative exon splicing. *Mol. Endocrinol.*, **4**, 920.
84. Waeber, G., Meyer, T. E., LeSieur, M., Hermann, H. L., Gérard, N., and Habener, J.

(1991) Developmental stage-specific expression of cyclic adenosine 3′, 5′ monophosphate response element-binding protein CREB during spermatogenesis involves alternative exon splicing. *Mol. Endocrinol.*, **5**, 1418.

85. Waeber, G. and Habener, J. (1992) Novel testis germ cell-specific transcript of the CREB gene contains an alternatively spliced exon with multiple in-frame stop codons. *Endocrinology*, **131**, 2010.
86. de Groot, R. P. and Sassone-Corsi, P. (1993) Hormonal control of gene expression: multiplicity and versatility of cyclic adenosine 3′, 5′-monophosphate-responsive nuclear regulators. *Mol. Endocrinol.*, **7**, 145.
87. Shaw, G. and Kamen, R. (1986) A conserved AU sequence from the 3′ untranslated region of GM-CSF mRNA mediates selective mRNA degradation. *Cell*, **46**, 659.
88. Raymond, V., Atwater, J. A., and Verma, I. M. (1989) Removal of an mRNA destabilizing element correlates with the increased oncogenicity of proto-oncogene *fos*. *Oncogene Res.*, **5**, 1.
89. Delmas, V., Laoide, B. M., Masquilier, D., de Groot, R. P., Foulkes, N. S., and Sassone-Corsi, P. (1992) Alternative usage of initiation codons in mRNA encoding the cAMP-responsive-element modulator (CREM) generates regulators with opposite functions. *Proc. Natl Acad. Sci. USA*, **89**, 4226.
90. Delmas, V., van der Hoorn, F., Mellström, B., Jégou, B., and Sassone-Corsi, P. (1993) Induction of CREM activator proteins in spermatids: downstream targets and implications for haploid germ cell differentiation. *Mol. Endocrinol.*, **7**, 1502.
91. de Kretser, D. M. and Kerr, J. B. (1988) The cytology of the testis. In *Physiology of reproduction* (ed. E. Knobil and J. Neill), Vol. I, p. 837. Raven.
92. Ewing, L. L. and Robert, B. (1989) Regulation of testicular function. *Ann. N. Y. Acad. Sci.*, **564**, 1.
93. Oakberg, J. (1956) A description of spermiogenesis in the mouse and its use in analysis of the cycle of the seminiferous epithelium and germ cell renewal. *Am. J. Anat.*,**99**, 391.
94. Parvinen, M. (1982) Regulation of the seminiferous epithelium. *Endocrinol. Rev.*, **3**, 404.
95. Jégou, B., Syed, V., Sourdaine, P. Byers, S., Gérard, N., Velez de la Calle, J., Pineau, C., Garnier, D. H., and Bauché, F. (1992) The dialogue between late spermatids and Sertoli cells in vertebrates: a century of research. In *Spermatogenesis, fertilization, contraception, molecular, cellular and endocrine events in male reproduction* (ed. E. Nieschlag and U.-F. Habenicht), p. 56. Schering Foundation, Springer-Verlag, New York.
96. Le Blond, C. P. and Clermont, Y. (1952) Definition of the stages of the cycle of the seminiferous epithelium in the rat. *Ann. N.Y. Acad. Sci.*, **55**, 548.
97. Perey, B., Clermont, Y., and Le Blond, C. P. (1961) The wave of the seminiferous epithelium in the rat. *Am. J. Anat.*, **47**, 108.
98. Willison, K. and Ashworth, A. (1987) Mammalian spermatogenic gene expression. *Trends Genet.*, **3**, 351.
99. Dym, M. (1983) The male reproductive system. In *Histology cell and tissue biology* (ed. L.Weiss). Elsevier Biomedical, New York.
100. Lyon, M. F. and Hawkes, S. G. (1970) X-linked gene for testicular feminization in the mouse. *Nature*, **227**, 1217.
101. Moutier, R. (1976) New mutations causing sterility restricted to the male in rats and mice. In *The laboratory animal in study of reproduction* (ed. T. Antikatzides, S. Erichsen, and A. Spiegel). Fischer Verlag, Stuttgart.
102. Kierszenbaum, A. L. and Tres, I. L. (1975) Structural and transcriptional features of the mouse spermatid genome. *J. Cell. Biol.*, **65**, 258.

103. Hecht, N. B. (1986) Regulation of gene expression during mammalian spermatogenesis. In *Experimental approaches to mammalian embryonic development* (ed. J. Rossant and R. A. Pedersen), p. 151. Cambridge University Press.
104. Kleene, K. C., Distel, R. J., and Hecht, N. B. (1984) Translational regulation and deadenylation of a protamine mRNA during spermiogenesis in the mouse. *Develop. Biol.*, **105**, 71.
105. Braun, R. E., Peschon, J. J., Behringer, R. R., Brinster, R. L., and Palmiter, R. D. (1989) Protamine 3′-untranslated sequences regulate temporal translation control and subcellular localization of growth hormone in spermatids of transgenic mice. *Genes Dev.*, **3**, 793.
106. Hecht, N. B. (1989) Mammalian protamines and their expression. In *Histones and other basic nuclear proteins* (ed. L. Hnilica, G. Stein, and J. Stein), p. 347. CRC Press, Boca Raton.
107. Hecht, N. B. (1990) Regulation of 'haploid expressed genes' in male germ cells. *J. Reprod. Fert.*, **88**, 679.
108. Van dar Hoorn, F. A. and Tarnasky, H. A. (1992) Factors involved in regulation of the RT7 promoter in a male germ cell-derived *in vitro* transcription system. *Proc. Natl Acad. Sci. USA*, **89**, 703.
109. Mali, P., Kaipia, A., Kangasniemi, M., Toppari, J., Sandberg, M., Hecht, N. B., and Parvinen, M. (1989) Stage-specific expression of nucleoprotein mRNAs during rat and mouse spermiogenesis. *Reprod. Fertil. Dev.*,**1**, 369.
110. Steinberger, E. (1971) Hormonal control of mammalian spermatogenesis. *Physiol. Rev.*, **51**, 1.
111. Lostroh, A. J. (1976) Hormonal control of spermatogenesis. In *Regulation mechanisms of male reproductive physiology* (ed. C. H. Spilman, T. J. Lobl and K. T. Kirton), p. 13. Excerpta Medica, Amsterdam.
112. Moore, R. Y. (1978) Neuroendocrine regulation of reproduction. In *Reproductive endocrinology* (ed. S. S. C. Yen and R. B. Jaffe). W. B. Saunders Company, Philadelphia.
113. Santen, R. J. (1987) The Testis. In *Endocrinology and metobolism* (ed. P. Felig, J. D. Baxter, A. E. Broadus, and L. A. Frohman). McGraw Hill Co., New York.
114. Feinberg, J., Pariset, C., Rondard, M., Loir, M., Lanneau, M., Weinman, S., and Demaille, J. (1983) Evolution of Ca^{2+}- and cAMP-dependent regulatory mechanisms during ram spermatogenesis. *Develop. Biol.*, **100**, 260.
115. Pariset, C., Feinberg, J., Dacheux, J. L., Oyen, O., Jahnsen, T., and Weinman, S. (1989) Differential expression and subcellular localization for subunits of cAMP-dependent protein kinase during ram spermatogenesis. *J. Cell Biol.*, **109**, 1195.
116. Oyen, O., Scott, J. D., Cadd, G. G., McKnight, G. S., Krebs, E. B., Hansson, V., and Jahnsen, T. (1988) A unique mRNA species for a regulatory subunit of cAMP-dependent protein kinase is specifically induced in haploid germ cells. *FEBS Lett.*, **229**, 391.
117. Oyen, O., Myklebust, F., Scott, J. D., Cadd, G. G., McKnight, S. G., Hansson, V., and Jahnsen, T. (1990) Subunits of cyclic adenosine 3′,5′-monophosphate-dependent protein kinase show differential and distinct expression patterns during germ cell differentiation: alternative polyadenylation in germ cells gives rise to unique smaller-sized mRNA species. *Biol. Reprod.*, **43**, 46.
118. Veldhuis, J. D. (1991) The hypothalamic-pituitary-testicular Axis. In *Reproductive Endocrinology* (ed. S. S. C. Yen and R. B. Jaffe), p. 409. W. B. Saunders Company, Philadelphia.

119. Toppari, J., Kangasniemi, M., Kaipia, A., Mali, P., Huhtaniemi, I., and Parvinen, M. (1991) Stage- and cell-specific gene expression and hormone regulation of the seminiferous epithelium. *J. Electr. Microsc. Tech.*, **19**, 203.
120. Lonnerberg, P., Parvinen, M., Jahnsen, T., Hansson, V., and Persson, H. (1992) Stage- and cell-specific expression of cyclic adenosine 3′,5′-monophosphate-dependent protein kinases in rat seminiferous epithelium. *Biol. Reprod.*, **46**, 1057.
121. Desjardins, C., Ewing, L. L., and Johnson, B. H. (1971) Effects of light deprivation upon the spermatogenic and steroidogenic elements of hamster testes. *Endocrinology*, **89**, 791.
122. Reiter, R. J. (1980) The pineal and its hormones in the control of reproduction in mammals. *Endocrin. Rev.*, **1**, 109 .
123. Pévet, P. (1988) The role of the pineal gland in the photoperiodic control of reproduction in different hamster species. *Reprod. Nutr. Dev.* **28**, 575.
124. Bartke, A. (1985) In *The hamster reproduction and behavior* (ed. M. I. Siegel), p. 73. Plenum Corp., New York.
125. Jetton, A. E., Fallest, P. C., Dahl, K. D., Schwartz, N. B., and Turek, F. W. (1991) Photoperiodic differences *in vitro*: gonadotropin basal secretion nad gonadotropin-releasing hormone responsiveness in the golden hamster. *Endocrinology*, **129**, 1025 .
126. Crowley, W. F., jun., Whitcomb, R. W., Jameson, J. L., Weiss, J., Finkelstein, J. S., and O'Dea, L. S. (1991) Neuroendocrine control of human reproduction in the male. *Rec. Progr. Horm. Res.* **47**, 27.
127. Russell, L. D. (1980) Sertoli-germ cell interrelations: a review. *Gamete Res.*, **3**, 179.
128. Bravo, R., Zerial, M., Toschi, L., Schurmann, M., Muller, R.,Hirai, S. I., Yaniv, M., Almendral, J. M., and Ryseck, R. P. (1988) Identification of growth-factor-inducible genes in mouse fibroblasts. *Cold Spring Harbor Symp. Quant. Biol.*, **53**, 901.
129. Nathans, D., Lau, L. F., Christy, B., Hartzell, S., Nakabeppu, Y., and Ryder, K. (1988) Genomic response to growth factors. *Cold Spring Harbor Symp. Quant. Biol.*, **53**, 893.
130. Meyer, T. E., Waeber, G., Lin, J., Beckman, W., and Habener, J. F. (1993) The promoter of the gene encoding 3′ 5′-cyclic adenosine monophopsphate (cAMP) response element binding protein contains cAMP response elements: evidence for positive autoregulation of gene transcription. *Endocrinology*, **132**, 770.
131. Delegeane, A., Ferland, L., and Mellon, P. L. (1987) Tissue-specific enhancer of the human glycoprotein hormone a subunit gene: dependence on cyclic AMP-inducible elements. *Mol. Cell. Biol.*, **7**, 3994.
132. McCormick, A., Brady, H., Theill, L., and Karin, M. (1990) Regulation of the pituitary-specific homeobox gene GHF-1 by cell autonomous and environmental cues. *Nature*, **345**, 829.
133. Hsu, J.-C., Laz, T., Mohn , K. L., and Taub, R. (1991) Identification of LRF-1, a leucine-zipper protein that is rapidly and highly induced in regenerating liver. *Proc. Natl Acad. Sci. USA*, **88**, 3511.
134. Tamarkin, L., Baird, C. J., and Almeida, O. F. X. (1985) Melatonin: a coordinating signal for mammalian reproduction? *Science*, **227**, 714.
135. Reiter, R. J. (1991) Pineal gland. Interface between the photoperiodic environment and the endocrine system. *Trends Endocr. Met.*, **1**, 13.
136. Klein, D. C., Moore, R. Y., and Reppert, S. M. (ed.) (1994) *Suprachiasmatic nucleus: the mind's clock.* Oxford University Press, Oxford.
137. Moore, R. Y. (1983) Organization and function of the central nervous system circadian oscillator; the suprachiasmatic hypothalamic nucleus. *Federation Proc.*, **42**, 2783.
138. Klein, D. C. (1985) Photoneural regulation of the mammalial pineal gland. In *Photoperiodism, melatonin and the pineal gland.* Pitman, London.

139. Sudgen, D., Vanecek, J., Klein, D. C., and Thomas, T. D. (1985) Activation of protein kinase C potentiates isoprenaline-induced cyclic AMP accumulation in rat pinealocytes. *Nature*, **314**, 359.
140. Vanecek, J., Sudgen, D., Weller, J., and Klein, D. C. (1985) Atypical synergistic α1 and β-adrenergic regulation of adenosine 3′,5′. *Endocrinology*, **116**, 2167.
141. Meijer, G. H.(1991) In *Suprachiasmatic nucleus: the mind's clock* (ed. D. C. Klein, R. Y. Moore, and S. M. Reppert). Oxford University Press, Oxford.
142. Tamarkin, L., Reppert, S. M., and Klein, D. C. (1979) Regulation of pineal melatonin in the syrian hamster. *Endocrinology*, **104**, 385.
143. Perlow, M. J., Reppert, S. M., Tamarkin, L., Wyatt, R. J., and Klein, D. C. (1980) Photic regulation of the melatonin rhythm: monkey and man are not the same. *Brain Res.*, **182**, 211.
144. Illnerova, H.,Vanecek, J., Wetterberg, L., and Sääf, J. J. (1979) Effect of one minute exposure to light at night on rat pineal serotonin N-acetyltransferase and melatonin. *Neurochem.*, **32**, 673.
145. Brownstein, M. and Axelrod, J. (1974) Pineal gland: 24-hour rhythm in norepinephrine turnover. *Science*, **184**, 163.
146. Craft, C. M., Morgan, W. W., and Reiter, R. J. (1984) 24 hour changes in catecholamine synthesis in rat and hamster pineal glands. *Neuroendocrinology*, **38**, 193.
147. Axelrod, J. (1988) The pineal gland: a neurochemical transducer. *Science*, **184**, 1341.
148. Takahashi, J. S. (1993) Circadian clocks à la CREM. *Nature*, **365**, 299.
149. Borrelli, E., Montmayeur, J. P., Foulkes, N. S., and Sassone-Corsi, P. (1992) Signal transduction and gene control: the cAMP pathway. *Crit. Rev. Oncogene*, **3**, 321.
150. Mellström, B., Naranjo, J. R., Foulkes, N. S., Lafarga, M., and Sassone-Corsi, P. (1993) Transcriptional response to cAMP in brain: specific ditribution and induction of CREM antagonists. *Neuron*, **10**, 655.
151. Busch, S. J. and Sassone-Corsi, P. (1990) Dimers, leucine zippers and DNA binding domains. *Trends Genet.*, **6**, 36.
152. Kara, C. J., Liou, H. C., Ivashkiv, L. B., and Glimcher, L. H. (1990) A cDNA for a human cyclic AMP response element-binding protein which is distinct from CREB and expressed preferentially in the brain. *Mol. Cell. Biol.*, **10**, 1347.
153. Maekawa, T., Sakura, H., Kanei-Ishii, C., Sudo, T., Yoshimura, T., Fujisawa, J., Yoshida, M., and Ishii, S. (1989) Leucine zipper structure of the protein CRE-BP1 binding to the cyclic AMP responsive element in brain. *EMBO J.*, **8**, 2023.
154. Comb, M., Birnberg, N. C., Seasholtz, A., Herbert, E., and Goodman, H. M. (1986) A cyclic-AMP and phorbol ester-inducible DNA element. *Nature*, **323**, 353.
155. Mizuki, N., Sarapata, D. E., Garcia-Sanz, J. A., and Kasahara, M. (1992) The mouse male germ cell-specific gene Tpx-1: molecular structure, mode of expression in spermatogenesis, and sequence similarity to two non-mammalian genes. *Mammalian Genome*, **3**, 274.
156. Van der Hoorn, F. A. and Tarnasky, H. A. (1992) Factors involved in regulation of the RT7 promoter in a male germ cell-derived *in vitro* transcription system. *Proc. Natl Acad. Sci. USA*, **89**, 703.
157. Johnson, P., Peschon, J. J., Yelick, P. C., Palmiter, R. D., and Hecht, N. B. (1988) Sequence homologies in the mouse protamine 1 and 2 genes. *Biochim. Biophys. Acta*, **950**, 45.
158. Lin, S. C. and Morrison-Bogorad, M. (1991) Cloning and characterization of a testis-specific thymosin b10 cDNA. *J. Biol. Chem.*, **266**, 23347.
159. Hummler, E., Cole, T. J., Blendy, J. A., Grauss, R., Aguzzi, A., Schmid, W., Beermann, F., and Schütz, G. (1994). Targeted mutation of the CREB gene: compensation within the CREB/ATF family of transcription factors. *Proc. Natl Acad. Sci. USA.*, **91**, 5647.

4 Regulation of Rel transcription complexes

THOMAS D. GILMORE

1. Introduction

Eukaryotic transcription factors play important roles in cellular decisions involving growth and differentiation. Techniques are readily available to define important domains within a given transcription factor, such as domains involved in site-specific DNA binding, the formation of oligomers, nuclear targeting, and transcriptional activation or repression. One of the next challenges is to understand the transcriptional, translational, and post-translational regulation of specific transcription factors.

Rel/NF-κB transcription factors control the expression of many cellular and viral genes. Furthermore, Rel transcription factors undergo various types of regulation: by combinatorial dimerization, by subcellular compartmentalization, by proteolysis, by protein–protein interactions, by transcriptional control, by expression of alternative mRNAs, by protein concentration, and probably by post-translational modifications. This has established Rel proteins as fertile ground for studying in detail the complex regulation of a family of related transcription factors. This chapter will attempt to cover critically what is known about the regulation of Rel proteins.

2. Rel proteins

The Rel family of eukaryotic transcription factors is comprised of several proteins that have been identified in vertebrates, insects, and one avian retrovirus (Table 1). Rel proteins and related topics have been the subject of numerous recent reviews (1–40), and therefore, only a modest degree of background information will be provided here.

In vertebrates, Rel DNA-binding complexes can consist of homodimers or heterodimers of most Rel family members. The different Rel complexes have differing affinities for DNA target sites (generically called κB binding sites or κB sites). Thus, the consensus κB site is a degenerate decamer (GGGRN/NYYCC, where R is any purine, N is any nucleotide, and Y is any pyrimidine) that contains two half-sites, which in part reflect the specificity of the various members that bind to each half-site. Therefore, Rel factors show combinatorial diversity based on the subunit com-

Table 1 *rel* family genes and proteins

Class I Rel proteins

Gene	Organism[a]	Product[b]	Expression[c]	References
nfkb1	Ch,R,M,H	p50/p105	Lymphoid	121,122,124, 166, 167
		p98/p84 (spliced variants)		128
nfkb2	Ch,M,H	p52/p100	Lymphoid	125,126, 165,189

Class II Rel proteins

Gene	Organism[a]	Product[b]	Expression[c]	References
v-*rel*	Rev-T virus	p59	Infection	190, 191
c-*rel*	Ch,T,M,H	p68–82	Haematopoietic, lymphoid, ovary	53, 80 191–196
relA	Ch, M,H	p65	Ubiquitous	197–199
relB	M,H	p68	Lymphoid	59, 200
Xrel1	Xe	p58	Oocyte, embryo	201
dorsal	Dr	p75	Embryo	81, 159, 160, 202
dif	Dr	p72	Larval fat body	66

[a] Source from which cDNA or peptide sequence is available. Dr, *D. melanogaster*, Xe, *Xenopus laevis*; Ch, chicken; T, turkey; R, rabbit; M, mouse; H, human.
[b] Size of protein product in kilodaltons.
[c] Primary tissue or developmental stage in which RNA expression has been described.

position of the Rel dimer. In general, Rel dimers bind to DNA with quite high affinity. In addition, there are other vertebrate non-Rel transcription factors that can bind to 'κB sites' (41–49; A. Israël, personal communication).

The structural essence of the Rel proteins is contained in a conserved domain of approximately 300 amino acids called the Rel homology (RH) domain. This domain contains sequences important for DNA binding, dimerization, inhibitor binding, and nuclear localization. A comparison of the RH domains of all known Rel proteins is shown in Fig. 1. Recently, the three-dimensional structure of the RH domain of a dimer of the Rel family p50 bound to a symmetrical κB site has been described (50, 51). There are multiple contacts between the RH domain and both sides of the κB site, and generally the p50 homodimer wraps itself around the DNA, giving the appearance of a 'butterfly' (52). The large number of contacts between the p50 dimer and DNA no doubt accounts for the high affinity with which most Rel dimers bind to DNA.

Sequences C-terminal to the RH domain can be used to divide Rel proteins into two classes (see Table 1 and Fig. 2). The two members of the first class of Rel proteins, p105 and p100, have long C-terminal domains that contain seven copies of a

```
     V  PYIEIFEQPRQRGTRFRYKCEGRSAGSIPCEHSTD-NNKTFPSIQILNYFGKVKIRTT--LVTKNEPYKPHPHDLVGKG----C-RDGYYEAEFGPERQ-VLSFQNLGIQCV
     T  PYIEIFEQPRQRGMRFRYKCEGRSAGSIPGEHSTD-NNKTFPSIQILNYFGKVKIRTT--LVTKNEPYKPHPHDLVGKD----C-RDGYYEAEFGPERR-VLSFQNLGIQCV
     C  PYIEIFEQPRQRGMRFRYKCEGRSAGSIPGEHSTD-NNKTFPSIQILNYFGKVKIRTT--LVTKNEPYKPHPHDLVGKD----C-RDGYYEAEFGPERR-VLSFQNLGIQCV
     M  PYVEIIEQPRQRGMRFRYKCEGRSAGSIPGERSTD-NNRTYPSVDIMNYYGKGKIRIT--LVTKNDPYKPHPHDLVGKD----C-RDGYYEAEFGPERR-PLFFQNLGIRCV
     H  PYIEIIEQPRQRGMRFRYKCEGRSAGSIPGEHSTD-NNRTYPSIQIMNYYGKGKVRIT--LVTKNDPYKPHPHDLVGKD----C-RDGYYEAEFGQERR-PLFFQNLGIRCV
   C65  PFVEILEQPKQRGMRFRYKCEGRSAGSIPGEHSTD-SARTHPTIRVNHYRGPGRVRVS--LVTKDPPHGPHPHELVGRH----C-QHGYYEAELSPERC-VHSFQNLGIQCV
   M65  PYVEIIEQPKQRGMRFRYKCEGRSAGSIPGERSTD-TTKTHPTIKINGYTGPGTVRIS--LVTKDPPHRPHPHELVGKD----C-RDGYYEADLCPDRS-IHSFQNLGIQCV
   H65  PYVEIIEQPKQRGMRFRYKCEGRSAGSIPGERSTD-TTKTHPTIKINGYTGPGTVRIS--LVTKDPPHRPHPHELVGKD----C-RDGFYEAELCPDRC-IHSFQNLGIQCV
 MRelB  PYLVITEQPKQRGMRFRYECEGRSAGSILGESSTE-ASKTQPAIELRDCGGLREVEVTACLVWKDWPHRVHPHSLVGKD----C-TDGVCRVRLRPHVSPRHSFNNLGIQCV
 HRelB  PHLVITEQPKQRGMPFRYECEGRSAGSILGESSTE-ASKTLPAIELRDCGGLREVEVTACLVWKDWPHRVHPHSLVGKD----C-TDGICRVRLRPHVSPRHSFNNLGIQCV
   C50  PYLQIIEQPKQRGFRFRYVCEGPSHGGLPGASSEK-NKKSYPQVKICNYVGPAKVIVQ--LVTNGKYRHLHAHSLVGKF----C-EDGVCTVNAGPKDMVVG-FANLGILHV
   M50  PYLQILEQPKQRGFRFRYVCEGPSHGGLPGASSEK-NKKSYPQVKICNYVGPAKVIVQ--LVTNGKNIHLHAHSLVGKH----C-EDGVCTVTAGPKDMVVG-FANLGILHV
   R50                RYVCEGPSHGGLPGASSEK                      KVIVQ--LVTDGKNIHLHA     KH----C-EDGICTVNAGPK
   H50  PYLQILEQPKQRGFRFRYVCEGPSHGGLPGASSEK-NKKSYPQVKICNYVGPAKVIVQ--LVTNGKNIHLHAHSLVGKH----C-EDGICTVTAGPKDMVVG-FANLGILHV
   C52  PYLVIIEQPKQRGFRFRYVCEGPSHGGLPGASSEK-GHKTYPTVKICNYEGMARIEVD--LVTHSDPPRVHAHSLVGKQ----CNEAGNCVAIVGPKDMTAQ-FSNLGVLHV
   H52  PYLVIVEQPKQRGFRFRYGCEGPSHGGLPGASSEK-GRKTYPTVKICNYEGPAKIEVD--LVTHSDPPRAHAHSLVGKQ----CSELGICAVSVGPKDMTAQ-FNNLGVLHV
  XRel  PPVEIIEQPKQRGMRFRYKCEGRSAGSIPGERSTD-TSKTHPTIKINNYQGPARIRIS--LVTKDSPHKPHPHELVGKD----C-KDGYYEAELSPDRSIHS-FQNLGIQCV
Dorsal  PYVKITEQPAGKALRFRYECEGRSAGSIPGVNSTPE-NKTYPTIEIVGYKGRAVVVVS--CVTKDTPYRPHPHNLVGKE---GC-KKGVCTLEINSETMRAV-FSNLGIQCV
   Dif  PHLRIVEEPTSNIIRFRYKCEGRTAGSIPGMNSSSETGKTFPTIEVCNYDGPVIIVVS--CVTSDEPFRQHPHWLVSKEEADAC-KSGIYQKKLPPEERRLVL-QKVGIQCA
    Co  P  I E P    FRY CEG   G   G S       P        G         V         H H LV K    C  G                    G
                 RxxRxRxxC

     V  KKKDLKESISLRISKKINPFNVP-EEQLHNIDEYDLNVVRLCFQAFLPDEH-GNYTLA---------------------------LPPLISNPIYDNRAPNTAELRICRVN
     T  KKKDLKESISLRISKKINPFNVP-EEQLHNIDEYDLNVVRLCFQAFLPDEH-GNYTLA---------------------------LPPLISNPIYDNRAPNTAELRICRVN
     C  KKKDLKESISLRISKKINPFNVP-EEQLHNIDEYDLNVVRLCFQAFLPDEH-GNYTLA---------------------------LPPLISNPIYDNRAPNTAELRICRVN
     M  KKKEVKEAIILRISAGINPFNVP-EQQLLDIEDCDLNVVRLCFQVFLPDEH-GNFTTA---------------------------LPPIVSNPIYDNRAPNTAELRICRVN
     H  KKKEVKEAIITRIKAGINPFNVP-EKQLNDIEDCDLNVVRLCFQVFLPDEH-GNLTTA---------------------------LPPVVSNPIYDNRAPNTAELRICRVN
   C65  KKRELEAAVAERIRTNNNPFNVPMEERGADY---DLSAVRLCFQVWVNGPGG-LCPLP-----------------------------PVLSQPIYDNRAPSTAELRILPGD
   M65  KKRDLEQAISQRIQTNNNPFHVPIEEQRGDY---DLNAVRLCFQVTVRDPAGRPLLLT-----------------------------PVLSHPIFDNRAPNTAELKICRVN
   H65  KKRDLEQAISQRIQTNNNPFQVPIEEQRGDY---DLNAVRLCFQVTVRDPSGRPLRLP-----------------------------PVLPHPIFDNRAPNTAELKICRVN
 MRelB  RKKEIEAAIERKIQLGIDPYNAGSLKNHQEV---DMNVVRICFQASYRDQQ-GHL----------------------------HRMDPILSEPVYDKKSTNTSELRICRIN
 HRelB  RKKEIEAAIERKIQLGIDPYNAGSLKNHQEV---DMNVVRICFQASYRDQQ-GQM----------------------------RRMDPVLSEPVYDKKSTNTSELRICRIN
   C50  TKKKVFETLETRMIDACKKGYNPGLLVHPELGYLQAEGCGDRQLTEREREIIRQAAVQQTKEMDLSVVRLMFTAFLPDSNGGFTRRLDPVISDAIYDSKAPNASNLKIVRMD
   M50  TKKKVFETLEARMTEACIRGYNPGLLVHSDLAYLQAEGGGDRQLTDREKEIIRQAAVQQTKEMDLSVVRLMFTAFLPDSTGSFTRRLEPVVSDAIYDSKAPNASNLKIVRMD
```

```
R50                     R YNPGLLVHPDLAYLQTEGGGD                                                     RLEPVVSDAIYDSK
H50     TKKKVFETLEARMTEACIRGYNPGLLVHPDLAYLQAEGGGDRQLGDREKELIRQAALQQTKEMDLSVVRLMFTAFLPDSTGSFTRRLEPVVSDAIYDSKAPNASNLKIVRMD
C52     TKKNMMEIMKEKLKKQKTRNT-NGLLTEAEL---------------REIEL---EAKELKKVMDLSIVRLRFTAYLRDSSGNFTLALQPVISDPIHDSKSPGASNLKISRMD
H52     TKKNMMGTMIQKLQRQRLRSRP-QGLTEAEQ---------------RELEQ---EAKELKKVMDLSIVRLRFSAFLRASDGSFSLPLKPVISQPIHDSKSPGASNLKISRMD
XRel    KKREVEDAVAHRIRTNNNPFNVSPEELKADY---DLNTVCLCFQVFIPD----QAA-----------------------GR-MLPLPFVVSQPIYDNRAPNTAELKICRVN
Dorsal  KKKDIEAALKAREEIRVDPFKTGFSHRFQPSSI-DLNSVRLCFQVFMESEQK---------------------------GRFTSPLPPVVSEPIFDKKA--MSDLVICRLC
Dif     KKLEMRDSLVERERRNIDPFNAKFDHKDQIDKI-NRYELRLCYQAFI---TV---------------------------GNSKVPLDPIVSSPIYGKS----SELTITRLC
Co      K                                                                                                          R

V       KNCGSVKGGDEIFLLCDKVQKDDIEVRFVLGN-------WEAKGSFSQADVHRQVAIVFRTPPFLG-DITEPITVKMQLRRPSDQAVSEPVDFRYLPDEEDPSGNKAKRQR
T       KNCGSVKGGDEIFLLCDKVQKDDIEVRFVLGN-------WEAKGSFSQADVHRQVAIVFRTPPFLR-DITEPITVKMQLRRPSDQEVSEPVDFRYLPDEEDSYGNKAKRQR
C       KNCGSVKGGDEIFILCDKVQKDDIEVRFVLDN-------WEAKGSFSQADVHRQVAIVFRTPPFLR-DITEPITVKMQLRRPSDQEVSEPMDFRYLPDEKDPYGNKAKRQR
M       KNCGSVRGGDEIFLLCDKVQKDDIEVRFVLND-------WEARGVFSQADVHRQVAIVFKTPPYCK-AILEPVTVKMQLRRPSDQEVSESMDFRYLPDEKDAYGNKSKKQK
H       KNCGSVRGGDEIFLLCDKVQKDDIEVRFVLND-------WEAKGIFSQADVHRQVAIVFKTPPYCK-AITEPVTVKMQLRRPSDQEVSESMDFRYLPDEKDTYGNKAKKQK
C65     RNSGSCQGGDEIFLLCDKVQKEDIEVRF------WA-EGWEAKGSFAAADVHRQVAIVFRTPPFRERSLRHPVTVRMELQRPSDRQRSPPLDFRYLPHQGDLQCIEEKRKR
M65     RNSGSCLGGDEIFLLCDKVQKEDIEVYF------TG-PGWEARGSFSQADVHRQVAIVFRTPPYADPSLQAPVRVSMQLRRPSDRELSEPMEFQYLPDTDDRHRIEEKRKR
H65     RNSGSCLGGDEIFLLCDKVQKEDIEVYF------TG-PGWEARGSFSQADVHRQVAIVFRTPPYADPSLQAPVRVSMQLRRPSDRELSEPMEFQYLPDTDDRHRIEEKRKR
MRelB   KESGPCTGGEELYLLCDKVQKEDISVVF------ST-ASWEGRADFSQADVHRQIAIVFKTPPYEDLEISEPVTVNVFLQRLTDGVCSEPLPFTYLPRDHDSYGVDKKRKR
HRelB   KESGPCTGGEELYLLCDKVQKEDISVVF------SR-ASWEGRADFSQADVHRQIAIVFKTPPYEDLEIVEPVTVNVFLQRLTDGVCSEPLPFTYLPRDHDSYGVDKKAKR
C50     RTAGCVTGGEEIYLLCDKVQKDDIQIRFYE-EDENG-GMWEGFGDFSPTDVHRQFAIVFKTPKYRDVNITKPASVFVQLRRKSDLETSEPKPFLYYPEIKDKEEVQRKRQK
M50     RTAGCVTGGEEIYLLCDKVQKDDIQIRFYE-EEENG-GVWEGFGDFSPTDVHRQFAIVFKTPKYKDVNITKPASVFVQLRRKSDLETSEPKPFLYYPEIKDKEEVQRKRQK
R50     RTAGCVTGGEEIYLLCDK K DIQIRFYE-EEENG-GIWEGFGDFSPTDVHR FAIVFK     KDVNITKPASVFVQLR KSDLETSEPKPFLYYPEIKD
H50     RTAGCVTGGEEIYLLCDKVQKDDIQIRFYE-EEENG-GVWEGFGDFSPTDVHRQFAIVFKTPKYKDINITKPASVFVQLRRKSDLETSEPKPFLYYPEIKDKEEVQRKRQK
C52     KTAGSVRGGDEVYLLCDKVQKDDIEVRFYE-DDENG---WQASGDFSPTDVHKQYAIVFRTPPYHKPKIDRPVTVFLQLKRKRGGDVSDSKQFTYYPVVEDKEEVERNRKK
H52     KTAGSVRGGDEVYLLCDKVQKDDIEVRFYE-DDENG---WQAFGDFSPTDVHKQYAIVFRTPPYHKMKIERPVTVFLQLKRKRGGDVSDSKQFTYYPLVEDKEEVQRKRRK
XRel    KNSGSCLGGDEIFLLCDKVQKEDIEVIFGL-------GNWEARGIFSQADVHRQVAIVFRTPAFQDTKIRQSVKVQMQLRRPSDKEVSEPMEFQYLPDEGDPHHIDEKRKR
Dorsal  SCSATVFGNTQIILLCEKVAKEDISVRFFE-EK-NGQSVWEAFGDFQHTDVHKQTAITFKTPRYHTLDITEPAKVFIQLRRPSDGVTSEALPFEYVPMDSDPAHLRRKRQK
Dif     SCAATANGGDEIIMLCEKIAKDDIEVRFYETDK-DGRETWFANAEFQPTDVFKQMAIAFKTPRYRNTEITQSVNVELKLVRPSDGATSAPLPFEYYPNPELLT-KHNRRVA
Co              G       LC K  K DI   F             W     F  DV  Q  AI F TP          V     R     S   F   P
                                                                                        ---                ------
                                                                                        PKA                 NLS
```

Fig. 1 Comparison of the sequences of the Rel homology domains of all known Rel proteins. V, v-Rel; c-Rel (Turkey; Chicken; Mouse; Human); RelA (p65); RelB; p50 (Rabbit); p52; XRel1; Dorsal; Dif. Only residues conserved (CO) in all proteins are indicated. In order, from the N-terminus, sequences involved in DNA binding (RxxRxRxxC), phosphorylation by protein kinase A (PKA), and nuclear targeting (NLS) are underlined.

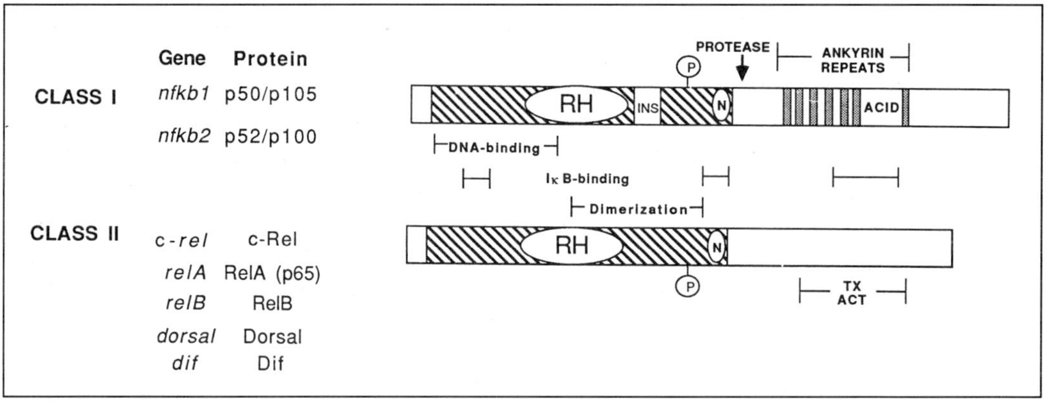

Fig. 2 Generalized structures of the two classes of Rel proteins. RH, Rel homology domain, in shading; INS, insertion of approximately 25 amino acids; N, nuclear targeting sequence; P, protein kinase A recognition sequence; TX ACT, transcriptional activation domain; Protease, proteolytic cleavage site. Adapted from Gilmore and Morin (25).

33 amino acid repeat called an ankyrin or cell-cycle repeat (called ANK repeat here). ANK repeats are known to mediate protein–protein interactions in other proteins (for review see 18). Members of the second class of Rel proteins, including RelA (also called p65), c-Rel, RelB, Dorsal, Dif, and v-Rel, have C-terminal sequences that do not contain ANK repeats and are generally unrelated amongst these Rel proteins. In addition, the RH domain of class II Rel proteins has an internal deletion of approximately 25 amino acids compared with class I Rel proteins. The C-terminal sequences from the class II Rel proteins all contain transcription activation domains (53–60). The strengths of these activation domains can vary greatly (e.g. the RelA activation domain is much stronger than that of c-Rel), and, at least in the case of c-Rel and RelA, these activation domains appear to be acidic in nature (like VP16; see Chapter 1) (35).

In addition to these two classes of Rel proteins, the NF-AT transcription factors (see Chapter 5) have a DNA-binding/dimerization domain that is highly related to the RH domain (61, 62). However, to date, there is no evidence that the Rel proteins covered in this review can form heterodimers with the NF-AT proteins.

Another distinguishing feature of Rel proteins is their varying subcellular location: that is, Rel proteins can be found in the cytoplasm or the nucleus of cells. In a few cell types, such as mature B cells, certain Rel complexes are constitutively active and present in the nucleus (63–65). However, in most cell types, Rel proteins are found in the cytoplasm as inactive proteins (i.e. proteins incapable of binding DNA) and, when induced, the Rel complex rapidly translocates to the nucleus where it can bind DNA and affect gene expression. Cytoplasmic sequestration and inhibition of DNA binding by Rel proteins is mediated by a family of Rel-binding proteins known as IκB proteins (for review see 22, 25; see also below).

Most vertebrate Rel proteins are expressed at highest levels in lymphoid cells and control the expression of genes, such as those encoding cytokines, cytokine receptors, and interferon, that are involved in various immunoregulatory responses (for review see 28, 35, 39). Interestingly, the Dif protein in *Drosophila melanogaster* can be induced by bacterial infection or lipopolysaccharide to enter the nucleus of cells in larval fat bodies, and Dif can directly activate genes involved in an immune-like response in flies, suggesting a vertical evolution between Dif and the vertebrate Rel proteins (66). NF-κB was the first vertebrate cellular Rel complex to be characterized, and the term 'NF-κB' is now generally used to refer to the p50–RelA (p50/p65) heterodimer. Although NF-κB was originally thought to be important for the developmental expression of the immunoglobulin κ light-chain gene in mouse B cells *in vitro* (63, 64), the *in vivo* relevance of this finding has recently been brought into question because transgenic mice that have a homozygous inactivation of their *nfkb1* gene (which encodes p50) have B cells that undergo apparently normal development (67); nevertheless, p50 knock-out mice do have a generally impaired immune response.

3. Protein–protein interactions involving Rel proteins

3.1 Interactions between Rel proteins and IκB proteins

The IκB proteins represent a family of proteins that interact with and modulate the activity of Rel protein complexes (for review see 22, 25). The IκB proteins include the vertebrate proteins IκB-α, IκB-β, IκB-γ, Bcl-3, and the *D. melanogaster cactus* gene product (see Fig. 3). IκB-γ comprises the C-terminal ANK repeat domain of p105, expressed from a subgenomic mRNA of p105 (68). In addition, full-length p105 and p100 can act as IκB proteins, since they can sequester other Rel proteins in the cytoplasm and inhibit DNA binding by certain Rel proteins *in vitro* (69–73).

All IκB proteins contain five to eight copies of the ANK repeat bounded by sequences of variable length (see Fig. 3). The ANK repeats appear to be sufficient for interaction of IκBs with Rel proteins and are likely to interact with and mask the nuclear localization sequences (NLS) of Rel proteins (74). For example, Inoue *et al.* (75) have shown that mutations in ANK repeats 1, 2, 4, or 5 of IκB-α abolish its ability to interact with and inhibit DNA binding by Rel proteins. In addition to the ANK repeats, the acidic sequences located C-terminal to the ANK repeats in most IκB proteins are also necessary for full inhibition of DNA binding by Rel proteins *in vitro* (75), and these IκB sequences probably interact directly with a conserved basic region in Rel proteins that is directly involved in DNA binding (74). It is likely that multiple sequences within Rel proteins interact with a given IκB; mutational analysis has indicated that the RXXRXRXXC sequence involved in DNA binding by Rel proteins (76), the NLS (77–79), and probably sequences C-terminal to the RH domain (78, 80, 81) all seem to be involved in interaction of Rel proteins with IκB proteins.

The various IκBs have distinct specificities for interacting with various Rel complexes. For example, IκB-α can interact with RelA of the NF-κB heterodimer and thus cause NF-κB to be sequestered in the cytoplasm and inhibit DNA binding by

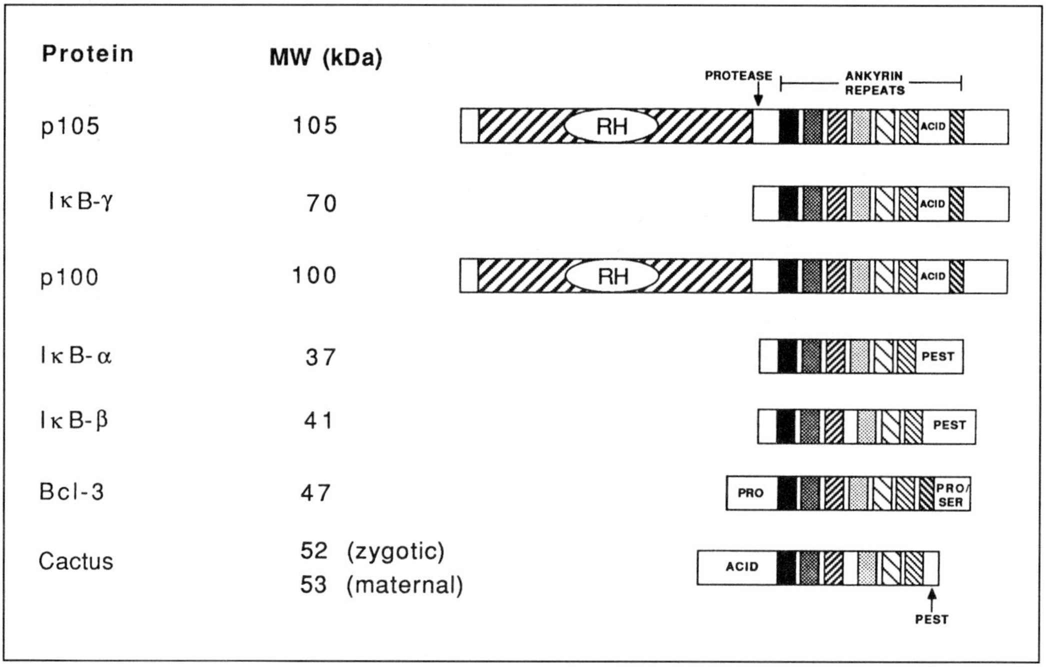

Fig. 3 Generalized structures of IκB proteins. The similar shadings indicate similarities between ANK repeats. PEST is a sequence rich in Proline, Serine, Threonine, and acidic amino acids that is found in many proteins that undergo rapid degradation. The Rel homology domain (RH) is described in Fig. 2. Adapted from Gilmore and Morin (25).

NF-κB *in vitro*. Interestingly, Hatada *et al.* (82) have shown that the specificity of IκB-γ can be altered by deleting three amino acids from its first ANK repeat; an IκB-γ molecule with this deletion inhibits DNA binding by RelA homodimers but not p50 homodimers (like IκB-α), whereas wild-type IκB-γ inhibits DNA binding by RelA and p50 homodimers. Isoforms of IκB-γ that are generated by alternate splicing have different inhibitory activities (83).

In addition to inhibiting Rel proteins from forming complexes on κB sites, most IκBs can remove pre-formed Rel complexes from DNA *in vitro* (84, 85). Most IκB molecules are small enough to enter the nucleus by passive diffusion. Thus, removal of Rel protein complexes from DNA by IκB could occur *in vivo* and would provide a mechanism whereby an induced effect on gene expression could be silenced.

The general model for regulation of the NF-κB p50–RelA complex by IκB-α is shown in Fig. 4 (86–92; for review see 22, 25). Upon receiving an appropriate signal, IκB-α is rapidly degraded on the NF-κB complex. The NF-κB complex is thus free to enter the nucleus and activate the expression of target genes. One of these target genes is that which encodes IκB-α (93); newly synthesized IκB-α can then act to remove NF-κB from DNA and re-sequester the NF-κB complex in the cytoplasm.

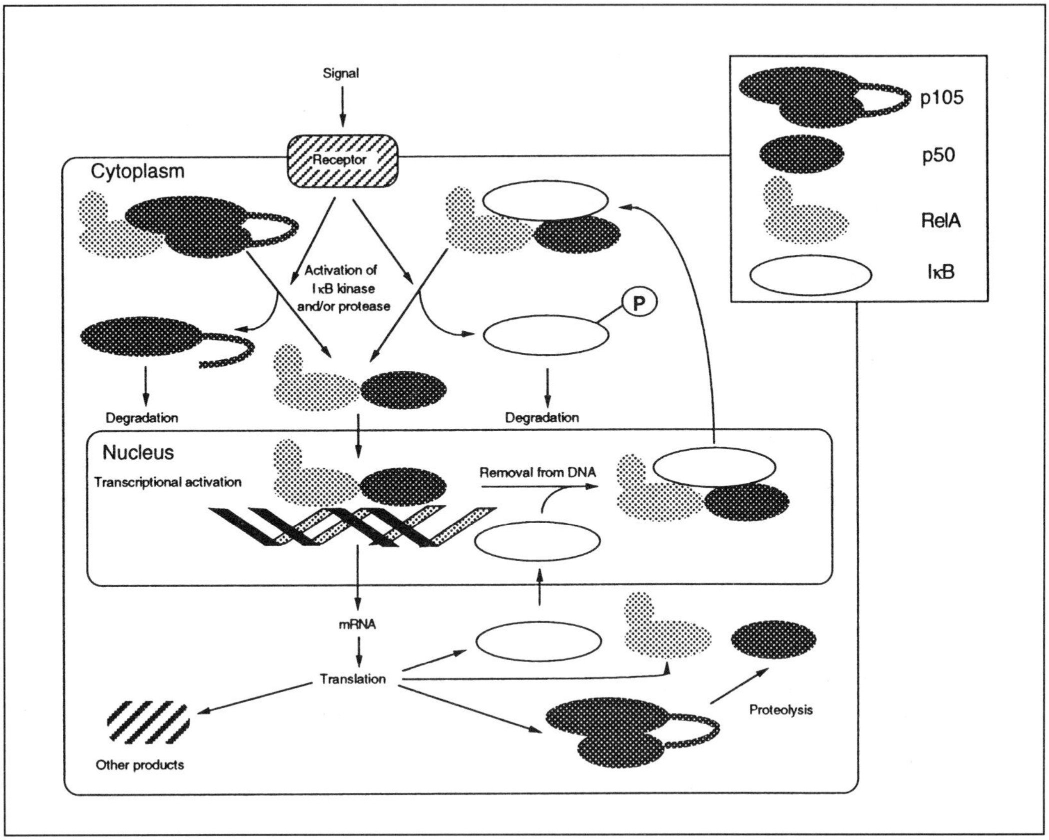

Fig. 4 Regulation of NF-κB by IκB. The two pathways for cytoplasmic retention of NF-κB complexes and nuclear induction of NF-κB DNA-binding activity are shown, and these pathways are described in the text. Largely adapted from Gilmore and Morin (25).

Recent experiments suggest that the interaction of NF-κB with IκB-α is a dynamic process even in non-activated cells. Rice and Ernst (94) have shown in co-precipitation experiments that all of the IκB-α in resting WEHI 231 mouse B cells is found in Rel complexes; that is, there is little if any free, stable IκB-α. However, pulse–chase experiments indicate that the half-life of IκB-α is much shorter than that of RelA or c-Rel. This suggests that there is continual cycling of IκB-α on to and off NF-κB complexes. Perhaps this IκB cycling assures that a small percentage of NF-κB can enter the nucleus even in non-stimulated cells to activate the transcription of genes required for general cell maintenance or cell-cycle control (95). Consistent with a general role for IκB-α in metabolism, mice with homozygous inactivation of their IκB-α gene die shortly after birth with a generalized atrophy of several organ systems (D. Baltimore, personal communication).

It must also be pointed out that induction of NF-κB complexes by dissociation of IκB-α results in nuclear translocation of only a small portion of the RelA protein in most cells. In other words, a significant portion of RelA and c-Rel within most resting cells is retained in the cytoplasm through association with p100 or p105 (69–72; see Fig. 4). Current estimates are that about 10–20% of RelA is retained in the cytoplasm as part of complexes that contain IκB-α, and about 80% of RelA is retained in complexes that contain p100 or p105. Therefore, it is thought that p52–RelA and p50–RelA complexes can also be induced by proteolysis of p100 and p105, respectively, in these complexes. In general, it is not known which stimuli activate this second pathway for Rel complex induction. Although phorbol esters can stimulate proteolysis of p105 and p100 in some cell types, the stimulated processing of p105 and p100 occurs rather slowly and is certainly not involved in the rapid appearance of active Rel complexes that is seen in many situations (70). In addition, there are also Rel complexes larger than dimers that exist in the cytoplasm of cells (96).

Cytoplasmic sequestration and inhibition of DNA binding are not the only effects of interaction of an IκB with a Rel complex. Bcl-3 has been shown to interact with p50 or p52 homodimers in the nucleus of some cells (97, 98), and this complex can still bind DNA. Moreover, it appears that the interaction of Bcl-3 with these Rel proteins increases their transcriptional activation function (97, 99). In these cases, the ANK repeats of Bcl-3 are needed for interaction with p50 or p52 and the Bcl-3 sequences N-terminal and C-terminal to the ANK repeats are needed for transcriptional activation. In addition, acidic sequences located between the last two ANK repeats of IκB-γ can efficiently activate transcription when fused to the heterologous DNA-binding domain of the yeast transcription factor GAL4 (100, 101). It is not known whether any IκB sequences that can act as transcriptional activation domains as GAL4 fusion proteins can interact with basal transcription factors such as TBP or TFIIB (see below and Chapter 1). Moreover, the *in vivo* relevance of transcriptional activation by IκB proteins is still unclear.

In summary, it is likely that IκB proteins have multiple effects on Rel protein complexes. It is not known whether IκB proteins can interact with proteins other than Rel proteins *in vivo*.

3.2 Interactions between Rel proteins and other cellular transcription factors

In addition to forming homodimers, heterodimers, and interacting with IκB inhibitors, Rel proteins have been demonstrated to interact with other cellular proteins (Table 2). These have usually been general (TBP and TFIIB; see Chapter 1) or enhancer-binding transcription factors, including those in the zinc finger, helix–loop–helix, and basic–leucine zipper families of DNA-binding proteins (102–110).

In general, associations between Rel proteins and transcription factors have been shown in two ways. First, Rel proteins can often influence transcription from reporter plasmids that contain no apparent κB sites. For example, v-Rel can activate transcription from reporter plasmids that contain Sp1 binding sites, and RelA can potentiate the effect of C/EBP on reporter plasmids containing C/EBP-binding sites (103, 105). Similarly, Fos and Jun, which cannot bind to κB sites, can increase

Table 2 Cellular transcription factors demonstrated to interact with Rel proteins

General factors

Transcription factor	Rel protein partners	References
TBP	c-Rel,v-Rel,RelA,Dorsal	111, 112, 148
TFIIB	p50,c-Rel,v-Rel	111, 112

Enhancer-binding proteins (DNA-binding motif)

Transcription factor	Rel protein partners	References
Sp1 (zinc finger)	v-Rel,c-Rel,RelA	105–108
Glucocorticoid receptor (zinc finger)	RelA	203, 204
Jun (bZIP)	p50, RelA	102, 104
Fos (bZIP)	RelA	104
C/EBPβ (bZIP)	p50, RelA	103, 205
GABPβ (bZIP)	RelA	206
ATFa, ATF2, ATF3 (b-ZIP)	p50, RelA	102
Elf-1(ets domain)	p50	109
IRF-1	p50	40
Twist (HLH)	Dorsal	161, 208
Snail (HLH)	Dorsal	(M. Levine, pers. commun.)

Other

Transcription factor	Rel protein partners	References
HMG I(Y) (HMG-box)	p50, RelA	113, 114
DSP-1	Dorsal, p50	116

Viral Protein

Transcription factor	Rel protein partners	References
HTLV-I Tax (non-DNA binding)	p100, p105, p52, p50	96, 115, 117
BZLF1	RelA	207

expression from a reporter plasmid containing κB sites (104). *In vitro* interactions between Rel proteins and these non-Rel transcription factors have generally been shown in co-precipitation assays using bacterially expressed or *in vitro*-synthesized proteins. The interactions of Rel proteins with heterologous enhancer-binding transcription factors are invariably mediated through the RH domain, and may reflect ways that Rel proteins can affect transcription without being directly bound to DNA. Alternatively, interactions between Rel proteins and heterologous transcription factors may occur on DNA *in vivo* when the factors are in proximity, either by occupying adjacent sites on DNA or by being brought into proximity due to alterations in the structure of chromatin (e.g. by transcription factor-induced DNA bending). Therefore, even though the interactions between Rel proteins and heterologous enhancer-binding proteins are often rather weak in solution, they may occur *in vivo* when the proteins are near one another on DNA.

Rel proteins can also interact with general transcription factors such as TATA-binding protein (TBP) and TFIIB (111, 112; S. Sarkar and T. D. Gilmore, unpublished work; see Chapter 1). These interactions may mediate the direct effect of Rel proteins on transcription when Rel complexes are bound to DNA regulatory elements. The identity of the Rel sequences needed for interaction with these general transcription factors is still somewhat controversial. Kerr *et al.* (111) have reported that the first 40 amino acids of mouse c-Rel mediate interaction with TBP and that sequences from the central region of the RH domain are important for interaction with TFIIB. In contrast, other results demonstrate that sequences that are C-terminal to the RH domain in v-Rel and chicken c-Rel interact with TBP and TFIIB (112; S. Sarkar and T. D. Gilmore, unpublished work). These latter results are more consistent with the general finding that transcription factors interact with TBP and TFIIB via their transcription activation domains (which are located in the C-terminal halves of v-Rel and c-Rel). However, certain TBP-associated factors (TAFs) may also be important for transcriptional activation by Rel proteins, and thus the experiments carried out to date may not describe the only interactions that occur *in vivo* between Rel proteins and general transcription factors.

Other DNA-binding proteins can also affect binding of Rel complexes to DNA and possibly influence their activity. This has been shown most clearly for the human interferon-β promoter (discussed in detail in Chapter 5). Binding of NF-κB to a κB site in the interferon-β promoter is enhanced by a DNA-binding protein called high mobility group protein I(Y) [HMG I(Y)] (113, 114). p50 and RelA can interact directly with HMG I(Y) while bound to DNA. This interaction is specific for the interferon-β promoter because HMG I(Y) does not affect binding of NF-κB to other κB sites. A similar HMG protein, DSP-1, appears to interact with Dorsal on the *zen* promoter and functions as a co-repressor for Dorsal (115).

It has been known for some time that the Tax protein of human T-cell leukaemia virus Type I can increase expression from several genes containing upstream κB sites, even though Tax does not bind directly to κB sites or DNA in general. There have now been several reports of an interaction between Tax and Rel family proteins (96, 116, 117), suggesting that Tax affects the DNA-binding activity or processing of certain Rel proteins. In addition, Tax has recently been shown to increase DNA binding by certain basic–leucine zipper-containing proteins (118). It is likely that other viral proteins will be shown to interact with cellular Rel proteins.

In summary, there may be multiple ways that interactions between Rel proteins and non-Rel transcription factors can affect gene expression (see Fig. 5). These interactions can turn Rel proteins into either activators or repressors. In some cases, Rel proteins may directly contact DNA, in others, Rel proteins may interact with non-Rel transcription factors without binding to DNA. The non-Rel factors may or may not be bound to DNA when they interact with Rel proteins.

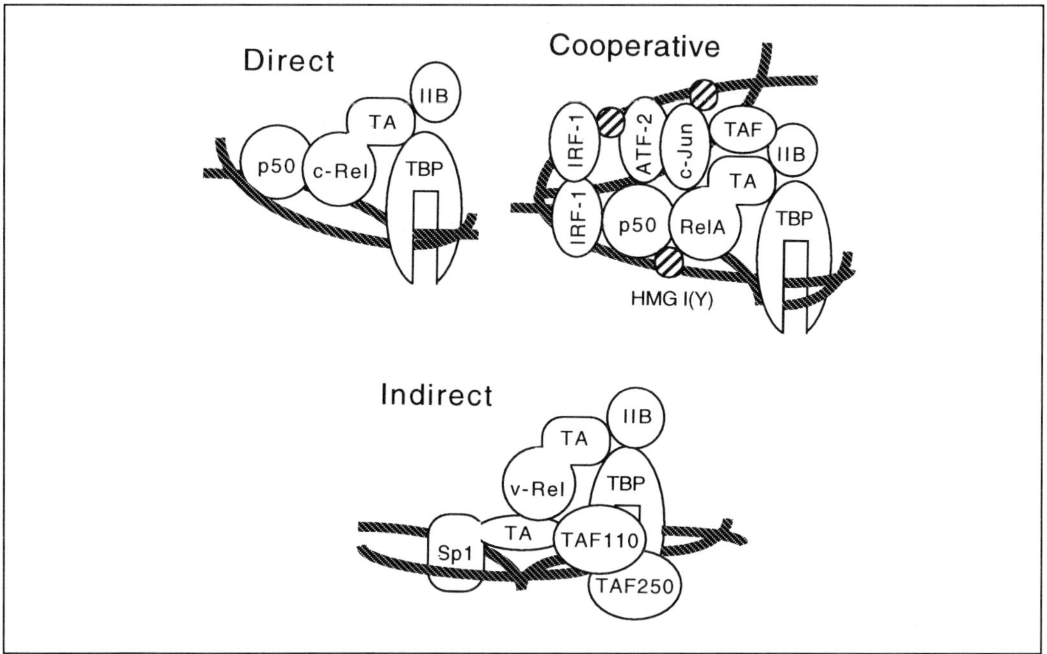

Fig. 5 Models for methods of transcriptional control by Rel proteins. In the example of the Direct model, a p50–c-Rel heterodimer binds to a κB site and the c-Rel C-terminal transcriptional activation domain (TA) directly contacts TBP and TFIIB. In the example of the Co-operative model (adapted from 40; see Chapter 5), the binding of NF-κB to DNA is facilitated by HMG I(Y) molecules and the p50–RelA heterodimer interacts with the components of the closely bound ATF-2–c-Jun heterodimer and IRF-1 (both of which may have been brought into proximity of bound NF-κB through effects on chromatin structure). The initiation of transcription of the interferon-β gene may also be potentiated through TAFs and direct contacts with basal transcription components. In the example of the Indirect model, non-DNA-bound v-Rel interacts with the TA domain of DNA-bound Sp1. This interaction may nucleate TAFs and general transcription factors near the start site of transcription and thereby increase transcriptional initiation. Pertinent to this model, the v-Rel TA domain has been shown to interact with TBP and TFIIB (112; S. Sarkar and T. D. Gilmore, unpublished work), the v-Rel RH domain can interact with the Sp1 TA domain (106), the Sp1 TA domain can interact with TAF110, and TAF110 can interact with TAF250, which can interact with TBP (see Chapter 1).

4. Regulation of Rel complexes by post-translational modification

4.1 Regulation of Rel complexes by proteolysis

Rel complexes can be regulated by proteolysis in three ways, although little is known about the proteases involved in any of these processes. First, p105 and p100 are precursor proteins that are cleaved to the mature DNA-binding proteins, p50 and p52, respectively. The C-terminal sequences of processed p105 and p100 are rapidly degraded. It has been shown that an ATP-dependent protease is responsible for the cleavage of p105 in COS cells (119), and the HIV-1 virally encoded protease can cleave p105 *in vitro* and *in vivo* (120). However, in many cells, the precursor proteins p105 and p100 are the predominant species, and it has been difficult, in pulse–chase experiments, to demonstrate a direct precursor–product relationship between p105 and p50. Furthermore, although proteolysis of C-terminal ANK repeats from *in vitro*-synthesized p100 and p105 is needed to generate a high affinity DNA-binding protein (121–126), full-length p100 and p105 when synthesized *in vivo* can bind DNA as well as their processed counterparts, p52 and p50, respectively (127; D. Baltimore, personal communication). The exact site of proteolysis in p100 and p105 *in vivo* is not known; however, it appears to occur within the glycine-rich hinge region between the RH and ANK domains. In addition, forms of p105 that are missing C-terminal sequences can also be generated by alternative splicing (128).

Secondly, as outlined above, upon activation of Rel complexes containing c-Rel and RelA, IκB-α is rapidly degraded (86–92). Like certain other proteins with short half-lives, IκB-α, IκB-β, and Cactus have C-terminal PEST sequences (129–131). It appears that IκB-α is subjected to proteolysis while still complexed with NF-κB (132–135). However, free IκB-α (like the C-terminal sequences of p105 discussed above) is also extremely unstable in cells (87). Recent experiments suggest that the protease that degrades IκB-α is part of the proteasome (136).

Many results suggest that phosphorylation of IκB-α initiates its degradation and release from NF-κB. The simplest observation that suggests this is the rapid appearance in stimulated cells of a more slowly migrating, phosphorylated form of IκB-α just prior to the disappearance of IκB-α on Western blots (86–92). Some experiments suggest that the ζ isoform of protein kinase C or the c-Raf kinase can phosphorylate IκB-α (88, 137–140), and IκB-α has been reported to be phosphorylated on serine and threonine (135) or tyrosine residues (141). Recent reports indicate that the activation-dependent phosphorylation of IκB-α occurs near its N-terminus (141a, 141b). Most current studies suggest that activation-dependent phosphorylation of IκB-α is not sufficient to cause its dissociation from NF-κB, but rather that this induced phosphorylation event acts as the signal for IκB-α to be ubiquitinated and then degraded by the proteasome while on the Rel complex in a process that also requires C-terminal PEST-containing sequences of IκB-α (141b, 141c, 141d). IκB-β, although it has a specificity like IκB-α for Rel complexes containing c-Rel and RelA, appears to be regulated in ways that are distinct from IκB-α (142).

However, phosphorylation also appears to be necessary for the inhibitory activity of some IκBs. It is likely that a C-terminal casein kinase II site in IκB-α must be phosphorylated for IκB-α to completely inhibit NF-κB (74; I. Verma, personal communication). Similarly, IκB-β, which also regulates complexes containing c-Rel and RelA, and Bcl-3, which regulates p50 homodimers, both appear to require phosphorylation for their inhibitory activity towards Rel complexes (98, 143).

Lastly, it has also been proposed that Dorsal and c-Rel might undergo proteolysis near their C-termini, which then allows them to enter the nucleus (80, 81). This proposal is based on the finding that deletion of C-terminal sequences from c-Rel and Dorsal can cause these proteins to enter the nucleus of tissue culture cells (80, 81). However, there is no evidence that c-Rel or Dorsal undergo proteolysis under any normal conditions *in vivo*. Although Evans *et al.* (144) have detected a smaller protein that cross-reacts with anti-c-Rel antiserum and enters the nucleus of chicken cells during S phase, the identity of this protein is not known.

4.2 Regulation of Rel proteins by phosphorylation

Rel proteins have been demonstrated to be phosphoproteins in every case where they have been analysed. However, surprisingly little is known about regulation of Rel proteins by phosphorylation. A consensus recognition site (Arg-Arg-X-Ser) for phosphorylation by protein kinase A (PKA) is conserved in most Rel proteins. This PKA site is located approximately 25 amino acids N-terminal to the nuclear targeting signal in the RH domain, and it lies in the heart of sequences needed for the formation of homodimers. This PKA site has been analysed in detail only in v-Rel, chicken c-Rel, RelA, and Dorsal. A peptide corresponding to this site in v-Rel can be phosphorylated *in vitro* by PKA, but this serine is only weakly phosphorylated in v-Rel in transformed spleen cells (145, 146). Furthermore, a serine to alanine change at this site does not affect the DNA-binding or transforming activity of v-Rel, and only weakly affects the DNA-binding activity of chicken c-Rel, indicating that this serine does not need to be phosphorylated for v-Rel or c-Rel to be active (145, 146). However, other changes at this site that introduce, for example, a constitutive negative charge (as would be provided by phosphorylation), greatly reduce the formation of homodimers, DNA binding and transformation by v-Rel, and DNA binding and homodimer formation by c-Rel (145, 146). In RelA, substitution of any of four amino acids for the serine in this PKA consensus sequence abolishes the ability of RelA to form homodimers, but does not affect its ability to form heterodimers with p50 and activate transcription (147). Thus, one might predict that phosphorylation of this serine residue would negatively regulate the formation of homodimers by Rel proteins, and therefore might promote heterodimer formation over homodimer formation. In addition, mutations at this serine residue induce nuclear translocation of c-Rel and Dorsal in tissue culture cells (145, 148). However, there is no direct *in vivo* evidence for regulation of Rel proteins by phosphorylation at this serine, and it may simply be that this site has been structurally conserved for other reasons. For example, an Arg to His change at the second residue of the recognition sequence (Arg-Arg-Pro-Ser) renders Dorsal and v-Rel temperature-sensitive *in vivo*, presumably by altering their ability to form stable homodimers and to bind DNA (149, 150).

DNA binding by p50 may be activated by phosphorylation within the RH domain; that is, hyperphosphorylated forms of p50 are seen in cells stimulated with phorbol ester and this phosphorylation correlates with increased DNA-binding activity (151).

Chicken c-Rel and v-Rel are phosphorylated at a large number of serine and threonine residues *in vivo*. Most of these sites are located C-terminal to the RH domain (145, 152). Therefore, one might suppose that some of these sites could affect the transcriptional activation domain in these proteins. In addition, there has also been one report of tyrosine phosphorylation of c-Rel and p105 in Jurkat T cells treated with phorbol esters (153); however, neither the sites nor the physiological relevance of tyrosine phosphorylation are known.

4.3 Regulation of Rel proteins by processes involving oxidation

Treatment of cells with micromolar concentrations of hydrogen peroxide, which increases the intracellular levels of reactive oxygen intermediates, can cause dissociation of IκB-α from NF-κB and the rapid appearance of nuclear NF-κB DNA-binding activity (154). In addition, a large number of inducers of NF-κB activity are also inducers of oxidative stress (for review see 21). Most interestingly, anti-oxidants such as *N*-acetyl-L-cysteine can block NF-κB induction by every agent tested so far, including such diverse agents as phorbol esters, the Tax protein of HTLV-I, the calcium ionophore A23187, and taxol. This suggests that oxidative stress might be an important, common step in the activation of NF-κB.

How might oxygen radicals activate NF-κB? Treatment of NF-κB with hydrogen peroxide *in vitro* does not cause IκB to be released from NF-κB, suggesting that oxygen radicals do not have a direct effect on the NF-κB–IκB complex (21, 154). Therefore, one might suppose that treatment of cells with hydrogen peroxide activates an IκB-specific kinase or protease. It is certain that there will be great interest in reactive oxygen intermediates as a second messenger system, especially concerned with the induction of NF-κB.

In addition, there is some evidence that oxidation–reduction might control the DNA-binding activity of Rel proteins. As mentioned above, a conserved RXXRXRXXC sequence, located towards the N-terminus of the RH domain in all Rel proteins appears to be directly involved in contacting DNA through the conserved cysteine (50, 51). *In vitro* experiments suggest that this conserved cysteine residue in this sequence is subject to redox regulation. In other words, oxidation of p50 and v-Rel reduces their DNA-binding activity *in vitro* and reducing agents such as dithiothreitol and β-mercaptoethanol can increase the DNA-binding activity of *in vitro*-synthesized p50 and v-Rel (155–157). Furthermore, a cysteine to serine mutation in v-Rel increases its DNA-binding activity *in vitro* and makes it resistant to redox regulation (155). These results are consistent with earlier findings that AP-1 (a Fos–Jun heterodimer) is also subject to redox regulation through cysteine residues (158). However, to date, there is no proof that Rel proteins are controlled *in vivo* by redox regulation through the conserved cysteine residue, and it is unlikely that cells would have reducing environments like those required to activate Rel proteins *in vitro*. Finally, as mentioned above, treatment of cells with oxidizing agents increases NF-κB activity (154).

5. Regulation of Rel protein complexes by concentration

Although rarely discussed, it is likely that intracellular or intranuclear concentration affects the activity of transcription factors. This has been elegantly shown in the case of Dorsal. Although the intracellular concentration of Dorsal protein is relatively constant within all cells of the *D. melanogaster* blastoderm, there is a nuclear gradient such that Dorsal is located almost exclusively in the nucleus of the ventral-most cells and exclusively in the cytoplasm of the dorsal-most cells (81, 159, 160). As a consequence of this nuclear gradient, the *twist* gene, which has low-affinity, upstream Dorsal-binding sites, is only expressed in the ventral-most cells (161). In contrast, the *rhomboid* gene, which has high-affinity Dorsal-binding sites, is expressed laterally where the nuclear concentration of Dorsal is lower (162); in ventral cells the *rhomboid* gene is specifically repressed by the Snail protein.

Are similar concentration effects operational in vertebrate cells? This is not precisely known. However, it has been shown that the v-Rel oncoprotein has a threshold level for transforming chicken spleen cells (163). Furthermore, RelA is located in the cytoplasm of cells when expressed at low levels due to association with IκB, but RelA is in the nucleus of cells when expressed at high levels (79). Therefore, Rel proteins expressed at high levels from expression plasmids may titrate out the endogenous IκB proteins and enter the nucleus in cells where their concentration would never otherwise be so high. Thus, as a cautionary note, many co-transfection experiments using κB site-containing reporter plasmids and Rel expression plasmids may not mimic physiologically relevant situations due to high levels of expression of the Rel proteins and/or concentration of target sites on the reporter plasmid. Determining whether the concentration of Rel factors or any transcription factor is normally involved in the control of transcription is a very difficult task.

6. Transcriptional regulation of *rel* family genes

rel family genes are expressed in all cell types, but generally appear to be expressed at highest levels in cells of lymphoid origin. In addition, expression of certain *rel* genes can be induced by mitogens (see Table 3). For example, mouse c-*rel* and *relB* can be induced in quiescent mouse 3T3 cells treated with serum (59, 164), and p105 and p100 cDNAs were isolated as mitogen-inducible genes in T cells (165–167).

The promoters for c-*rel*, *nfkb1*, *nfkb2*, and *relA* have been cloned. The c-*rel*, *nfkb1*, and *nfkb2* promoters lack a TATA box and contain κB sites (168–172). Each of these promoters has been shown to undergo autoregulation: c-*rel* can be repressed by over-expression of c-Rel or v-Rel (168, 169); and in different cell types, *nfkb1* can be

Table 3 Promoters of Rel and IκB family genes

Gene	Promoter type	Inducible expression
c-*rel*	GC-rich (κB sites)	Serum, T-cell mitogens
nfkb1	GC-rich (κB sites)	T-cell mitogens
nfkb2	GC-rich (κB sites)	T-cell mitogens
relA	TATA box (no κB sites)	No
relB	Unknown	Serum
Ikba	GC-rich (κB sites)	Monocyte adherence, serum, liver regeneration

repressed or activated by either v-Rel or p50 homodimers (170, 171; P. Enrietto, personal communication). However, it should be noted that mutation of the κB site in the chicken c-*rel* promoter does not alter its ability to be repressed by c-Rel or v-Rel (169), suggesting that c-Rel might act through other factors or sites to affect its expression. The *relA* gene has a typical TATA box-containing promoter, which does not contain κB sites (173), and does not appear to be inducible by mitogens. In general, the RelA protein is expressed at low levels in cells; this may reflect the fact that RelA has an extremely potent C-terminal transcriptional activation domain, which needs to be tightly controlled.

7. Oncogenic Rel proteins appear to be unregulated versions of their normal counterparts

The v-Rel protein of the highly oncogenic Rev-T retrovirus causes lethal lymphomas *in vivo* and can transform a variety of haematopoietic chicken cells and chicken embryo fibroblasts (CEF) *in vitro* (for review see 1, 2, 7, 8, 10, 14, 174). The most significant structural difference between v-Rel and c-Rel is a deletion of 118 C-terminal amino acids in v-Rel compared with c-Rel, and this deletion has several functional consequences. First, when expressed from a retroviral vector, v-Rel and a c-Rel protein, with its C-terminal residues deleted, are located in the nucleus of CEF, whereas wild-type c-Rel is located in the cytoplasm (80). It is not clear if removal of these C-terminal amino acids is responsible for a decreased *in vivo* affinity of IκB for v-Rel compared with c-Rel (175) or if the deletion alters an intramolecular interaction in c-Rel that masks the nuclear targeting sequence (78, 80, 145). Moreover, most of the v-Rel protein is actually located in the cytoplasm of transformed chicken spleen cells (176, 177). Secondly, the deletion in v-Rel removes much of a strong C-terminal c-Rel transcription activation domain (57). However, v-Rel still retains a weak ability to activate transcription, and this activity appears to be essential for its transforming function (178, 179). Thirdly, v-Rel and a C-terminally deleted c-Rel protein are at least ten times more transforming than wild-type c-Rel (54, 152), indicating that C-terminal sequences of c-Rel inhibit its oncogenicity. Although there are several models for how v-Rel malignantly transforms cells, the preponderance of evidence indicates that v-Rel is transforming because it binds to promoter elements as a homodimer and constitutively activates certain genes normally activated in a regulated fashion by c-Rel (178–180; D.W.White, G.Pitoc, and T.D.Gilmore, unpublished observations). However, one recent report has suggested that v-Rel might also have direct effects on DNA replication (181).

rel family genes may be rearranged in up to 10–15% of human lymphoid cancers (reviewed in 15); in addition, the IκB family gene, *bcl-3*, was originally identified as a rearranged gene in a human B-cell cancer (182). To date, the rearrangements in human *rel* family genes include alterations in c-*rel* and *nfkb2* (123, 183–185). Similarly to v-Rel, in each case the rearrangement removes sequences encoding residues C-terminal to the RH domain of human c-Rel or p100. The removal of ANK repeat sequences from p100 creates a p100 protein that is now located in the nucleus, can bind DNA with high affinity *in vitro* (as a homodimer or heterodimer), and can activate transcription (184). Thus, the oncogenic activation of p100 in human tumours may mimic the activation seen with v-Rel and c-Rel in the avian system. However, it should be noted that deleted human p100 and c-Rel proteins have not been shown to be transforming *in vitro*. In fact, there has only been one report of a Rel protein, an alternatively spliced form of human RelA, transforming mammalian cells in cell culture (186), and this finding has recently been disputed (187).

8. Conclusions

The Rel family of eukaryotic transcription factors controls the expression of a number of important cellular and viral genes. As described in this chapter, Rel factors are regulated at myriad levels (see Table 4). Understanding precisely how they are regulated *in vivo* will be a daunting task. However, an understanding of the regulation of Rel complexes is likely to have therapeutic relevance, due to the large number of cellular and viral genes controlled by specific Rel complexes. For example, it is likely that the identification of inhibitors of IκB release from NF-κB and/or inhibitors of degradation of IκB will be important reagents for blocking undesirable activation of Rel complexes. As such, anti-oxidants may have therapeutic effects; for example, during the course of HIV infection or in certain lymphoid cancers. Furthermore, antisense oligonucleotides to *relA* have been shown to have potent anti-tumour effects in a mouse model system (188). In any event, it is certain that Rel proteins will continue to provide guide-posts for understanding the regulation of transcription factors in general.

Table 4 Regulation of Rel and IκB factors

Form of regulation	**rel gene or protein**
Transcriptional induction	*c-rel, nfkb1, nfkb2, relB, Ikba, dorsal*
Alternative promoter usage	*nfkb1* (for IκB-γ), *nfkb2*
Alternative splicing	*nfkb1, nfkb2, relA*, IkB-g
mRNA stability	*Ikba*
Combinatorial dimerization	All
Subcellular location (nucleus vs. cytoplasm)	All
Proteolysis	50/p105, p52/p100, IκB-α, IκB-β, Cactus
Phosphorylation	Unknown (Possibly v-Rel, c-Rel, p50, RelA, Dorsal, IκB-α, IκB-β)
Redox regulation	v-Rel, p50, IκB-α (?)
Nuclear concentration	Dorsal

Acknowledgements

I thank the members of my lab for help with the figures and for critical reading of the manuscript. I also thank several investigators for communicating unpublished data. The work in my lab has recently been sponsored by grants from the National Cancer Institute, the American Cancer Society, the Children's Leukemia Research Association, and the Council for Tobacco Research USA, Inc.

References

1. Moore, B. E. and Bose, H. R., jun. (1988) Transformation of avian lymphoid cells by reticuloendotheliosis virus. *Mutation Res.*, **195**, 79.
2. Rice, N. R. and Gilden, R. V., (1988) The *rel* oncogene. In *The oncogene handbook* (ed. E. P. Reddy, A. M. Skalka, and T. Curran), p. 495. Elsevier Science Publishers, New York.
3. Lenardo, M. and Baltimore, D. (1989) NF-κB: a pleiotropic mediator of inducible and tissue-specific gene control. *Cell*, **58**, 227.
4. Gilmore, T. D. (1990) NF-κB, KBF1, *dorsal*, and *rel*ated matters. *Cell*, **62**, 841.
5. Baeuerle, P. A. (1991) The inducible transcription activator NF-κB: regulation by distinct protein subunits. *Biochim. Biophys. Acta*, **1072**, 63.
6. Baeuerle, P. A. and Baltimore, D. (1991) The physiology of the NF-κB transcription factor. *Mol. Asp. Cell. Regul.*, **6**, 409.
7. Enrietto, P. J. (1991) The *rel* gene and its role in oncogenesis. *Sem. Canc. Biol.*, **1**, 399.
8. Gilmore, T. D. (1991) Malignant transformation by mutant Rel proteins. *Trends Genet.*, **7**, 318.
9. Govind, S. and Steward, R. (1991) Dorsoventral pattern formation in *Drosophila*: signal transduction and nuclear targeting. *Trends Genet.*, **7**, 119.
10. Hannink, M. and Temin, H. M. (1991) Molecular mechanisms of transformation by the v-*rel* oncogene. *Crit. Rev. Oncogen.*, **2**, 293.
11. Schmitz, M. L., Henkel, T., and Baeuerle, P. A. (1991) Proteins controlling the nuclear uptake of NF-κB, Rel and dorsal. *Trends Cell Biol.*, **1**, 130.
12. Blank, V., Kourilsky, P., and Israël, A. (1992) The NF-κB family: Rel/Dorsal homology and ankyrin-like repeats. *Trends Biochem. Sci.*, **17**, 135.
13. Bours, V., Franzoso, G., Brown, K., Park, S., Azarenko, V., Tomita-Yamaguchi, M., Kelly, K., and Siebenlist, U. (1992) Lymphocyte activation and the family of NF-κB transcription complexes. *Curr. Topics Microbiol. Immunol.*, **182**, 411.
14. Bose, H. R., jun. (1992) The Rel family: models for transcriptional regulation and oncogenic transformation. *Biochim. Biophys. Acta*, **1114**, 1.
15. Gilmore, T. D. (1992) Role of *rel* family genes in normal and malignant lymphoid cell growth. *Canc. Surveys*, **15**, 69.
16. Inoue, J. (1992) The regulation of transcription factor Rel/NF-κB. *Tanpakushitsu Kakusan Koso–Protein, Nucleic Acid, Enzyme*, **37**, 2947.
17. Israël, A. (1992) La famille de facteurs de transcription rel/NF-κB: un méchanisme original de contrôle de l'expression génique. *Pathol. Biol.*, **40**, 212.
18. Michaely, P. and Bennett, V. (1992) The ANK repeat: a ubiquitous motif involved in macromolecular recognition. *Trends Cell Biol.*, **2**, 127.
19. Nolan, G. P. and Baltimore, D. (1992) The inhibitory ankyrin and activator Rel proteins. *Curr. Opin. Genet. Dev.*, **2**, 211.

20. Rushlow, C. and Warrior, R. (1992) The rel family of proteins. *BioEssays*, **14**, 89.
21. Schreck, R., Albermann, K., and Baeuerle, P. A. (1992) Nuclear factor κB: an oxidative stress-responsive transcription factor of eukaryotic cells. *Free Radical Res. Commun.*, **17**, 221.
22. Beg, A. A. and Baldwin, A. S., jun. (1993) The IκB proteins: multifunctional regulators of Rel/NF-κB transcription factors. *Genes Dev.*, **7**, 2064.
23. Collins, T. (1993) Biology of disease. Endothelial nuclear factor-κB and the initiation of the atherosclerotic lesion. *Laboratory Invest.*, **68**, 499.
24. Duckett, C. S. (1993) Structure and function of the NF-κB family of transcription factors. *Promega Notes*, **44**, 40.
25. Gilmore, T. D. and Morin, P. J. (1993) The IκB proteins: members of a multifunctional family. *Trends Genet.*, **9**, 427.
26. Govind, S. and Steward, R. (1993) Coming to grips with cactus. *Curr. Biol.*, **3**, 351.
27. Greene, W. C., Sun, S.-C., Ganchi, P., Doerre, S., and Ballard, D. W. (1993) The NF-κB/IκB connection: a novel strategy for transcriptional control. In *Molecular basis of the immune response* (ed. H. Nariuchi *et al.*), p. 73. Academic Press, New York.
28. Grilli, M., Chiu, J. J.-S., and Lenardo, M. J . (1993) NF-κB and Rel—participants in a multiform transcriptional regulatory system. *Internatl. Rev. Cytology*, **143**, 1.
29. Grimm, S. and Baeuerle, P. A. (1993) The inducible transcription factor NF-κB: structure-function relationship of its protein subunits. *Biochem. J.*, **290**, 297.
30. Hay, R. T. (1993) Control of nuclear factor κB DNA-binding activity by inhibitory proteins containing ankrin repeats. *Biochem. Soc. Trans.* **21**, 926.
31. Liou, H.-C. and Baltimore, D. (1993) Regulation of NF-κB/rel transcription factor and IκB inhibitor system. *Curr. Opin. Cell Biol.*, **5**, 477,
32. Müller, J. M., Ziegler-Heitbrock, J. H. W., and Baeuerle P. A. (1993) Nuclear factor κB, a mediator of lipopolysaccharide effects. *Immunobiol.*, **187**, 233.
33. Steward, R. and Govind, S. (1993) Dorsal-ventral polarity in the *Drosophila* embryo. *Curr. Opin. Genet. Dev.*, **3**, 556.
34. Wasserman, S. (1993) A conserved signal transduction pathway regulating the activity of the Rel-like proteins Dorsal and NF-κB. *Mol. Biol. Cell*, **4**, 767.
35. Baeuerle, P. A. and Henkel, T. (1994) Function and activation of NF-κB in the immune system. *Ann. Rev. Immunol.*, **12**, 141.
36. Kopp, E. and Ghosh, S. (1994) NF-κB and Rel proteins in innate immunity. *Adv. Immunol.* **58**, 1.
37. Lenardo, M. and Siebenlist, U. (1994) Bcl-3-mediated nuclear regulation of the NF-κB transactivating factor. *Immunol. Today*, **15**, 145.
38. Siebenlist, U., Brown, K., and Franzoso, G. (1994) NF-κB: a mediator of pathogen and stress responses. In *Advances in research on neurodegeneration, II* (ed. Y. Mizuno *et al.*), p. 93. Birkhäuser, Boston.
39. Siebenlist, U., Franzoso, G., and Brown, K. (1994) Structure, regulation and function of NF-κB. *Ann. Rev. Cell Biol.* **10**, 405.
40. Thanos, D. and Maniatis, T. (1995) NF-κB: a lesson in family values. *Cell*, **80**, 529.
41. Baldwin, A. S., jun., and Sharp, P. A. (1989) Binding of a nuclear factor to a regulatory sequence in the promoter of the mouse $H\text{-}2K^b$ class I histocompatibility gene. *Mol. Cell. Biol.*, **7**, 305.
42. Baldwin, A. S., jun., LeClair, K. P., Singh, H., and Sharp, P. A. (1990) A large protein containing zinc finger domains binds to related sequence elements in the enhancers of the class I major histocompatibility complex and kappa immunoglobulin genes. *Mol. Cell. Biol.*, **10**, 1406.

43. Fan, C.-M. and Maniatis, T. (1990) A DNA-binding protein containing two widely separated zinc finger motifs that recognize the same DNA sequence. *Genes Dev.*, **4**, 29.
44. Rustgi, A. K., Van't Veer, L. J., and Bernards, R. (1990) Two genes encode factors with NF-κB- and H2TF1-like DNA-binding properties. *Proc. Natl Acad. Sci. USA*, **87**, 8707.
45. Adams, B. S., Leung, K., Hanley, E. W., and Nabel, G. J. (1991) Cloning of RκB, a novel DNA-binding protein that recognizes the interleukin-2 receptor a chain κB site. *New Biol.*, **3**, 1063.
46. Ron, D., Brasier, A. R., and Habener, J. F. (1991) Angiotensin gene-inducible enhancer-binding protein 1, a member of a new family of large nuclear proteins that recognize nuclear factor κB-binding sites through a zinc finger motif. *Mol. Cell. Biol.*, **11**, 2887.
47. Lattion, A.-L., Espel, E., Reichenbach, P., Fromental, C., Bucher, P., Israël, A., Baeuerle, P., Rice, N. R., and Nabholz, M. (1992) Characterization of a new tissue-specific transcription factor binding to the simian virus 40 enhancer TC-II (NF-κB) element. *Mol. Cell. Biol.*, **12**, 5217.
48. Korner, M., Rattner, A., Mauxion, F., Sen, R., and Citri, Y. (1989) A brain-specific transcription activator. *Neuron*, **3**, 563.
49. Rattner, A., Korner, M., Walker, M. D., and Citri, Y. (1993) NF-κB activates the HIV promoter in neurons. *EMBO J.*, **12**, 4261.
50. Ghosh, G., van Duyne, G., Ghosh, S., and Sigler, P. B. (1995) Structure of NF-κB p50 homodimers bound to a κB site. *Nature*, **373**, 303.
51. Muller, C. W., Rey, F. A., Sodeoka, M., Verdine, G. L., and Harrison, S. C. (1995). Structure of the NF-κB p50 homodimer bound to DNA. *Nature*, **373**, 311.
52. Baltimore, D. and Beg, A. A. (1995) A butterfly flutters by. *Nature*, **373**, 287.
53. Bull, P., Morley, K. L., Hoekstra, M. F., Hunter, T., and Verma, I. M. (1990) The mouse c-*rel* protein has an N-terminal regulatory domain and a C-terminal transcriptional transactivation domain. *Mol. Cell. Biol.*, **10**, 5473.
54. Kamens, J., Richardson, P., Mosialos, G., Brent, R., and Gilmore, T. (1990) Oncogenic transformation by v-Rel requires an amino-terminal activation domain. *Mol. Cell. Biol.*, **10**, 2840.
55. Ballard, D. W., Dixon, E. P., Peffer, N. J., Bogerd, H., Doerre, S., Stein, B., and Greene, W. C. (1992) The 65-kD subunit of human NF-κB functions as a potent transcriptional activator and a target for v-Rel mediated repression. *Proc. Natl Acad. Sci. USA*, **89**, 1875.
56. Dobrzanski, P., Ryseck, R.-P., and Bravo, R. (1993) The N- and C-terminal domains of RelB are required for full transactivation: role of the N-terminal leucine zipper-like motif. *Mol. Cell. Biol.*, **13**, 1572.
57. Richardson, P. M. and Gilmore, T. (1991) v-Rel is an inactive member of the Rel family of transcriptional activating proteins. *J. Virol.*, **65**, 3122.
58. Ruben, S. M., Narayanan, R., Klement, J. F., Chen, C.-H., and Rosen, C. A. (1992) An alternatively spliced form of NF-κB p65 defines essential functional domains. *Mol. Cell. Biol.*, **12**, 444.
59. Ryseck, R.-P., Bull, P., Takamiya, M., Bours, V., Siebenlist, U., Dobrzanski, P., and Bravo, R. (1992) RelB, a new Rel family transcription activator that can interact with p50–NF-κB. *Mol. Cell. Biol.*, **12**, 674.
60. Schmitz, M. L. and Baeuerle, P. A. (1991) The p65 subunit is responsible for the strong transcription activating potential of NF-κB by p65. *EMBO J.*, **10**, 3805.
61. McCaffrey, P. G., Luo, C., Kerpolla, T. K., Jain, J., Badalien, A. N., Ho, E., Burgin, W. S., Lane, J. N., Lambert, T., Vedine, G. L., Rao, A., and Hogan, P. G. (1993). Isolation of the cyclosporin-sensitive T cell transcription factor NF-ATp. *Science*, **262**, 750.

62. Northrop, J. P., Ullman, K. S., and Crabtree, G. (1993) Characterisation of the nuclear and cytoplasmic components of the lymphoid-specific nuclear factor of activated T cells (NF-AT) complex. *J. Biol. Chem.*, **268**, 2917.
63. Sen, R. and Baltimore, D. (1986) Multiple nuclear factors interact with the immunoglobulin enhancer sequences. *Cell*, **46**, 705.
64. Sen, R. and Baltimore, D. (1986) Inducibility of κ immunoglobulin enhancer-binding protein NF-κB by a post-translational mechanism. *Cell*, **47**, 921.
65. Lernbecher, T., Müller, U., and Wirth, T. (1993) Distinct NF-κB/Rel transcription factors are responsible for tissue-specific and inducible gene activation. *Nature*, **365**, 767.
66. Ip, Y. T., Reach, M., Engstrom, Y., Kadalayil, L., Cai, H., González-Crespo, S., Tatei, K., and Levine, M. (1993) *dif*, a *dorsal*-related gene that mediates an immune response in Drosophila. *Cell*, **75**, 753.
67. Sha, W. C., Liou, H.-C., Tuomanen, E. I., and Baltimore, D. (1995) Targeted disruption of the p50 subunit of NF-κB leads to multifocal defects in immune responses. *Cell*, **80**, 321.
68. Inoue, J., Kerr, L. D., Kakizuka, A., and Verma, I. (1992) IκBγ, a 70 kd protein identical to the C-terminal half of p110 NF-κB: a new member of the IκB family. *Cell*, **68**, 1109.
69. Rice, N. R., MacKichan, M. L., and Israël, A. (1992) The precursor of NF-κB p50 has IκB-like functions. *Cell*, **71**, 243.
70. Mercurio, F., DiDonato, J. A., Rosette, C., and Karin, M. (1993) p105 and p98 precursor proteins play an active role in NF-κB-mediated signal transduction. *Genes Dev.*, **7**, 705.
71. Naumann, M., Nieters, A., Hatada, E. N., and Scheidereit, C. (1993) NF-κB precursor p100 inhibits nuclear translocation and DNA binding of NF-κB/*rel*-factors. *Oncogene*, **8**, 2275.
72. Naumann, M., Wulczyn, F. G., and Scheidereit, C. (1993) The NF-κB precursor p105 and the proto-oncogene product Bcl-3 are IκB molecules and control nuclear translocation of NF-κB. *EMBO J.*, **12**, 213.
73. Scheinman, R. I., Beg, A. A., and Baldwin, A. S., jun. (1993) NF-κB p100 (Lyt-10) is a component of H2TF1 and can function as an IκB-like molecule. *Mol. Cell. Biol.*, **13**, 6089.
74. Ernst, M. K., Dunn, L. L., and Rice, N. R. (1995) The PEST-like sequence of IκBα is responsible for inhibition of DNA binding but not for cytoplsamic retention of c-Rel or RelA homodimers. *Mol. Cell. Biol.*, **15**, 872.
75. Inoue, J., Kerr, L. D., Rashid, D., Davis, N., Bose, H. R., jun., and Verma, I. M. (1992) Direct association of pp40/IκBβ with rel/NF-κB transcription factors: role of ankyrin repeats in the inhibition of DNA binding activity. *Proc. Natl Acad. Sci. USA*, **89**, 4333.
76. Kumar, S. and Gélinas, C. (1993) IκBα-mediated inhibition of v-Rel DNA-binding requires direct interaction with the RxxRxRxxC Rel/κB DNA-binding motif. *Proc. Natl Acad. Sci. USA*, **90**, 8962.
77. Beg, A. A., Ruben, S. M., Scheinman, R. I., Haskill, S., Rosen, C. A., and Baldwin, A. S., jun. (1992) IκB interacts with the nuclear localization sequences of the subunits of NF-κB: a mechanism for cytoplasmic retention. *Genes Dev.*, **6**, 1899.
78. Ganchi, P. A., Sun, S.-C., Greene, W. C., and Ballard, D. W. (1992) IκB/MAD-3 masks the nuclear localization signal of NF-κB p65 and requires the transactivation domain to inhibit NF-p65 DNA binding. *Mol. Biol. Cell*, **3**, 1339.
79. Zabel, U., Henkel, T., dos Santos Silva, M., and Baeuerle, P. A. (1993) Nuclear uptake control of NF-κB by MAD-3, an IκB protein present in the nucleus. *EMBO J.*, **12**, 201.
80. Capobianco, A. J., Simmons, D. L., and Gilmore, T. D. (1990) Cloning and expression of

a chicken c-*rel* cDNA: unlike $p59^{v\text{-}rel}$, $p68^{c\text{-}rel}$ is a cytoplasmic protein in chicken embryo fibroblasts. *Oncogene*, **5**, 257.

81. Rushlow, C. A., Han, K., Manley, J. L., and Levine, M. (1989) The graded distribution of the *dorsal* morphogen is initiated by selective nuclear transport in Drosophila. *Cell*, **59**, 1165.
82. Hatada, E. N., Naumann, M., and Scheidereit, C. (1993) Common structural constituents confer IκB activity to NF-κB p105 and IκB/MAD-3. *EMBO J.*, **12**, 2781.
83. Grumont, R. J. and Gerondakis, S. (1994) Alternative splicing of RNA transcripts encoded by the murine p105 NF-κB gene generates IκBγ isoforms with different inhbitory activities. *Proc. Natl Acad. Sci. USA*, **91**, 4367.
84. Zabel, U. and Baeuerle, P. A. (1990) Purified human IκB can rapidly dissociate the complex of the NF-κB transcription factor with its cognate DNA. *Cell*, **61**, 255.
85. Léveillard, T. and Verma, I. M. (1993) Diverse molecular mechanisms of inhibition of NF-κB/DNA binding complexes by IκB proteins. *Gene Expression*, **3**, 135.
86. Beg, A. A., Finco, T. S., Nantermet, P. V., and Baldwin, A. S., jun. (1993) Tumor necrosis factor and interleukin-1 lead to phosphorylation and loss of IκBα: a mechanism for NF-κB activation. *Mol. Cell. Biol.*, **13**, 3301.
87. Brown, K., Park, S., Kanno, T., Franzoso, G., and Siebenlist, U. (1993) Mutual regulation of the transcriptional activator NF-κB and its inhibitor, IκB-α. *Proc. Natl Acad. Sci. USA*, **90**, 2532.
88. Cordle, S. R., Donald, R., Read, M. A., and Hawiger, J. (1993) Lipopolysaccharide induces phosphorylation of MAD3 and activation of c-Rel and related NF-κB proteins in human monocytic THP-1 cells. *J. Biol. Chem.*, **268**, 11803.
89. Henkel, T., Machleidt, T., Alkalay, I., Krönke, M., Ben-Neriah, Y., and Baeuerle, P. A. (1993) Rapid proteolysis of IκB-α in response to phorbol ester, cytokines and lipopolysaccharide is a necessary step in the activation of NF-κB. *Nature*, **365**, 182.
90. Mellits, K. H., Hay, R. T., and Goodbourn, S. (1993) Proteolytic processing of MAD3 (IκBα) and enhanced processing of the NF-κB precursor p105 are obligatory steps in the activation of NF-κB. *Nucl. Acids Res.*, **21**, 5059.
91. Scott, M. L., Fujita, T., Liou, H.-C., Nolan, G. P., and Baltimore, D. (1993) The p65 subunit of NF-κB regulates IκB by two distinct mechanisms. *Genes Dev.*, **7**, 1266.
92. Sun, S.-C., Ganchi, P. A., Ballard, D. W., and Greene, W. C. (1993) NF-κB controls expression of inhibitor IκBα: evidence for an inducible autoregulatory pathway. *Science*, **259**, 1912.
93. de Martin, R., Vanhove, B., Cheng, Q., Hofer, E., Csizmadia, V., Winkler, H., and Bach, F. H. (1993) Cytokine-inducible expression in endothelial cells of an IκBα-like gene is regulated by NFκB. *EMBO J.*, **12**, 2773.
94. Rice, N. R. and Ernst, M. K. (1993) *In vivo* control of NF-κB activation by IκBα. *EMBO J.*, **12**, 4685.
95. Baldwin, A. S., jun., Azizkhan, J. C., Jensen, D. E., Beg, A. A., and Coodly, L. R. (1991) Induction of NF-κB DNA-binding activity during the G_0-to-G_1 transition in mouse fibroblasts. *Mol. Cell. Biol.*, **11**, 4943.
96. Watanabe, M., Muramatsu, M.-A., Hirai, H., Suzuki, T., Fujisawa, J., Yoshida, M., Arai, K.-I., and Arai, N. (1993) HTLV-I encoded Tax in association with NF-κB precursor p105 enhances nuclear localization of NF-κB p50 and p65 in transfected cells. *Oncogene*, **8**, 2949.
97. Fujita, T., Nolan, G. P., Liou, H.-S., Scott, M. L., and Baltimore, D. (1993) The candidate proto-oncogene *bcl-3* encodes a transcriptional coactivator that activates through NF-κB p50 homodimers. *Genes Dev.*, **7**, 1354.

98. Nolan, G. P., Fujita, T., Bhatia, K., Huppi, C., Liou, H.-C., Scott, M., and Baltimore, D. (1993) The *bcl-3* proto-oncogene encodes a nuclear IκB-like molecule that preferentially interacts with NF-κB p50 and p52 in a phosphorylation-dependent manner. *Mol. Cell. Biol.*, **13**, 3557.
99. Bours, V., Franzoso, G., Azarenko, V., Park, S., Kanno, T., Brown, K., and Siebenlist, U. (1993) The oncoprotein Bcl-3 directly transactivates through κB motifs via association with DNA-binding p50B homodimers. *Cell*, **72**, 729.
100. Morin, P. J. and Gilmore, T. D. (1992) The C terminus of the NF-κB precursor and an IκB isoform contain transcription activation domains. *Nucl. Acids Res.*, **20**, 2453.
101. Morin, P. J., Subramanian, G. S., and Gilmore, T. D. (1993) GAL4-IκBα and GAL4-IκBγ activate transcription by different mechanisms. *Nucl. Acids Res.*, **21**, 2157.
102. Kaszubska, W., Hooft van Huijsduijnen, R., Ghersa, P., DeRaemy-Schenk, A.-M., Chen, B. P. C., Hai, T., DeLarmarter, J. F., and Whelan, J. (1993) Cyclic AMP-independent ATF family members interact with NF-κB and function in the activation of the E-selectin promoter in response to cytokines. *Mol. Cell. Biol.*, **13**, 7180.
103. Stein, B., Cogswell, P. C., and Baldwin, A. S., jun. (1993) Functional and physical associations between NF-κB and C/EBP family members: a Rel domain-bZIP interaction. *Mol. Cell. Biol.*, **13**, 3964.
104. Stein, B., Baldwin, A. S., jun., Ballard, D. W., Greene, W. C., Angel, P., and Herrlich, P. (1993) Cross-coupling of the NF-κB p65 and Fos/Jun transcription factors produces potentiated biological function. *EMBO J.*, **12**, 3879.
105. Sif, S., Capobianco, A. J., and Gilmore, T. D. (1993) The v-Rel oncoprotein increases expression from Sp1 site-containing promoters in chicken embryo fibroblasts. *Oncogene*, **8**, 2501.
106. Sif, S. and Gilmore, T. D. (1994) Interaction of the v-Rel oncoprotein with cellular transcription factor Sp1. *J. Virol.*, **68**, 7131.
107. Perkins, N. D., Edwards, N. L., Duckett, C. S., Agranoff, A. B., Schmid, R. M., and Nabel, G. J. (1993) A cooperative interaction between NF-κB and Sp1 is required for HIV-1 enhancer activation. *EMBO J.*, **12**, 3551.
108. Majello, B., DeLuca, P., Hagen, G., Suske, G., and Lania, L. (1994) Different members of the Sp1 multigene family exert opposite transcriptional regulation of the long terminal repeat of HIV-1. *Nucl. Acids Res.*, **22**, 4914.
109. John, S., Reeves, R. B., Lin, J.-X., Child, R., Leiden, J. M., Thompson, C. A., and Leonard, W. J. (1995) Regulation of cell-type-specific interleukin-2 receptor alpha-chain gene expression: potential role of physical interactions between Elf-1, HMGI/Y, and NF-κB family proteins. *Mol. Cell. Biol.*, **15**, 1786.
110. Stein, B. and Baldwin, A. S., jun. (1993) Distinct mechanisms for regulation of the interleukin-8 gene involve synergism and cooperativity between C/EBP and NF-κB. *Mol. Cell. Biol.*, **13**, 7191.
111. Kerr, L. D., Ransone, L. J., Wamsley, P., Schmitt, M. J., Boyer, T. G., Zhou, Q., Berk, A. J., and Verma, I. M. (1993) Association between proto-oncoprotein Rel and TATA-binding protein mediates activation by NF-κB. *Nature*, **365**, 412.
112. Xu, X., Proprock, C., Ishikawa, H., Maldonado, E., Ito, Y., and Gélinas, C. (1993) Functional interaction of the v-Rel and c-Rel oncoproteins with the TATA-binding protein and association with transcription factor IIB. *Mol. Cell. Biol.*, **13**, 6733.
113. Thanos, D. and Maniatis, T. (1992) High mobility group protein HMG I(Y) is required for NF-κB-dependent virus induction of the human IFN-β gene. *Cell*, **71**, 777.
114. Du, W., Thanos, D., and Maniatis, T. (1993) Mechanisms of transcription synergism between distinct virus-inducible enhancer elements. *Cell*, **74**, 887.

115. Lehming, N., Thanos, D., Brickman, J. M., Ma, J., Maniatis, T., and Ptashne, M. (1994) An HMG-like protein that can switch a transcriptional activator to a repressor. *Nature*, **371**, 175.
116. Hirai, H., Fujisawa, J., Suzuki, T., Ueda, K., Muramatsu, M., Tsuboi, A., Arai, N., and Yoshida, M. (1992) Transcriptional activator Tax of HTLV-1 binds to the NF-κB precursor p105. *Oncogene*, **7**, 1737.
117. Beraud, C., Sun, S.-C., Ganchi, P., Ballard, D. W., and Greene, W. C. (1994) Human T-cell leukemia virus type I associates with and is negatively regulated by the NF-κB2 p100 gene product: implications for viral latency. *Mol. Cell. Biol.*, **14**, 1374.
118. Wagner, S. and Green, M. (1993) HTLV-I Tax protein stimulation of DNA binding by bZIP proteins by enhancing dimerization. *Science*, **262**, 395.
119. Fan, C.-M. and Maniatis, T. (1991) Generation of p50 subunit of NF-κB by processing of p105 through an ATP-dependent pathway. *Nature*, **354**, 395.
120. Rivière, Y., Blank, V., Kourilsky, P., and Israël, A. (1991) Processing of the precursor of NF-κB by the HIV-1 protease during acute infection. *Nature*, **350**, 625.
121. Ghosh, S., Gifford, A. M., Riviere, L. R., Tempst, P., Nolan, G. P., and Baltimore, D. (1990) Cloning of the p50 DNA binding subunit of NF-κB: homology to *rel* and *dorsal*. *Cell*, **62**, 1019.
122. Kieran, M., Blank, V., Logeat, F., Vandekerckhove, J., Lottspeich, F., Le Bail, O., Urban, M. B., Kourilsky, P., Baeuerle, P. A., and Israël, A. (1990) The DNA binding subunit of NF-κB is identical to factor KBF1 and homologous to the *rel* oncogene product. *Cell*, **62**, 1007.
123. Neri, A., Chang, C.-C., Lombardi, L., Salina, M., Corradini, P., Maiolo, A. T., Chaganti, R. S. K., and Dalla-Favera, R. (1991) B cell lymphoma-associated chromosomal translocation involves candidate oncogene *lyt*-10, homologous to NF-κB p50. *Cell*, **67**, 1075.
124. Capobianco, A. J., Chang, D., Mosialos, G., and Gilmore, T. D. (1992) p105, the NF-κB p50 precursor, is one of the cellular proteins complexed with the v-Rel oncoprotein in transformed chicken cells. *J. Virol.*, **66**, 3758.
125. Mercurio, F., DiDonato, J., Rosette, C., and Karin, M. (1992) Molecular cloning and characterization of a novel Rel/NF-κB family member displaying structural and functional homology to NF-κB p50/p105. *DNA Cell Biol.*, **11**, 523.
126. Sif, S. and Gilmore, T. D. (1993) NF-κB p100 is one of the high-molecular-weight proteins associated with the v-Rel oncoprotein in transformed chicken spleen cells. *J. Virol.*, **67**, 7612.
127. Potter, D. A., Larson, C. J., Eckes, P., Schmid, R. M., Nabel, G. J., Verdine, G. L., and Sharp, P. A. (1993) Purification of the MHC class I transcription factor H2TF1: the full-length product of the *nfkb2* gene. *J. Biol. Chem.*, **268**, 18882.
128. Grumont, R. J., Fecondo, J., and Gerondakis, S. (1994) Alternate splicing of murine of *nfkb1* generates a nuclear isoform of the p50 precursor NF-κB1 that can function as a transactivator of NF-κB-regulated transcription. *Mol. Cell. Biol.*, **14**, 8460.
129. Haskill, S., Beg, A. A., Tompkins, S. M., Morris, J. S., Yurochko, A. D., Sampson-Johannes, A., Mondal, K., Ralph, P., and Baldwin, A. S., jun. (1991) Characterization of an immediate-early gene induced in adherent monocytes that encodes IκB-like activity. *Cell*, **65**, 1281.
130. Geisler, R., Bergmann, A., Hiromi, Y., and Nüsslein-Volhard, C. (1992) *cactus*, a gene involved in dorsoventral pattern formation of Drosophila, is related to the IκB gene family of vertebrates. *Cell*, **71**, 613.
131. Kidd, S. (1992) Characterization of the Drosophila *cactus* locus and analysis of interactions between cactus and dorsal proteins. *Cell*, **71**, 623.

132. Sun, S.-C., Elwood, J., Beraud, C., and Greene, W. C. (1994) Human T cell leukemia virus type I Tax activation of NF-κB/Rel involves phosphorylation and degradation of IκBα and RelA (p65)-mediated induction of the *c-rel* gene. *Mol. Cell. Biol.*, **14**, 7377.
133. Traenkner, E. B.-M., Wilk, S., and Baeuerle, P. A. (1994) A proteasome inhibitor prevents activation of NF-κB and stabilises a newly phopshorylated form of IκBα that is still bound to NF-κB. *EMBO J.*, **13**, 5433.
134. Alkalay, I., Yaron, A., Hatzubai, A., Jung, S., Avraham, A., Gerlitz, O., Paschut-Lavon, I., and Ben-Neriah, Y. (1995) In vivo stimulation of IκB phosphorylation is not sufficient to activate NF-κB. *Mol. Cell. Biol.*, **15**, 1294.
135. DiDonato, J. A., Mercurio, F., and Karin, M (1995) Phosphorylation of IκBα precedes but is not sufficient for its dissociation from NF-κB. *Mol. Cell. Biol.*, **15**, 1302.
136. Palombella, V. J., Rando, O. J., Goldberg, A. L., and Maniatis, T. (1994) The ubiquitin-proteasome pathway is required for processing the NF-κB1 precursor protein and the activation of NF-κB. *Cell*, **78**, 773.
137. Ghosh, S. and Baltimore, D. (1990) Activation *in vitro* of NF-κB by phosphorylation of its inhibitor IκB. *Nature*, **344**, 678.
138. Dominguez, I., Sanz, L., Arenzana-Seisdedos, F., Diaz-Meco, M. T., Virelizier, J.-L., and Moscat, J. (1992) Inhibition of protein kinase C ξ subspecies blocks the activation of an NF-kB-like activity in *Xenopus laevis* oocytes. *Mol. Cell. Biol.*, **13**, 1290.
139. Diaz-Meco, M. T., Berra, E., Municio, M. M., Sanz, L., Lozano, J., Domingues, I., Diaz-Golpe, V., Lain de Lera, M. T., Alcamí, J., Paya, C. V., Arenzana-Seisdedos, F., Virelizier, J.-L., and Moscat, J. (1993) A dominant negative protein kinase C ξ subspecies blocks NF-κB activation. *Mol. Cell. Biol.*, **13**, 4770.
140. Li, S. and Sedivy, J. M. (1993) Raf-1 protein kinase activates the NF-κB transcription factor by dissociating the cytoplasmic NF-κB-IκB complex. *Proc. Natl Acad. Sci. USA*, **90**, 9247.
141. Koong, A. C., Chen, E. Y., and Giaccia, A. J. (1994). Hypoxia casues the activation of nuclear factor κB through phosphorylation of IκB-α on tyrosine residues. *Cancer Res.*, **54**, 1425.
141a. Brown, K., Gerstberger, S., Carlson, L., Franzoso, G., and Siebenlist, U. (1995) Control of IκB-α proteolysis by site-specific, signal-induced phosphorylation. Science, **267**, 1485.
141b. Brockman, J. A., Scherer, D. C., McKinsey, T. A., Hall, S. M., Qi, X., Lee, W. Y., and Ballard, D. W. (1995) Coupling of a signal response domain in IκBα to multiple pathways for NF-κB activation. *Mol. Cell. Biol.*, **15**, 2809.
141c. Lin, Y.-C., Brown, K., and Siebenlist, U. (1995) Activation of NF-kB requires proteolysis of the inhibitor IκB-α: signal-induced phosphorylation of IκB-α alone does not release active NF-κB. *Proc. Natl Acad. Sci. USA*, 92, 552.
141d. Chen, Z., Hagler, J., Palombella, V. J., Melandri, F., Scherer, D., Ballard, D., and Maniatis, T. (1995) Signal-induced site-specific phosphorylation targets IκBα to the ubiquitin-proteasome pathway. *Genes Dev.*, **9**, in press.
142. Thompson, J. E., Phillips, R. J., Erdjument-Bromage, H., Tempst, P., and Ghosh, S. (1995) IκB-β regulated the persistent response in a biphasic activation of NF-κB. *Cell*, **80**, 573.
143. Link, E., Kerr, L. D., Schreck, R., Zabel, U., Verma, I., and Baeuerle, P. A. (1992) Purified IκB-β is inactivated upon dephosphorylation. *J. Biol. Chem.*, **267**, 239.
144. Evans, R. B., Gottleib, P. D., and Bose, H. R., jun. (1993) Identification of a Rel-related protein in the nucleus during S phase of the cell cycle. *Mol. Cell. Biol.*, **13**, 6147.

145. Mosialos, G., Hamer, P., Capobianco, A. J., Laursen, R. A., and Gilmore, T. D. (1991) A protein kinase A recognition sequence is structurally linked to transformation by p59$^{v\text{-}rel}$ and cytoplasmic retention of p68$^{c\text{-}rel}$. *Mol. Cell. Biol.*, **11**, 5877.
146. Mosialos, G. and Gilmore, T. D. (1993) v-Rel and c-Rel are differentially affected by mutations at a consensus protein kinase recognition sequence. *Oncogene*, **8**, 721.
147. Ganchi, P. A., Sun, S.-C., Greene, W. C., and Ballard, D. W. (1993) A novel NF-κB complex containing p65 homodimers: implications for transcriptional control at the level of subunit dimerization. *Mol. Cell. Biol.*, **13**, 7826.
148. Norris, J. L. and Manley, J. L. (1992) Selective nuclear transport of the *Drosophila* morphogen *dorsal* can be established by a signaling pathway involving the transmembrane protein *Toll* and protein kinase A. *Genes Dev.*, **6**, 1654.
149. Isoda, K., Roth, S., and Nüsslein-Volhard, C. (1992) The functional domains of the *Drosophila* morphogen *dorsal:* evidence from the analysis of mutants. *Genes Dev.*, **6**, 619.
150. White, D. W. and Gilmore, T. D. (1993) Temperature-sensitive transforming mutants of the *v-rel* oncogene. *J. Virol.*, **67**, 6776.
151. Li, C.-C. H., Dai, R.-M., Chen, E., and Longo, D. L. (1994) Phosphorylation of NF-κB1/p50 is involved in NF-κB activation and stable DNA binding. *J. Biol. Chem.*, **269**, 30089.
152. Mosialos, G. (1993) Mutational analysis of a conserved consensus phosphorylation site in the retroviral oncoprotein vRel and in chicken cRel. Ph.D. Thesis, Boston University.
153. Neumann, M., Tsapos, K., Schleppler, J. A., Ross, J., and Franza, R., jun. (1992) Identification of complex formation between two intracellular tyrosine kinase substrates: human c-Rel and the p105 precursor of p50 NF-κB. *Oncogene*, **7**, 2095.
154. Schreck, R., Rieber, P., and Baeuerle, P. A. (1991) Reactive oxygen intermediates as apparently widely used messengers in the activation of the NF-κB transcription factor and HIV-1. *EMBO J.*, **10**, 2247.
155. Kumar, S., Rabson, A. B., and Gélinas, C. (1992) The RxxRxRxxC motif conserved in all rel/κB proteins is essential for the DNA-binding activity and redox regulation of the v-Rel oncoprotein. *Mol. Cell. Biol.*, **12**, 3094.
156. Matthews, J. R., Kaszubska, W., Turcatti, G., Wells, T. N. C., and Hay, R. T. (1993) Role of cysteine 62 in DNA recognition by the P50 subunit of NF-κB. *Nucl. Acids Res.*, **21**, 1727.
157. Toledano, M. B., Ghosh, D., Trinh, F., and Leonard, W. J. (1993) N-terminal DNA-binding domains contribute to differential DNA-binding specificities of NF-κB p50 and p65. *Mol. Cell. Biol.*, **13**, 852.
158. Abate, C., Patel, L., Rauscher, III, F. J., and Curran, T. (1990) Redox regulation of Fos and Jun DNA-binding activity in vitro. *Science*, **249**, 1157.
159. Roth, S., Stein, D., and Nüsslein-Volhard, C. (1989) A gradient of nuclear localization of the *dorsal* protein determines dorsoventral pattern in the *Drosophila* embryo. *Cell*, **59**, 1189.
160. Steward, R. (1989) Relocalization of the *dorsal* protein from the cytoplasm to the nucleus correlates with its function. *Cell*, **59**, 1179.
161. Ip, Y. T., Park, R. E., Kosman, D., Bier, E., and Levine, M. (1992) The *dorsal* gradient morphogen regulates stripes of *rhomboid* expression in the presumptive neuroectoderm of the *Drosophila* embryo. *Genes Dev.*, **6**, 1728.
162. Jiang, J., Kosman, D., Ip, Y. T., and Levine, M. (1991) The *dorsal* morphogen gradient regulates the mesoderm determinant *twist* in early *Drosophila* embryos. *Genes Dev.*, **5**, 1881.

163. Miller, C. K., Embretson, J. E., and Temin, H. M. (1988) Transforming viruses spontaneously arise from nontransforming reticuloendotheliosis virus strain T-derived viruses as a result of increased accumulation of spliced viral RNA. *J. Virol.*, **62**, 1219.
164. Bull, P., Hunter, T., and Verma, I. M. (1989) Transcriptional induction of the murine c-*rel* gene with serum and phorbol-12-myristate-13-acetate in fibroblasts. *Mol. Cell. Biol.*, **9**, 5239.
165. Bours, V., Burd, P. R., Brown, K., Villalobos, J., Park, S., Ryseck, R.-P., Bravo, R., Kelly, K., and Siebenlist, U. (1992) A novel mitogen-inducible gene product related to p50/p105-NF-κB participates in transactivation through a κB site. *Mol. Cell. Biol.*, **12**, 685.
166. Bours, V., Villalobos, J., Burd, P. R., Kelly, K., and Siebenlist, U. (1990) Cloning of a mitogen-inducible gene encoding a κB DNA-binding protein with homology to the *rel* oncogene and to cell-cycle motifs. *Nature*, **348**, 76.
167. Meyer, R., Hatada, E. N., Hohmann, H.-P., Haiker, M., Bartsch, C., Röthlisberger, U., Lahm, H.-W., Schlaeger, E. J., van Loon, A. P. G. M., and Scheidereit, C. (1991) Cloning of the DNA-binding subunit of human NF-κB: the level of its mRNA is strongly regulated by phorbol ester and tumor necrosis factor α. *Proc. Natl Acad. Sci. USA*, **88**, 966.
168. Hannink, M. and Temin, H. M. (1990) Structure and autoregulation of the c-*rel* promoter. *Oncogene*, **5**, 1843.
169. Capobianco, A. J. and Gilmore, T. D. (1991) Repression of the chicken c-*rel* promoter by vRel in chicken embryo fibroblasts is not mediated through a consensus NF-κB binding site. *Oncogene*, **6**, 2203.
170. Ten, R. M., Paya, C. V., Israël, A., Le Bail, O., Mattei, M.-G., Virelizier, J.-L., Kourilsky, P., and Israël, A. (1992) The characterization of the promoter of the gene encoding the p50 subunit of NF-κB indicates that it participates in its own regulation. *EMBO J.*, **11**, 195.
171. Cogswell, P. C., Scheiman, R. I., and Baldwin, A. S., jun. (1993) Promoter of the human NF-κB p50/p105 gene. Regulation by NF-κB subunits and by c-REL. *J. Immunol.*, **150**, 2794.
172. Liptay, S., Schmid, R. M., Nabel, E. G., and Nabel, G. J. (1994) Transcriptional regulation of NF-κB2: evidence for κB-mediated postive and negative autoregulation. *Mol. Cell. Biol.*, **14**, 7695.
173. Ueberla, K., Lu, Y., Chung, E., and Haseltine, W.A. (1993) The NF-κB p65 promoter. *J. AIDS*, **6**, 227.
174. Boehmelt, G., Mdruga, J., Dorfler, P., Briegel, K., Schwartz, H., Enrietto, P., and Zenke, M. (1995) Dendritic cell progenitor is transformed by a conditional v-Rel oestrogen receptor fusion protein v-RelER. *Cell*, **80**, 341.
175. Diehl, J. A., McKinsey, T. A., and Hannink, M. (1993) Differential $pp40^{I\kappa B\text{-}\beta}$ inhibition of DNA binding by *rel* proteins. *Mol. Cell. Biol.*, **13**, 1769.
176. Gilmore, T. D. and Temin, H. M. (1986) Different localization of the product of the v-*rel* oncogene in chicken fibroblasts and spleen cells correlates with transformation by REV-T. *Cell*, **44**, 791.
177. Gilmore, T. D. and Temin, H. M. (1988) v-*rel* oncoproteins in the nucleus and in the cytoplasm transform chicken spleen cells. *J. Virol.*, **62**, 703.
178. Sarkar, S. and Gilmore, T. D. (1993) Transformation by the vRel oncoprotein requires sequences carboxy-terminal to the Rel homology domain. *Oncogene*, **8**, 2245.
179. Walker, W. H., Stein, B., Ganchi, P. A., Hoffman, J. A., Kaufmann, P. A., Ballard, D. W., Hannink, M., and Greene, W. C. (1992) The v-*rel* oncogene: insights into the mechanism of transcriptional activation, repression and transformation. *J. Virol.*, **66**, 5018.

180. Hrdlicková, R., Nehyba, J., and Humphries, E. H. (1993) v-*rel* induces expression of three avian immunoregulatory surface receptors more efficiently than c-*rel*. *J. Virol.*, **68**, 308.
181. Ishikawa, H., Asano, M., Kanda, T., Kumar, S., Gélinas, C., and Ito, Y. (1990) Two novel functions associated with the Rel oncoproteins: DNA replication and cell-specific transcriptional activation. *Oncogene*, **8**, 2889.
182. Ohno, H., Takimoto, G., and McKeithan, T. W. (1990) The candidate proto-oncogene *bcl*-3 is related to genes implicated in cell lineage determination and cell cycle control. *Cell*, **60**, 991.
183. Lu, D., Thompson, J. D., Gorski, G. K., Rice, N. R., Mayer, M. G., and Yunis, J. J. (1991) Alterations at the *rel* locus in human lymphoma. *Oncogene*, **6**, 1235.
184. Fracchiolla, N. S., Lombardi, L., Salina, M., Migliazza, A., Baldini, L., Berti, E., Cro, L., Polli, E., Maiolo, A. T., and Neri, A. (1993) Structural alterations of the NF-κB transcription factor *lyt*-10 in lymphoid malignancies. *Oncogene*, **8**, 2839.
185. Zhang, J., Chang, C.-C., Lombardi, L., and Dalla-Faverra, R. (1994) Rearranged NF-κB2 gene in the HUT78 T-lymphoma cell line codes for a constitutively nuclear factor lacking transcriptional repressor functions. *Oncogene*, **9**, 1931.
186. Narayanan, R., Klement, J. F., Rubin, S. M., Higgins, K. A., and Rosen, C. A. (1992) Identification of a naturally occurring transforming variant of the p65 subunit of NF-κB. *Science*, **256**, 367.
187. Grimm, S. and Baeuerle, P. A. (1994) Failure of the splicing variant p65Δ of the NF-κB subunit p65 to transform fibroblasts. *Oncogene*, **9**, 2391.
188. Kitajima, I., Shinohara, T., Bilakovics, J., Brown, D. A., Xu, X., and Nereneberg, M. (1992) Ablation of transplanted HTLV-I Tax-transformed tumors in mice by antisense inhibition of NF-κB. *Science*, **258**, 1792.
189. Schmid, R. M., Perkins, N. D., Duckett, C. S., Andrews, P. C., and Nabel, G. J. (1991) Cloning of an NF-κB subunit which stimulates HIV transcription in synergy with p65. *Nature*, **352**, 733.
190. Stephens, R. M., Rice, N. R., Hiebsch, R. R., Bose, H. R., jun., and Gilden, R. V. (1983) Nucleotide sequence of v-*rel*: the oncogene of reticuloendotheliosis virus. *Proc. Natl Acad. Sci. USA*, **80**, 6229.
191. Wilhelmsen, K. C., Eggleton, K., and Temin, H. M. (1984) Nucleic acid sequences of the oncogene v-*rel* in reticuloendotheliosis virus strain T and its cellular homolog, the proto-oncogene c-*rel*. *J. Virol.*, **52**, 172.
192. Hannink, M. and Temin, H. M. (1989) Transactivation of gene expression by nuclear and cytoplasmic *rel* proteins. *Mol. Cell. Biol.*, **9**, 4323.
193. Grumont, R. J. and Gerondakis, S. (1989) Structure of mammalian c-*rel* protein deduced from the nucleotide sequence of murine cDNA clones. *Oncogene Res.*, **4**, 1.
194. Brownell, E., Mittereder, N., and Rice, N. R. (1989) A human *rel* proto-oncogene cDNA containing an *Alu* fragment as a potential coding exon. *Oncogene*, **4**, 935.
195. Brownell, E., Mathieson, B., Young, H. A., Keller, J., Ihle, J. N., and Rice, N. R. (1987) Detection of c-*rel*-related transcripts in mouse hematopoietic tissues, fractionated lymphocyte populations, and cell lines. *Mol. Cell. Biol.*, **7**, 1304.
196. Brownell, E., Ruscetti, F. W., Smith, R. G., and Rice, N. R. (1988) Detection of *rel*-related RNA and protein in human lymphoid cells. *Oncogene*, **3**, 93.
197. Ruben, S. M., Dillon, P. J., Schreck, R., Henkel, T., Chen, C.-H., Maher, M., Baeuerle, P. A., and Rosen, C. A. (1991) Isolation of a *rel*-related human cDNA that potentially encodes the 65-kD subunit of NF-κB. *Science*, **251**, 1490.

198. Nolan, G. P., Ghosh, S., Liou, H.-C., Tempst, P., and Baltimore, D. (1991) DNA binding and IκB inhibition of the cloned p65 subunit of NF-κB, a *rel*-related polypeptide. *Cell,* **64**, 961.
199. Ikeda, T., Honjo, K., Hirota, Y., and Onodera, T. (1993) Isolation of the chicken NF-κB p65 subunit-encoding cDNA and characterisation of its products. *Gene,* **133**, 237.
200. Ruben, S. M., Klement, J. F., Coleman, T. A., Maher, M., Chen, C.-H., and Rosen, C. A. (1992) I-Rel: a novel *rel*-related protein that inhibits NF-κB transcriptional activity. *Genes Dev.,* **6**, 745.
201. Kao, K. R. and Hopwood, N. D. (1991) Expression of a mRNA related to c-*rel* and dorsal in early *Xenopus laevis* embryos.*Proc. Natl Acad. Sci. USA,* **88**, 2697.
202. Steward, R. (1987) *Dorsal,* an embryonic polarity gene in *Drosophila,* is homologous to the vertebrate proto-oncogene, c-*rel*. *Science,* **238**, 692.
203. Ray, A. and Prefontaine, K. E. (1994) Physical association and functional antagonsim between the p65 subunit of transcription factor NF-κB and the glucocorticoid receptor. *Proc. Natl Acad. Sci. USA,* **91**, 752.
204. Scheinman, R. I., Gualberto, A., Jewell, C. M., Cidlowski, J., and Baldwin, A. S., jun. (1995) Characterisation of mechanisms involved in trans-repression of NF-κB by activated glucorticoid receptors. *Mol. Cell. Biol.* **15**, 943.
205. LeClair, K. P., Blanar, M. A., and Sharp, P. A. (1992) The p50 subunit of NF-κB associates with the NF-IL6 transcription factor. *Proc. Natl Acad. Sci. USA,* **89**, 8145.
206. Bruder, J. T., Weske, J., Brown, T., Kyriakis, J. M., Avruch, J., and Rapp, U. R. (1993) GABP binds to the NF-κB sites in the HIV LTR and is activated by Raf-1. *Abstracts of the ninth meeting on oncogenes,* p. 416.
207. Gutsch, D. E., Holley-Guthrie, E. A., Zhang, Q., Stein, B., Blanar, M. A., Baldwin, A. S., and Kenny, S. C. (1994) The bZIP transactivator of Epstein Barr virus, BZLF1, functionally and physically interacts with the p65 subunit of NF-κB. *Mol. Cell. Biol.,* **14**, 1939.
208. Jiang, J. and Levine, M. (1993) Binding affinities and co-operative interactions with bHLH activators delimit threshold responses to the dorsal gradient morphogen. *Cell,* **72**, 741.

5 | Multi-factorial and combinatorial models for specific transcriptional regulation

STEPHEN GOODBOURN and PETER KING

1. Introduction

A simple model for specific transcriptional regulation proposes that a promoter contains binding sites for a single transcription factor. This deterministic factor would be produced in only a restricted cell type, for example during development, or would be activated in response to a specific stimulus. The nature of the activation could operate at several levels, and examples of these are discussed elsewhere in this book. For example, CREB-1 and CREM are constitutive DNA-binding proteins whose ability to interact with the basal transcriptional apparatus is regulated by post-translational modification (see Chapters 1 and 3), while NF-κB is a protein whose ability to bind DNA is modulated by regulating partitioning between the nucleus and cytoplasm (see Chapter 4). However, it has long been clear that gene regulation cannot always be explained in such simple terms. For example, although NF-κB activation is a prerequisite for β-interferon induction by double-stranded RNA in many cell types, it is insufficient for activation. Similarly, over-expression of another transcriptional activator, IRF-1, cannot bring about β-interferon activation on its own. Rather, it is the combination of several such factors that achieves the specificity and magnitude of this activation. There are also cases in which specific transcription requires the integration of two discrete signals; the IL-2 promoter is activated in resting T cells by a combination of mitogenic stimulation and changes in the calcium ion concentration—either signal alone can activate other responses in these cells. Finally, there are cases in which promoters can discriminate between different signals to give alternative transcriptional responses; in yeast cells, mating type-specific genes can be activated or repressed according to the types of promoter-binding factors present (some other examples of this are given in Chapters 6 and 7). The purpose of this chapter is to highlight examples of the types of interactions between transcription factors that can explain some of these complex patterns of regulatory behaviour.

2. The octamer-binding factors: regulation by recruitment of co-activators

One of the very earliest studies on tissue-specific transcriptional regulation centred on the production of immunoglobulin genes in B-lymphocytes (reviewed in 1). These studies defined the proto-typical tissue-specific enhancer and promoter, and provided one of the first examples of using gel-retardation assays to identify sequence-specific DNA-binding proteins. Although there appear to be a multitude of specific regulatory elements and transcription factors that are required for efficient expression, one sequence, the so-called octamer motif (ATGCAAAT) is found in every immunoglobulin promoter and in most immunoglobulin enhancers; promoter dissection experiments have established that this motif is essential for function. A DNA-binding protein that recognizes this motif has been identified in B-cells (referred to as Oct-2; 2–4). Oct-2 is a member of a group of related transcription factors, referred to as the POU family, that share a common homeodomain and POU-specific domain (5). Several of the POU family in addition to Oct-2 can bind strongly to the octamer motif and these show different tissue distributions; for example, Oct-4 is found in germ-line tissue and embryonic stem cells (6), while Oct-1 is ubiquitous (7). The latter observations suggest that the Oct family may play roles in the regulation of genes that are not activated in a B-cell-specific manner.

Evidence to support the notion that Oct-2 was a specific transcriptional activator of immunoglobulin genes came from transfection studies in which it was shown that Oct-2 could support immunoglobulin gene transcription in HeLa cells (4, 8). However, it has subsequently been shown that mice bearing targeted disruptions of both Oct-2 alleles can produce immunoglobulin normally, at least in the early stages of B-cell development (9). These experiments show that Oct-2 is not required until late in the maturation of B-cells. One possible explanation for the continued expression in early B-cells could be that Oct-2 is a redundant factor, and the obvious replacement factor would be the ubiquitous factor, Oct-1. However, Oct-1 does not appear to be an efficient transactivator of either immunoglobulin (4, 8) or other octamer-motif-containing (10, 11) promoters in HeLa cells. This dilemma appears to have a simple solution. In *in vitro* transcription assays it had been noted that immunoglobulin promoters are much more active in extracts from B-cells than from HeLa cells (12, 13), and biochemical experiments have allowed the identification of an activity that appears to be responsible for this effect. The factor, called OCA-B, can confer strong activation on either Oct-1 or Oct-2 (13). Recently, a cDNA that appears to encode OCA-B has been isolated (14, 15); this clone (called OBF-1) encodes a B-cell-specific factor that can form a ternary complex on an octamer motif with either Oct-1 or Oct-2 (but not Oct-4) through interactions with the POU domains, although OBF-1 does not itself bind to DNA. When OBF-1 is expressed in HeLa cells (which have abundant levels of Oct-1) it allows immunoglobulin gene transcription. Thus OBF-1 appears to have all of the properties of a tissue-specific co-activator, which is recruited to the DNA by protein–protein interactions (see

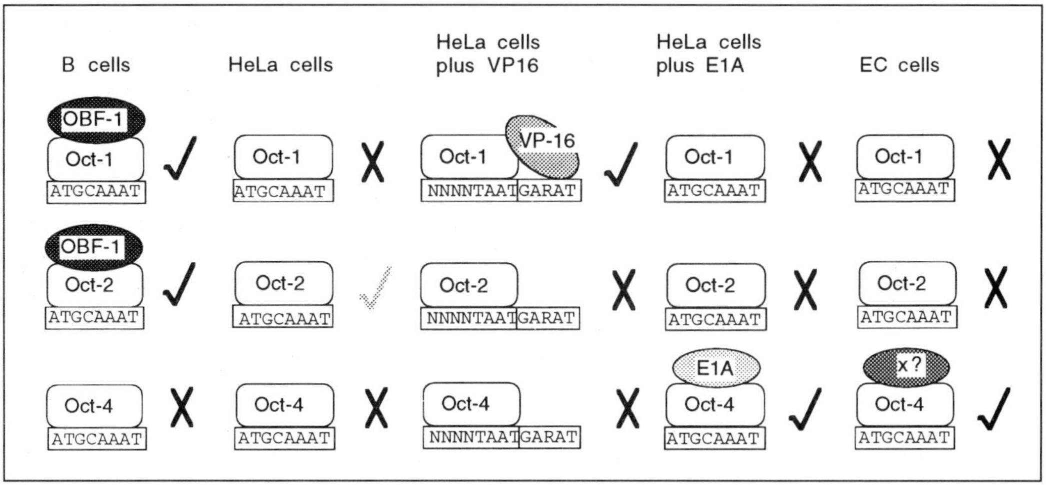

Fig. 1 Cell-type-specific regulation of Octamer-binding protein (Oct) activity. B-lymphocytes contain Oct-1 and Oct-2, but not Oct-4. Both Oct-1 and Oct-2 are capable of interacting with a B-cell-specific factor called OBF-1 that enables them to stimulate transcription from an octamer motif (ATGCAAAT); Oct-4 cannot interact with OBF-1. HeLa cells only contain Oct-1. However, in transfection experiments Oct-2, but not Oct-4, can stimulate transcription in these cells. The degree of stimulation is not as great as that achieved by Oct-2 in the presence of OBF-1. When the herpes virus VP16 product is expressed in HeLa cells, Oct-1 (but not Oct-2 or Oct-4) can support high levels of transcription from the TAATGARAT motif. The adenoviral E1A product has a similar selective effect on Oct-4; it is assumed that in embryonal carcinoma (EC) cells, the effect of E1A can be mimicked by a cellular E1A-like product.

Fig. 1). In the absence of Oct-2 in B-cells, this co-activator could still be recruited to the DNA via Oct-1.

Notwithstanding the enhancement of transcription mediated by OBF-1, it is clear that Oct-2 can stimulate transcription in HeLa cells in a much more efficient manner than Oct-1 (4, 8, 10, 11). This poses the question of what is the role of Oct-1 in other cells? At least part of the answer lies in the observations that octamer motifs are found in some 'housekeeping' genes such as snRNAs (both RNA polymerase II- and RNA polymerase III-dependent; 16) and histone H2B genes (17). In the former case, there is good evidence from transfection experiments that Oct-1 is required for transcriptional activation (10, 11). In the case of histone H2B gene expression, it has been shown that Oct-1 can stimulate transcription *in vitro* and that this property is varied throughout the cell cycle such that transcription ceases during mitosis, in accord with the known properties of histone H2B genes *in vivo* (18). Oct-1 prepared from cells shows a cyclic variation in phosphorylation, and when Oct-1 is modified by phosphorylation *in vitro*, it shows a marked loss of affinity for the octamer motif (19), giving rise to a model in which the ability of Oct-1 to stimulate histone H2B transcription is regulated by changes in its affinity for the octamer motif.

It is not yet known whether the stimulatory effects of Oct-1 for transcription of snRNAs and histone H2B genes require co-activators of the OBF-1 type. However, it is clear that Oct-1 does not act as a general stimulator of RNA polymerase II-dependent transcription when bound to octamer motifs alone. The reasons for this are not clear, although there are many potential explanations, such as the need to recruit other factors that are themselves template-dependent, or are tissue-specific. An example of this may be the stimulation of pituitary-specific transcription, in which Oct-1 and another related POU family member, Pit-1, form a heterodimer and can bind to an element found in pituitary-specific promoters (20). Similarly, Oct-1 can bind to the interleukin-2 promoter, but to do so efficiently requires the binding of AP-1 to an adjacent site (21). Oct-1 and Oct-2 can also co-operate with each other by stimulating the binding of a second molecule to adjacent non-cognate sites that, on their own, are not capable of binding these proteins (22–24); these effects seem to involve the formation of protein dimers formed through POU domain interactions (25). In each case, Oct-1 is being recruited to DNA through motifs other than the octamer. It is characteristic of homeodomain proteins that they can tolerate variations in their binding sites (reviewed in 26), and it may be that in binding to these motifs Oct-1 adopts an altered conformation that might allow it to function as an activator—this general idea appears to hold in other cases (see, for example, the yeast transactivator, MCM1, and its mammalian homologue, SRF, discussed below).

An interesting example of the use of Oct-1 to stimulate RNA polymerase II-dependent transcription is seen in the case of immediate early gene expression in herpes simplex virus (HSV; see Fig. 1). Transcription of these genes does not require *de novo* host protein synthesis, and yet, unusually, does require a virion protein, the product of the VP16 gene (reviewed in 27). Analysis of the sequences of the immediate early gene promoters revealed the presence of a common motif, the so-called TAATGARAT element (the R is a purine residue). However, the VP16 protein does not appear to show any specific binding to this sequence. The host-cell proteins required to recruit VP16 to the TAATGARAT motif have been identified as Oct-1 and an additional host-cell factor (HCF) of unknown function (28, 29). The recruitment of VP16 to DNA strongly stimulates RNA polymerase II-mediated transcription of the HSV immediate early promoters. The means by which VP16 stimulates transcription are discussed in detail in Chapter 1.

Curiously, the TAATGARAT motif is quite different from the octamer motif (ATGCAAAT). Nevertheless, Oct-1 can bind to the former, a property that is typical of the relaxed specificities shown by homeobox proteins. Extensive structural studies on the interactions of Oct-1 with the octamer motif have indicated that the POU-specific part of the protein can interact with the ATGC part of the motif, while the homeodomain can interact with the AAAT part (30–34). The TAATGARAT motif contains a sequence (TAAT) that is very similar to the AAAT recognized by the homeodomain, and indeed many homeodomain proteins can interact with a TAAT motif. Interestingly, the presence of GARAT at the 3′ flank of TAAT causes the POU domain of Oct-1 to adopt a novel conformation that is directly responsible for recruiting VP16 (35).

Although the VP16 protein can be recruited to DNA by Oct-1, it is incapable of mediating its effects through the closely related Oct-2 (28). A single amino acid difference in the helix 1 of the POU homeodomains of Oct-1 and Oct-2 allows the interacting proteins to discriminate between them (36–38).

Oct-4 also appears to require helper factors to stimulate transcription. In this case, it has been shown that Oct-4 can form a complex with the adenovirus E1A protein such that the viral transactivator is recruited to DNA (39; see Fig.1). It has long been hypothesized that embryonal carcinoma cells contain a cellular E1A-like activity (40), although the nature of this remains unknown, and it is probable that this activity can also be recruited to DNA by Oct-4. Again, E1A interacts with the POU domain of Oct-4, and cannot interact with Oct-1 or Oct-2 (39).

Although the Oct factors may be able to stimulate transcription to varying degrees on their own, it is clear from the above that in most well-characterized cases they act by recruiting other proteins, a strategy that has been adopted by certain viral proteins. These interactions are summarized in Fig. 1. It may be appropriate to consider the Oct proteins as 'scaffold' factors that allow the assembly of other regulatory molecules on the promoter. Interestingly, the Oct proteins are large (Oct-1 is >100 kDa), implying that they may have multiple functions. However, interactions characterized to date all map to the relatively small POU domains. The Oct proteins are examples of homeodomain proteins that often function as key mediators of developmental decisions (see Chapter 6); it is intriguing to speculate that they are primarily DNA-binding templates that allow other factors to be recruited to promoters. In this role, their tissue-specificity is a key component whereby they would allow common activator functions to be assembled in a very defined manner.

3. Yeast mating types: a paradigm for cell-type-specific regulation of transcription using common components

The determination of yeast mating type comprises perhaps the simplest example of a differentiation decision made by a eukaryotic cell (reviewed in 41). Haploid yeast cells are of two types, designated **a** and α, which can mate only with cells of the opposite type to produce **a**/α diploids. The mating type is determined by the constitutive expression of a set of genes that are either **a**- or α-specific (for example the mating pheromones themselves). When cells are exposed to the mating pheromone from cells of the other mating type, the expression of these genes is up-regulated and cells arrest in G1; these events lead to cell fusion and the formation of a diploid cell. Diploid cells do not need to mate and therefore repress the regulation of the mating type and other haploid-specific genes. The expression of the **a**- and α-specific gene sets are governed by 'master regulators', which are located at a specific genomic position called the mating type locus (*MAT*). All yeast cells possess silent copies of the regulator genes for both **a** and α mating types stored at transcriptionally inactive loci, called *HMR***a** and *HML*α respectively. Yeast are able to switch mating types by a gene-conversion event catalysed by the HO endonuclease, which replaces the information at the active *MAT* locus with information from either *HMR***a** or *HMLα*. When cells are of the α mating type, the *MAT* locus is occupied by the *MATα1* and *MATα2* genes, which are thus constitutively expressed; **a** mating type cells express the *MAT***a***1* gene from the *MAT* locus.

The α- and **a**-specific genes are regulated via transcriptional events that are properties of their upstream activator sequences (UASs). Transcriptional activation of **a**-specific genes (for example the *STE2* gene, which encodes the receptor for α factor) in **a** cells requires a nearly palindromic sequence element called a P-box that binds to an essential protein called MCM1 (42–48; see Fig. 2). Basal-level expression from **a**-specific promoters also requires the activities of a second protein called STE12 (47, 49–52) that binds to a distinct sequence called a PRE (53, 54; see Fig. 2). P-Boxes and PREs are physically close in the promoters of **a**-specific genes. Alone, STE12 binds to the PRE in an inefficient manner, and it appears that it can occupy this site by co-operatively binding with MCM1 to the P-box (54). This co-operativity is mediated by protein–protein interactions between STE12 and MCM1 (54, 55). Exposure to pheromone causes the activation of a protein kinase cascade that results in the phosphorylation of STE12 (reviewed in 56); the modification of STE12 leads to an increase in expression of the mating type genes.

STE12 and MCM1 are expressed in both mating types. Nevertheless, **a**-specific genes are not expressed in α cells; this is a consequence of specific repression, which requires sequences that flank the P-boxes. These sequences bind to the product of the *MATα2* gene (MATα2; see Fig. 2; 42, 47, 57–60). In this situation, a molecule of MATα2 can bind to each side of the P-box and can co-operatively bind with an MCM1 molecule located at the P-box. Again, co-operativity is mediated by protein–protein interactions, using a domain of MATα2 that is separable from the DNA-binding domain (59, 61). Although considerable plasticity is observed in the spacing requirements for binding *in vitro* to the two MATα2 sites, the spacing is rigidly conserved in **a**-specific UASs, presumably as a consequence of the requirement to interact with MCM1 (59). There is evidence to suggest that the co-operative binding by MATα2 requires not only a motif to interact with MCM1 and a DNA-binding domain, but also an intact N-terminus (57, 60). Thus, MCM1 can serve to activate or repress transcription of **a**-specific promoters depending upon the presence of MATα2. The mechanism by which the stable MCM1–MATα2 complex can mediate repression has been extensively studied, and it is believed that it can phase nucleosomes and thus exclude the transcription apparatus from the adjacent **a**-specific promoters (see Chapter 2); furthermore, it can recruit the TUP1–SSN6 complex to the DNA template, which can act as a general repressor of transcription (see Chapter 1).

α-Specific promoters (e.g. the *STE6* gene, which is required for the secretion of the α factor pheromone) also contain a P-box, but unlike the site in **a**-specific promoters, this is not a palindromic sequence (indicated as P* in Fig. 2) and is a very weak binding site for MCM1 (45, 62–66). Proper function as a transcriptional activator sequence requires an adjacent sequence called a Q-box (43, 67). Although the

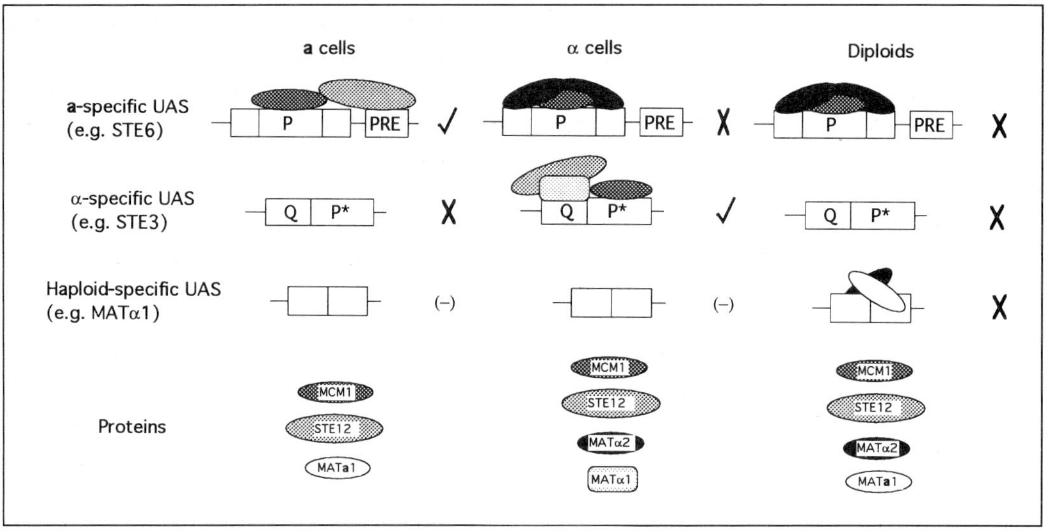

Fig. 2 Combinatorial regulation of yeast cell type-specific gene expression. In cells of the **a** mating type, **a**-specific gene regulation requires an upstream activator sequence (UAS) that contains binding sites for MCM1 (the P-box) and STE12 (the PRE-box). α-Specific UASs cannot function in these cells because of the lack of a strong P-box. In cells of the α mating type, α-specific gene regulation requires the *MAT*α1 gene product, that can bind to the Q-box and facilitate co-operative binding of MCM1 to the weak P-box (P*); STE12 can also bind to MATα1 to stimulate transcription. α Mating type cells also contain MATα2, which can bind co-operatively with MCM1 to the UASs of **a**-specific genes to repress their expression. In diploid cells, **a**-specific UASs remain repressed by the MATα2–MCM1 complex; these cells also express the product of the *MAT***a**1 gene, which can interact with MATα2 to generate a repressor of a set of haploid-specific genes. The *MAT*α1 gene is one of these genes, and the loss of its gene product ensures that α-specific genes are not expressed in diploid cells.

product of the *MAT*α1 gene (MATα1) cannot bind to the Q-box on its own, it can show co-operativity of binding with the MCM1 product so that a ternary complex is formed (48, 67, 68). The ternary complex appears to stabilize MCM1 in a conformation that enhances transcription (43, 62, 67, 69, 70). Thus, MCM1 can be a cell type-specific transcriptional activator depending on the type of co-activator protein expressed. Like **a**-specific promoters, α-specific promoters also require STE12 for efficient basal function, and to respond to pheromone exposure; however, they do not contain a PRE binding site for STE12. In this case, it appears that STE12 can bind to MATα1 (68, 71), and perhaps also to MCM1, in the absence of DNA (see Fig. 2).

The results detailed above are sufficient to describe the specific activation of genes in both **a** cells and α cells. Each of the components described above appears to play other roles in transcriptional regulation. The SWI5 gene is an example of a gene that is transcribed in a cell-cycle-dependent manner (see Chapter 8); using gel retardation assays, Lydall et al. (72) have determined that MCM1 binds to a crucial promoter element in the presence of another unidentified protein called SFF. As well as the MCM1 protein, the components of the mating type determinants are used to regulate other genes. Other promoters, for example that of the haploid-specific (expressed in both **a** and α cells, but not in diploids) FUS1 gene, contain multiple binding sites for STE12 but lack an MCM1 site; in these cases, basal expression and pheromonal response only require these sequences and STE12 (73, 74). STE12 is also required for the expression of another set of genes that encode components of the invasive response in haploid cells; in this case, the activity of STE12 appears to be regulated by a kinase cascade that is distinct from the pheromone-response pathway (75; reviewed in 56).

So far, we have not discussed a role for the *MAT***a**1 gene product expressed at the *MAT* locus in **a** cells. Since the activation of **a**-specific genes does not require cell type-specific factors, there would appear to be no requirement for any **a**-specific regulators. However, it should be remembered that in diploid cells, the products of both *MAT* loci might be expected to be expressed; the consequence of this would be that MATα1 would combine with MCM1 to stimulate α-specific genes, while MATα2 would combine with MCM1 to repress **a**-specific genes. These cells would therefore behave as an α mating type, whereas diploid cells do not in fact mate. This is prevented by the product of the *MAT***a**1 gene (MAT**a**1), which combines with MATα2 to give a heterodimer that can bind to a specific sequence element present in the promoter of the *MAT*α1 gene and repress its activity (reviewed in 41; see Fig. 2). This ensures that α-specific genes cannot be transcribed in diploids. Thus genes of both **a** and α cell types are inactive in diploid cells, as are other haploid-specific genes (such as the HO endonuclease needed to initiate the mating type gene-conversion event) that are also repressed by the MAT**a**1–MATα2 complex. The sequences responsible for the repression in diploid cells of the *MAT*α1 gene have been mapped to a 28-bp element that is related to sequences found in the 5′ flanking regions of other haploid-specific genes. This element has a binding site for MATα2 that is quite distinct from that seen in **a**-specific UASs. Thus, in the presence of MAT**a**1, MATα2 acquires an additional DNA-binding specificity, extending the range of genes that can be repressed. Both MAT**a**1 and MATα2 are members of the homeodomain proteins; the target specificity appears to derive from the different constraints imposed by different protein–protein interactions on the homeodomain proteins—as a result the preferred target sites for the two distinct dimeric homeodomain complexes contain different spacings between the two actual homeodomain contact points. Similarly, in both nematodes and *Drosophila*, interactions between homeodomain proteins have been shown to affect their specificities (see above and Chapter 6).

3.1 Induction of c-Fos gene expression: a mammalian equivalent of the yeast mating type paradigm?

The induction of c-Fos gene expression in response to growth factors has many parallels to the yeast mating type loci (see Section 4.2 and Fig. 4 below). The serum response element (SRE), which is related to the P-box, was first identified in the promoter region of the c-Fos gene as an essential sequence element that mediated the rapid protein synthesis-independent induction in response to release from serum starvation (reviewed in 76). Similar elements are found in a number of growth factor-inducible genes, and are related to, but distinct from, an element found in muscle-specific genes (see Chapter 7). A ubiquitous protein referred to as serum response factor (SRF), which is related to the yeast MCM-1 protein, has been shown to bind to the SRE sequence (77). Related cDNAs encode factors that can bind to the muscle-specific elements (reviewed in 78, 79, see Chapter 7).

The SRE is occupied *in vivo* before serum treatment in a manner that is indistinguishable from SRF binding *in vitro* (80–82); furthermore, binding activity of SRF does not increase in response to release from serum starvation (83). These observations would be consistent with the SRE being permanently occupied by SRF. SRF appears to be essential for the induction by serum, as micro-injection of antibodies against SRF blocks induction (84). However, it is not yet clear whether SRF can mediate induction in the absence of co-factors. Mutational analyses indicate that sequences that flank the SRE can influence the degree of serum response in some cell lines (but not all), and are essential for the induction by some growth factors (notably those that work through protein kinase C; 85). These sequences are required for SRF to form a ternary complex with the SRE and a second protein called TCF (ternary complex factor; 86), which alone can bind weakly if at all to these sequences (82, 85, 86; reviewed in 79). A number of cDNAs have been identified that encode molecules with the ability to form such a ternary complex. These molecules called SRF-accessory proteins 1 and 2 (87) and Elk-1 (88, 89) are related, and are members of the ets family of transcription factors. Strikingly, in response to serum induction, a marked change in the mobility of the SRF–Elk-1 ternary complex can be observed in gel retardation assays; this change can be mimicked *in vitro* in response to phosphorylation of Elk-1 by mammalian MAP kinase, a kinase that is known to be activated by serum induction (90). Elk-1 can act as a strong transcriptional activator when fused to a DNA-binding domain, and the ability to transactivate depends upon the integrity of the MAP kinase phosphorylation sites (91). These experiments suggest that serum induction may be brought about by phosphorylating Elk-1, which is stabilized in its ability to bind to DNA by association with SRF. Genomic footprints suggest that Elk-1 is stably associated with SRF before serum stimulation (80–82); this is not a problem, since activated mammalian MAP kinase is known to localize to the nucleus. The recruitment of Elk-1 by SRF to mediate some growth factor responses is similar to the MCM1–MATα1 interaction discussed above in connection with the specific stimulation of α-specific gene expression in yeast; not only is the P-box related to the SRE, and MCM1 to SRF, but Elk-1 and MATα1 share some homology, at least in the regions required to interact with the DNA-binding protein (reviewed in 78, 79).

Despite its lack of ability to stimulate transcription in transfection experiments, SRF acts *in vitro* as a strongly positive SRE-dependent transcription factor (77). It is thus possible that it can contribute to the magnitude of induction *in vivo*. This is consistent with gene fusion experiments in which the C-terminus of SRF has been shown to be important for maximal activation *in vivo* (91). It is interesting to note that some serum responsive elements do not appear to have the binding site to permit TCF formation (92); furthermore, point mutants that disrupt the sequences that flank the SRF binding site do not lead to loss of serum inducibility in all situations (83, 93, 94). It is thus possible that some other currently uncharacterized pathway exists to activate SRF function. This could be a direct effect on SRF, or could involve additional factors. It is interesting to speculate that such an additional factor might be a mammalian functional equivalent of STE12, which can interact with MCM1 and MATα1 in a DNA-independent manner to mediate a response to a MAP kinase pathway.

4. Discriminating between signals, and signal enhancement

4.1 The molecular basis of the differential response to type I and type II interferons: differential activation of signal transducing activators of transcription (STATs)

The interferons (IFNs) are a class of cytokines first identified by their ability to induce antiviral activity in target cells. They also have profound immunomodulatory effects and growth- and differentiation-altering properties (95). There are two types of IFN, type I (α- and β-IFN) and type II (γ-IFN), which, although they are unrelated polypeptides that are induced by different mechanisms and bind distinct receptors, induce an overlapping set of genes in a protein synthesis-independent manner (96), suggesting that treatment of cells with types I and II IFNs activates common signal transduction components. Much recent work has elucidated the mechanisms by which the IFNs stimulate transcription, and a novel signal transduction pathway has been revealed, which is also shared by many other cytokines and growth factors. The involvement of common signal transducing kinases and DNA-binding protein complexes in the activation of transcription by cytokines raises many interesting questions about how specificity in these systems is maintained.

The isolation of genes activated by types I and II IFNs and analysis of their promoter regions led to the identification of two DNA sequence elements necessary and sufficient for their induction by IFNs, namely, the type I IFN stimulatable response element (ISRE; 97, 98), and the type II gamma activation site (GAS; 99). These are both highly conserved sequence elements, which can confer responsiveness on to heterologous promoters. Their consensus sequences are shown below:

ISRE	A/GNGAAANNGAAACT
GAS	TTNCNNNAA

These response elements are unrelated, and at first gave no insight into the spectrum of type I and II IFN inducibility of the genes controlled by these elements: although some genes are exclusively activated by one type of IFN or the other, others are inducible by both (96). In the case of the guanylate binding protein (GBP) gene, which is inducible by type I and type II IFNs (100), its promoter was found to contain both an ISRE and a GAS (99). However, other genes inducible by both types of IFN appeared to have only one class of response element, leading to the speculation that, in certain cases, type I IFN could act through a type II IFN response element and vice versa.

Gel retardation analyses using ISREs as probes identified a complex, called ISGF3, that could be produced in a protein synthesis-independent manner within minutes of α-IFN treatment (98, 101). The complex was known to be activated in the cytoplasm and contain at least two different components, an α-IFN inducible component (ISGF3α), and a constitutively active DNA-binding component, which could also be induced by γ-IFN (ISGF3γ; 102). Purification of these components showed that ISGF3α itself was composed of three factors of molecular weights 113, 91, and 84 kDa, whereas ISGF3γ was found to be a single factor of 48 kDa (103). Once the cDNAs encoding these factors were obtained, it became clear that p113 and p91 are novel related proteins, and that p84 is a C-terminally truncated splice product of p91, lacking the final 38 amino acids (104, 105). They each have a Src homology 2 (SH2) and a Src homology 3 (SH3) domain and are tyrosine phosphorylated in response to α-IFN treatment (106). p48 is a member of the IRF-1 transcription factor family, which have also been shown to bind to ISREs (107, 108). Further analysis of the ISGF3 complex determined that upon treatment with α-IFN, the cytoplasmically localized ISGF3α factors are phosphorylated on a single tyrosine residue towards their C-termini, and p113 together with either p91 or the less abundant p84 form a complex with the ISGF3γ subunit, p48. The mature ISGF3 then translocates to the nucleus to bind to an ISRE and activate transcription (106).

Similar studies led to the identification of a 91-kDa DNA-binding protein that, upon γ-IFN treatment, bound rapidly to the GAS element upstream of the GBP promoter. This factor was named gamma activated factor (GAF), and was shown to be phosphorylated on tyrosine following γ-IFN treatment (109). Intriguingly, it cross-reacted with antibodies raised against p91, which itself was shown to become tyrosine phosphorylated when cells were treated with γ-IFN (106, 110).

The identity of these two proteins, p91 and GAF, was confirmed by complementation of a number of cell lines that are defective in their response to IFN treatment. In one such line, transfection of the cDNA for p91 was found to restore both α- and γ-IFN responses, although, interestingly, only the type I response could be restored using the p84 cDNA (111). Other complementation groups were found to be mutant in members of the family of cytoplasmic tyrosine kinases called JAKs (originally called 'just another kinase' because they were randomly isolated by homology to other kinases; 112).

The analysis of these mutant cell lines and the identification of the different ISGF3 components and JAKs required to restore their responses to both α- and γ-IFN enabled the derivation of the signal transduction pathways for both type I and type II IFNs: following type I IFN binding to the type I receptor, Tyk2 and JAK1 kinases are activated, leading to the tyrosine phosphorylation of p113, p91, and p84 and the formation of mature ISGF3 as previously described (113). Type II IFN binding to its receptor causes activation of JAK1 and JAK2 and the phosphorylation of p91 on the same tyrosine as type I IFN; p91 then dimerizes via its SH2 domains and phosphorylated tyrosine residues and is translocated to the nucleus to bind GAS elements (114, 115; for review see 116).

The pathways leading to both type I and type II IFN-dependent transcriptional activation are summarized in Fig. 3. Because of their ability to carry signals from the cell membrane directly to the nucleus and activate transcription, either alone or with other factors, the ISGF3α components have been renamed STATs (signal transducers and activators of transcription): p91 and p84 are known as STAT1α and STAT1β respectively, and p113 is known as STAT2 (117).

The fact that STAT1 can be activated by both α- and γ-IFN suggests a mechanism for how the two ligands can directly induce overlapping patterns of gene expression. As well as the production of ISGF3, α-IFN activation might also be expected to lead to the formation of STAT1α homodimers, which should be identical to GAF. This should lead to the stimulation of type II IFN-activated genes through their GAS elements. This has indeed been demonstrated for the IRF-1 promoter, where induction by α-IFN utilizes a GAS site (118). The magnitude of this induction is smaller than that obtained by γ-IFN, perhaps because most of the STAT1α activated by α-IFN is incorporated into ISGF3.

Since one consequence of activation of GAF by α-IFN would be the production of IRF-1 (which is itself a transcriptional activator of ISRE-like sequences: see below), ISREs should be indirectly activated by α-IFN, as well as directly activated by ISGF3. Similarly, γ-IFN activation of IRF-1 might also be expected to indirectly activate ISREs. It is possible that γ-IFN can have some direct effects on ISREs; although STAT1 homodimers cannot bind ISRE sequences, STAT1 might recruit other cellular factors to form a complex similar to ISGF3 (119, 120). The ISRE of the 9-27 gene has been reported to be inducible by type II IFN (121), although it is not known if this effect is indirect; interestingly, it has been reported that the γ-IFN response of this promoter is impaired in cell lines that lack functional p48 (116).

4.2 STATs are activated by a multitude of extracellular signals: induction of c-Fos by PDGF and EGF

Recently, it has become clear that the type of signal transduction pathway used to activate transcription in response to type I and type II IFNs also plays a major role in the stimulation of cells by many other cytokines and some growth factors (for reviews see 116, 122–124). In common with the IFNs, binding of these ligands to their receptors leads to the activation of a JAK complex that tyrosine phosphorylates members of the STAT family to activate transcription through GAS-related elements. However, different ligands recruit different JAKs into their receptor–kinase complexes, with the result that a subset of different STAT molecules becomes activated, possibly by recognition of the complexes by the SH2 domains of the STATs (125, 126). So far, a third JAK, JAK3, and at least six STATs, have been reported (123, 127).

The GAS elements bound by these different STATs are very similar, once again

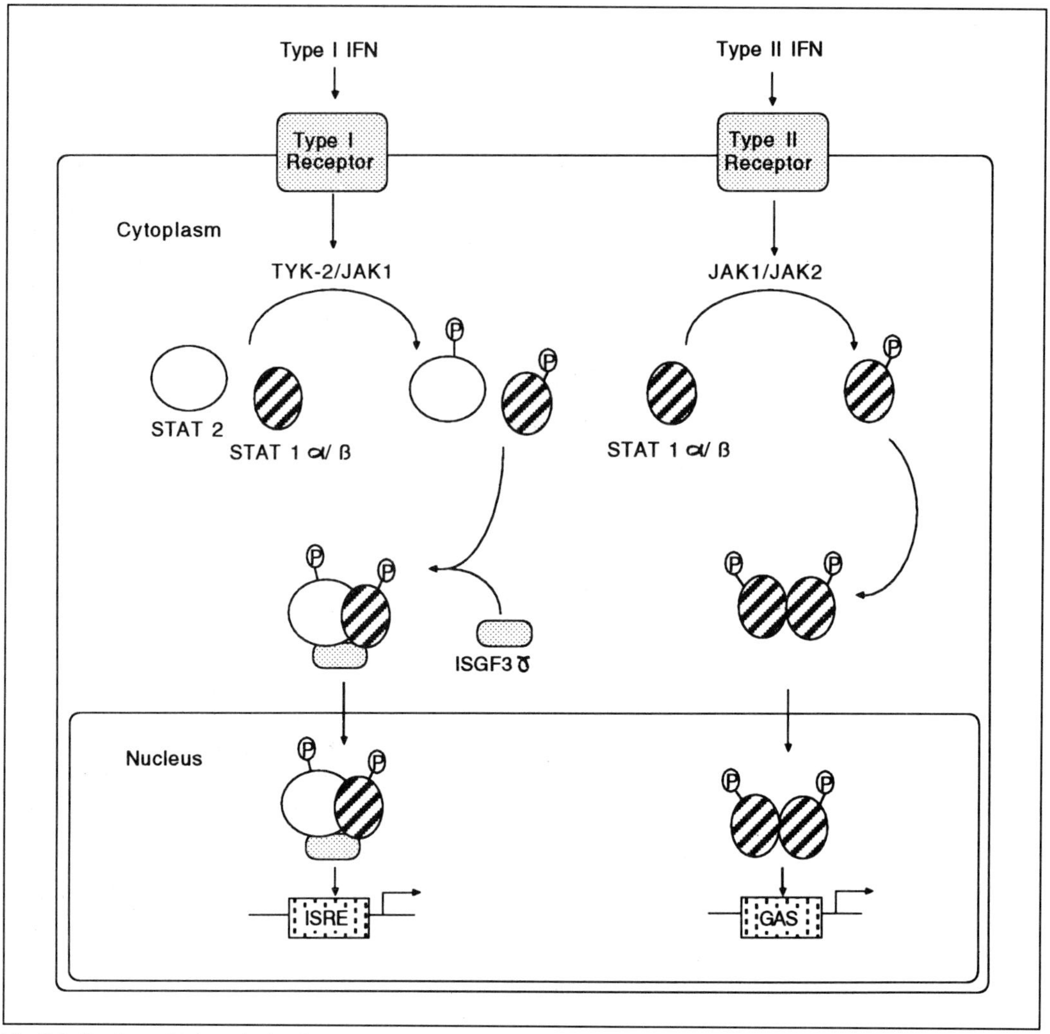

Fig. 3 Activation of transcription in response to type I and type II interferons (IFNs). Type I and type II IFNs bind to distinct receptor proteins, and as a consequence cause activation of specific JAKs and phosphorylation of specific STATs. In the case of type I IFN, STAT1α/β and STAT 2 are tyrosine phosphorylated and become associated with ISGF3γ to form a heterotrimeric complex called ISGF3. This complex translocates to the nucleus to activate transcription through ISREs. In the case of type II IFN, only STAT1α/β are tyrosine phosphorylated, leading to the formation of homodimers (called GAF) that translocate to the nucleus to activate transcription through GAS elements.

raising questions of how specificity in cytokine induction is achieved. Studies on the c-Fos promoter may help to understand the mechanism of transcriptional specificity. Although specific growth factors such as epidermal growth factor (EGF) and platelet-derived growth factor (PDGF) can activate c-Fos expression through the SRE as discussed above, full activation requires an additional element called the sis-inducible element (SIE), which resides about 30 base pairs upstream from the SRE

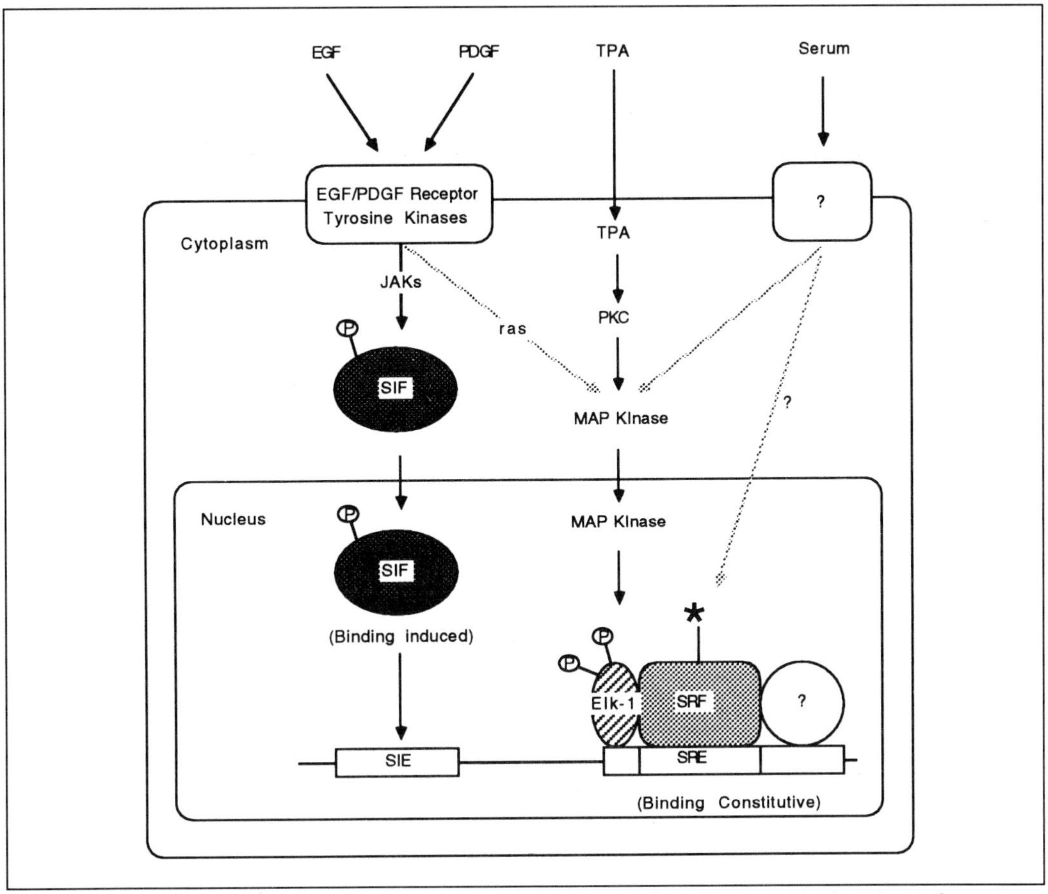

Fig. 4 Model for induction of the c-Fos promoter. The response of the c-Fos promoter to release from serum starvation utilizes the serum response element (SRE) as a minimal inducible element. Serum response factor (SRF) appears to be constitutively bound to the SRE, but can recruit a helper factor to adjacent sequences to form a ternary complex. In this figure, the helper factor is shown as Elk-1. Elk-1 is a target for MAP kinases, which can be activated by several signal transduction pathways, including growth factor receptor tyrosine kinases and protein kinase C (PKC); activated MAP kinase can translocate to the nucleus and directly phosphorylate Elk-1. Since some SREs appear to lack binding sites for Elk-1 and equivalent factors, it is possible that serum stimulation can also activate SRF in a ternary complex-independent manner. Growth factor receptor tyrosine kinases can also contribute to c-Fos induction through a second regulatory element called the SIE. This is a binding site for members of the STAT family of factors that are activated in the cytoplasm and subsequently translocate to the nucleus; STAT1 homodimers, STAT3 homodimers, and STAT1/STAT3 heterodimers have all been identified as potential regulatory factors through the SIE. Maximal c-Fos induction is probably brought about by the interaction of events at both the SIE and SRE. Adapted from Treisman (79).

(128; see Fig. 4). Unlike the SRE, treatment with PDGF or EGF causes an increase in DNA binding of a sequence-specific factor, which is called sis-inducible factor, or SIF (128). The SIE is closely related to the GAS element and SIF has recently been shown to be a member of the STAT family of transcription factors (see below).

Detailed analyses of SIF have revealed that there appear to be three forms of complex, referred to as SIF-A, SIF-B, and SIF-C, suggesting that more than one factor can be involved in complex formation (129). An analysis of the composition of these complexes indicated that SIF-C contained STAT1, and by comparison with the effects of γ-IFN, this complex would appear to be a STAT1 homodimer (129); SIF-B also contains STAT1, but in association with another component. By screening cDNA libraries for clones that are homologous to STAT1, a related clone encoding a protein called STAT3 has been isolated (130). Antibodies raised against STAT3 specifically react with the SIF-A complex in EGF-treated A431 cells, and partially affect the SIF-B complex (130). It thus appears that EGF treatment (and probably PDGF treatment) can induce the formation of SIE-binding complexes containing STAT1 homodimers, STAT3 homodimers, and STAT1/3 heterodimers.

Since there appears to be very little difference between the sequences of the SIE and GAS sites, it is of interest to determine whether other extracellular stimuli that activate STATs can induce SIF complex formation. In fact, both γ- and α-IFN can induce a homodimeric STAT1 complex (118, 131); however, neither of these is an inducer of c-Fos expression. It is therefore possible that the activation of c-Fos requires the presence of the STAT3 component, either as a homodimer, or as a heterodimer with STAT1. This simple view is challenged by the observation that IL-6, a poor or ineffective inducer of c-Fos, can also induce STAT3 homodimers, and in some circumstances can also induce STAT1/3 heterodimers (129, 130). The most likely explanation of the differences between EGF, γ-IFN, and IL-6 is that activation of a particular STAT complex is not sufficient for transcriptional induction of the genes. Full induction would require an additional element (an analogous situation has been discussed above for the effect of STE12 on mating type expression in yeast). In the case of the c-Fos promoter, the proteins bound at the SIE would interact with the SRE components, perhaps in a direct manner. Since SRE activation appears to require MAP kinase-mediated events, IL-6 and γ-IFN would not be able to stimulate through this sequence. Whether there are any consequences of the production of only a restricted range of STATs without other signals is an interesting question; γ-IFN is a known down-regulator of activated c-Fos expression (reviewed in 132), and it is interesting to speculate that STAT-1 homodimers could mediate this, perhaps by sterically occluding the access of positive factors to the SIE, or even by interfering with the ability of SRF–Elk-1 complexes to function.

As the signal transduction pathways of other cytokines and growth factors, which use similar mechanisms to those described above, are investigated, a clearer picture should emerge as to precisely how the interactions between ligands and receptors affect the recruitment of particular JAKs, which in turn activate specific subsets of STAT family members. The array of homo- and heterodimers produced would presumably bind as yet unrecognized subclasses of GAS elements with different affinities, which might also help to explain differential gene regulation by these factors.

4.3 The β-interferon promoter

β-Interferon (β-IFN) is a member of the type I IFNs that function as antiviral compounds. They achieve this by a variety of mechanisms, including the establishment of an antiviral state in target cells, and the activation of immune response functions. Type I IFNs also negatively regulate cell growth in some target cells. A consequence of lack of expression or response to type I IFNs is manifested as an inability to survive many types of viral infection, while inappropriate expression of type I IFNs can lead to autoimmune disease (reviewed in 95, 132). Thus it is essential that these genes are tightly regulated. β-IFN has been studied in much more detail than the 20 or so α-IFN genes. The expression of this gene can be induced in many cell types by viral infection, a process that can be mimicked in fibroblast-like cells by synthetic double-stranded RNA (dsRNA), such as poly(I)–poly(C), which is thought to resemble viral replicative intermediates. The induction of gene expression occurs at the level of transcriptional activation. Interestingly, there are other inducers of β-IFN expression, but, significantly, the magnitude of these events is much smaller than that achieved by virus infection or dsRNA treatment; since type I IFNs are very potent molecules, and these inducers include immune modulators such as lipopolysaccharide (LPS), tumour necrosis factor (TNF), and γ-IFN, these levels of induction may be physiologically significant in terms of local activations, but are clearly in a different class of events to that achieved by dsRNA or virus infection (reviewed in 132). The induction of β-IFN transcription by dsRNA does not require *de novo* protein synthesis, and in many cases, the post-transcriptional shut-off is also protein synthesis-independent; thus, the entire induction/turn-off cycle can be initiated and completed by the response of cellular factors to the presence of increased levels of dsRNA. The means by which this is achieved in a specific manner is an interesting example of concerted regulation through a relatively small promoter target.

The β-IFN gene contains all of the information for high level expression in response to dsRNA within a fragment of about 70 base pairs (see Fig. 5). The promoter contains four regions (called positive regulatory domains, or PRDs) that contribute in a co-operative manner to the strength of expression (reviewed in 133, 134). The sequences of the individual PRDs are distinct, implying that a number of different DNA-binding proteins may be required to bring about induction. An interesting property of some PRD-driven heterologous promoters is that they exhibit significant basal activity (135–138), and can be efficiently induced by signals other than dsRNA. For example, PRD II multimers are active in HeLa cells, and can be induced to high levels by TNF (139). In contrast, the lack of basal activity of the intact β-IFN promoter suggests that the PRDs are repressed in their natural context. This has been shown to be the case by a number of observations on the transcriptional superinduction of β-IFN by protein synthesis inhibitors (140, 141), and the identification of sequences that act as negative regulatory domains (NRDs; 142–146; see Fig. 5). The poor inducibility of the intact β-IFN promoter by TNF (147) also suggests that the PRDs are unresponsive in their natural context, and we have shown that removal of the NRD I sequence that flanks PRD II does allow the β-IFN promoter to be induced by TNF (unpublished observations). Thus at least part of the specificity of induction by dsRNA operates at the level of a dominant inhibitor that can only be inactivated by dsRNA.

Although the nature of the β-IFN repressor(s) remains unknown, a considerable

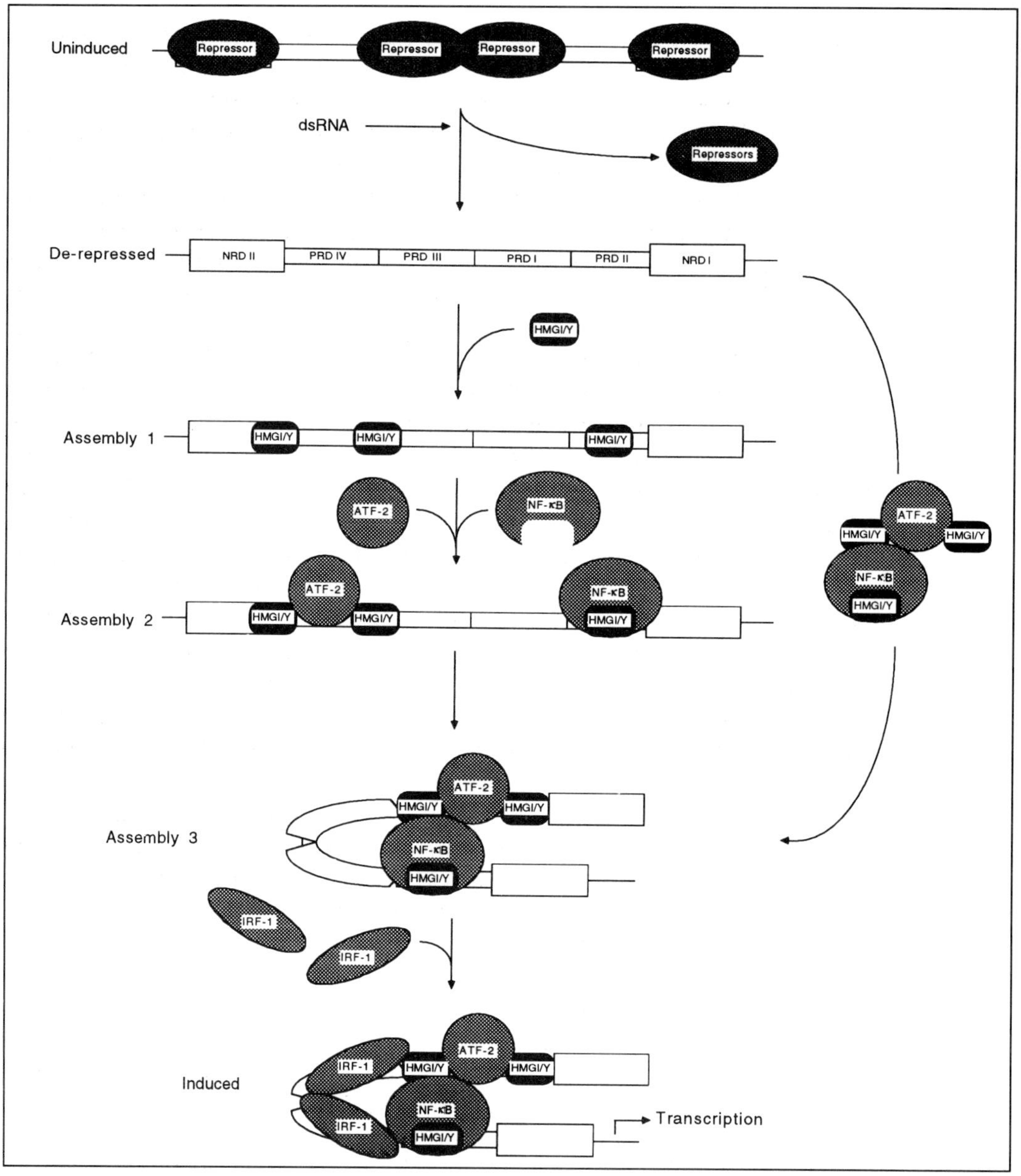

Fig. 5 Model for induction of the β-interferon (β-IFN) promoter by dsRNA. Prior to induction, the β-IFN promoter is bound at multiple sites by repressor molecules. Exposure of cells to double-stranded RNA (dsRNA) causes the loss of repression, and the activation of transcription through four distinct positive regulatory domains (PRDs). Although the details of the activation pathway have not been established, the work of Maniatis and his colleagues (150, 151, 164, 165) has led to the proposal that the HMGI/Y protein can occupy several sites in the β-IFN promoter. Once bound, this factor can co-operatively recruit NF-κB and ATF-2 to PRD II and PRD IV, respectively. NF-κB and ATF-2 interact with each other, and this could cause the template to bend; such a bend may allow IRF-1 to bind to PRD I and PRD III, and thus lead to transcriptional activation. In an alternative model, the mutual interactions between HMGI/Y, NF-κB, and ATF-2 could allow a multi-protein complex to preform, which could considerably enhance the efficiency of activation.

amount of information is available about the positive activators. Induction through PRD II involves the activation of NF-κB (136, 148, 149). Interestingly, as discussed in Chapter 4, this transcription factor can be activated by a number of stimuli (e.g. IL-1, TNF, dsRNA, cAMP, bacterial lipopolysaccharides, viral transactivators) that are not efficient inducers of β-IFN. A DNA-binding protein that can specifically recognize PRD IV was isolated and shown to be the bZIP protein ATF-2 (150), and it has been subsequently shown that PRD IV-specific complexes that contain ATF-2, as a homodimer or as a heterodimer with c-Jun, can be detected in extracts from induced cells, although the mechanism of induction is unknown (151). ATF-2 is a member of the CREB family of transcription factors that mediate a transcriptional response to cAMP (see Chapter 3). Induction through PRD I and PRD III is probably mediated by IRF-1 (152–158), although the results of biochemical analyses (108, 138) and targeted disruption experiments (158–160) suggest that there may be other factors that can substitute for IRF-1. The PRD I and PRD III elements may also be negatively regulated by an IRF-1-related protein called IRF-2 (154, 155, 158).

Although there are clear cell-specific variations in the requirements for different PRDs (see, for example, 161; reviewed in 134), most analyses are consistent with each of the four PRDs contributing to the magnitude of induction. At a very simple level, the strong transcriptional activation observed in response to dsRNA might be explained by the cumulative effect of multiple transcription factors. The nature of the signal transduction pathway activated by dsRNA is unclear, although it has recently been shown that the dsRNA-dependent protein kinase R can phosphorylate IκB and lead to NF-κB activation (162; see Chapter 4). As we have discussed above, some of the transcription factors involved in β-IFN induction are also activated by other stimuli, most notably NF-κB, which can be activated by TNF, although the synthesis of IRF-1 can also be induced by other signals (see above and 163). The failure of other stimuli such as TNF to efficiently induce β-IFN might be due to a simple inability to activate enough individual positive regulatory factors. However, de-repression appears to be obligatory, and is specific to dsRNA, since in the absence of functional negative regulatory regions, the β-IFN gene can respond to other inducers (unpublished observations).

This simple model has been challenged of late by the implication of a role for a chromatin-associated factor in β-IFN regulation (reviewed in 164). Although the binding of NF-κB to PRD II *in vitro* is insensitive to mutations in the central A-T-rich core (136) these mutations are deleterious to induction (144). A resolution of this was provided by the experiments of Thanos *et al.* (165) who isolated a cDNA that encoded a protein for the chromatin-associated high mobility group protein, HMGI/Y, that could bind to the central A-T-rich core of PRD II, but not to the NF-κB binding site from the immunoglobulin κ chain gene enhancer, which lacks a central A-T-rich core. HMGI/Y appears to be essential for the activation of PRD II and β-IFN induction, but is not required for the activation of an NF-κB binding site derived from the immunoglobulin κ chain gene enhancer (165). *In vitro*, HMGI/Y could enhance the binding of NF-κB to PRD II; it is thought that the HMGI/Y protein can bind to the minor groove of DNA and bend it such that

NF-κB binding is facilitated. The recent solution of the crystal structure of the DNA-bound complex of the p50 subunit of NF-κB (166–168) shows that NF-κB is unlikely to prevent HMGI/Y from co-occupying the DNA. It is clear from these observations that HMGI/Y would act as a factor that aided the binding of NF-κB to DNA. However, HMGI/Y does not appear to be required for the activation of NF-κB, and it is not obvious at this point why there is a need for the HMGI/Y protein for specific induction by dsRNA (see below).

The HMGI/Y protein also appears to be required for activation through PRD IV (151). The finding of ATF-2 as a PRD IV-binding protein is surprising given that PRD IV is not very similar to a typical cAMP response element (CRE; see Chapter 3). However, in the case of PRD IV, it appears that binding of ATF-2 to the central core site requires the presence of A-T-rich flanking sequences, which also act to bind HMGI/Y (151). Like NF-κB, the DNA-binding activity of the PRD IV complex is inducible by dsRNA; unlike NF-κB, the inducible complex seems to require HMGI/Y for its formation, although the molecular basis for this is not known.

A finding that may be of key relevance in understanding the role of HMGI/Y in induction is the observation that not only does this factor aid the individual binding of NF-κB and ATF-2 to DNA, but also that it appears to aid the binding of these factors in a co-operative manner. It is capable of binding to each of NF-κB and ATF-2 in solution (151, 164); the consequence of this would be the generation of a complex that had enhanced specificity for DNA, such that when one molecule has bound, the other would be presented to the template at effectively infinite concentration. Furthermore, NF-κB and several members of the bZIP family, including ATF-2, can interact with each other in solution (reviewed in 169); this would extend the potential for co-operative binding such that the binding of any single component would be followed by binding of all of these components. Even if the affinities of the individual protein components are not high enough to form a stable pre-assembled complex in solution, once one component had bound to DNA others could show enhanced recruitment since the potential interaction sites now include both the DNA template and the protein partner. While the importance of this type of enhanced assembly model for co-operative transcriptional stimulation is clear, it may not be obvious how it can influence promoter specificity. In fact, in the case of the β-IFN promoter, PRD II has a relatively low affinity for NF-κB; on its own NF-κB may not be efficiently recruited to this site. If, however, NF-κB can be brought to the promoter in association with other sequence-specific DNA-binding proteins that are also specifically inducible by dsRNA, it would significantly enhance activation. In fact, the use of a low affinity site would allow discrimination against activation by other NF-κB inducers.

A model for the activation of the β-IFN promoter is shown in Fig. 5. In addition to the interactions described above, it is also possible that NF-κB can interact with IRF-1 and aid its recruitment to the DNA (see 164).

The transcriptional organizational role for HMGI/Y may not be restricted to β-IFN. A similar arrangement of bZIP and NF-κB binding sites is found in the promoter of the E-selectin gene, and there is evidence to support a role for HMGI/Y in the recruitment and assembly of these factors to the promoter (170–172). Induction of class II MHC by γ-IFN can also be impaired by specific antisense oligonucleotides against HMGI/Y (165). Furthermore, HMGI/Y has been shown to bind to the promoters of the lymphotoxin gene (173) and the IL-2α receptor (174). In the latter case, there is evidence that HMGI/Y-binding activity is enhanced by external stimuli (this has not been shown to be the case for dsRNA in the induction of the β-IFN gene), and that it can bind to Elf-1, which also binds to the IL-2α receptor promoter (although there is no evidence that HMGI/Y can enhance binding of Elf-1 to DNA). Since a second region of the IL-2α receptor promoter binds to NF-κB, and this factor can directly interact with Elf-1, it is possible that specific regulation of this gene is also mediated by the assembly of a multi-protein complex whose individual components are each regulated by a common signal. Other proteins may play similar organizational roles, such as LEF-1, which binds to the T-cell receptor promoter and induces a sharp bend (175). It is believed that this bend is required to induce a conformational change in the promoter such that other transcription factors are brought into close proximity to permit maximal transcriptional stimulation (176, 177; reviewed in 178).

5. Integrating divergent signals: NF-AT and the IL-2 promoter

Interleukin-2 (IL-2) is one of an array of cytokines whose transcription is activated upon stimulation of T cells. Analysis of the IL-2 promoter has revealed a specific motif that can mediate activation, and this element has been shown to bind to a nuclear factor that is present in activated T cells, called NF-AT (reviewed in 179, 180). T-cell activation requires the generation of two intracellular signals, namely the activation of protein kinase C (PKC) and the release of Ca^{2+} from intracellular stores. The analysis of how distinct signals become integrated into a unified transcriptional response provides a striking example of specificity.

Biochemical analysis showed that NF-AT is composed of a ubiquitous nuclear component (NF-ATn) and a T-cell-specific component that becomes re-localized from the cytoplasm to the nucleus upon activation (NF-ATc; see Fig. 6): Mature NF-AT can be reconstituted by simply mixing cytoplasmic and nuclear fractions (181). The presence of the nuclear component in T cells depends upon protein synthesis in response to stimulation. This process is the PKC-dependent arm of the pathway, and since the activation of T cells is accompanied by a shift from G0 to G1, is a classical example of an immediate early gene activation. The Ca^{2+} release is required to activate the uptake of NF-ATc to the nucleus. The immunosuppressant drugs cyclosporin A (CsA) and FK506 inhibit T-cell activation by blocking this step, achieving this by inhibiting the activity of a Ca^{2+}-activated phosphatase called calcineurin (reviewed in 182). Cells that over-express a constitutively active form of calcineurin require only the PKC signal to allow NF-AT-dependent transcription (183, 184).

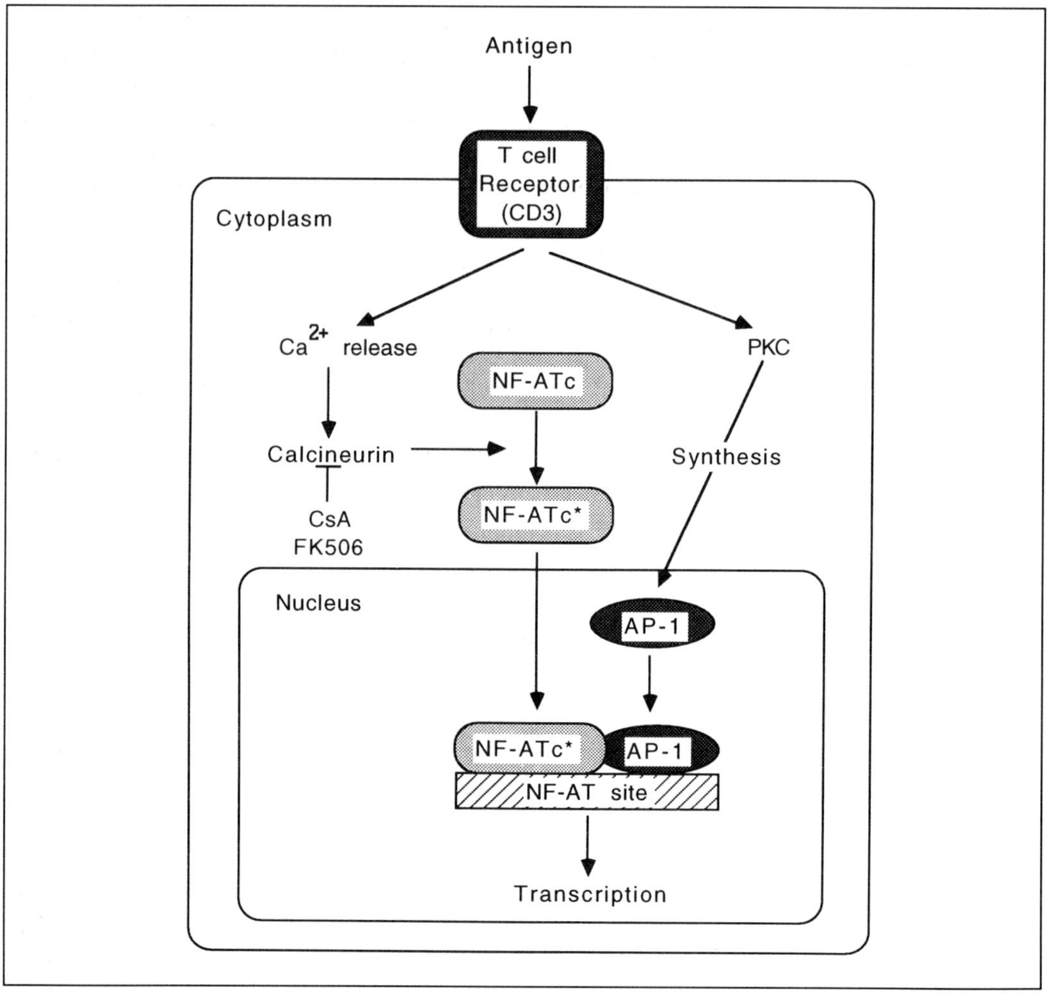

Fig. 6 Integration of signals at the interleukin-2 promoter. T lymphocytes are activated by the binding of antigen to the T-cell receptor, which leads to the production of two distinct signals (intracellular release of Ca^{2+} and the activation of PKC); the generation of these signals from alternative receptor interactions can also trigger activation. PKC induces the synthesis of c-Fos, which is a component of the AP-1 transcription factor. Ca^{2+} release activates calmodulin, which in turn activates the calmodulin-dependent phosphatase, calcineurin. The activity of this enzyme (which can be blocked by the immunosuppressant drugs, FK506 and cyclosporin A; CsA) causes the translocation of the NF-ATc component to the nucleus as a result of dephosphorylation (NF-ATc*). After translocation, NF-ATc can interact with AP-1 on DNA to stimulate transcription. Adapted from Goodbourn (179).

Two candidate genes for NF-ATc have been isolated (185, 186). These genes encode related products that may well perform similar roles at different stages of T-cell development (reviewed in 179). The NF-ATc genes show some similarity to the DNA-binding domain of c-Rel, a member of the NF-κB family of transcription factors (see Chapter 4). The observation of homology is interesting, since the NF-κB family of transcription factors are also cytoplasmic prior to the receipt of activating stimuli. NF-κB is held in the cytoplasm by association with an inhibitor molecule called IκB. Although it is not yet known whether the cytoplasmic location of NF-ATc is ensured by association with IκB or a related molecule, it is interesting to note that nuclear uptake of NF-κB can be blocked by CsA in response to some signals in stimulated T cells, suggesting that IκB may indeed be a target for calcineurin (187).

Consistent with its designation as an immediate early gene product, NF-ATn contains c-Fos and c-Jun (188). These factors normally combine to give a transcription factor called AP-1. However, AP-1 is not able to bind to the NF-AT site in the IL-2 promoter, unless NF-ATc is also present. Similarly, although NF-ATc can bind on its own, it cannot bind in a stable manner in the absence of AP-1. AP-1 binding requires intact leucine zippers, suggesting that both proteins may contact the DNA. This binding interdependence ensures that IL-2 activation can only occur when two distinct signals are detected. Intriguingly, it has been previously reported that NF-κB can interact with proteins that are related to the AP-1 factor (see above), and this suggests that combinatorial interactions between proteins of different families may considerably enhance the degree of inducer specificity (reviewed in 169).

NF-AT may also play a role in the activation of other T-cell-specific genes (reviewed in 180). In contrast to the protein synthesis-dependent activation of IL-2 transcription, induction of tumour necrosis factor (TNFα) expression is a primary event. This clearly indicates that pre-existing factors are sufficient for the activation of this gene, and that AP-1 is not required. Accordingly, the activation of TNFα appears to involve NF-ATc as part of a novel DNA-binding complex that does not contain AP-1 (189). Since a large number of genes have been demonstrated to be activated by AP-1, there are clearly many outcomes of the combinations of stimulating AP-1 and NF-AT activities.

6. Conclusions

The systems described above suggest that the naïve view that every promoter is activated by a single sequence-specific transcription factor that is produced under defined circumstances, is incorrect. Rather, specific transcriptional regulation is brought about by the interaction of a number of different transcription factors. These factors can bind to non-overlapping sequences on the promoter such as seen in the interactions between the SRF and SIE elements in the c-Fos promoter and the PRE- and P-box in the promoters of yeast **a**-specific genes, or to sequences that are at least partially overlapping such as the SRF–Elk-1 or NF-ATc–AP-1 interactions. It is becoming increasingly clear that the occupation of these sites can be co-operative, and in some cases it is probable that such co-operativity may be essential to ensure efficient function. This is certainly the case with the yeast **a**-specific genes and the SRF–Elk-1 or NF-ATc–AP-1 assemblies; the use of HMGI/Y in protein-complex assembly on the β-IFN promoter is an extreme example of such co-operativity. Protein–protein interactions also seem to be required to specifically recruit factors that cannot bind by themselves to DNA, but which are essential to modulate transcription (for example, the Oct protein co-factors such as VP16). Given the overwhelming evidence that co-operative protein–protein interactions are important to ensure promoter occupation, it is interesting to speculate that the concentrations of specific transcription factors would be of critical importance in determining whether activation is brought about; insufficient amounts of any one factor would lead to a failure to activate. In this light the concentration of the array of STAT homo- and heterodimers activated by cytokines will be important in specifying the response achieved by binding to GAS sites of different affinities.

Chapter 1 describes our current knowledge of the basal transcriptional machinery; in the case of the complex promoters described here, it appears that more than one transcriptional activation domain would be presented to this machinery, and would presumably lead to synergistic activation of transcription. Although the targets in the basal machinery for interactions of many factors are now known, it is not yet clear how the properties of transcription factors can change in response to cellular signals, and this promises to be an exciting area for future research.

References

1. Staudt, L. M. and Lenardo, M. J. (1991) Immunoglobulin gene transcription. *Annu. Rev. Immunol.*, **9**, 373.
2. Staudt, L. M., Singh, H., Sen, R., Wirth, T., Sharp, P. A., and Baltimore, D. (1986) A lymphoid-specific protein binding to the octamer motif of immunoglobulin genes. *Nature*, **323**, 640.
3. Clerc, R. G., Corcoran, L. M., LeBowitz, J. H., Baltimore, D., and Sharp, P. A. (1988) The B-cell-specific Oct-2 protein contains POU box- and homeo box-type domains. *Genes Dev.*, **2**, 1570.
4. Muller, M. M., Ruppert, S., Schaffner, W., and Matthias, P. (1988) A cloned octamer transcription factor stimulates transcription from lymphoid-specific promoters in non-B-cells. *Nature*, **336**, 544.
5. Herr, W., Sturm, R. A., Clerc, R. G., Corcoran, L. M., Baltimore, D., Sharp, P. A., Ingraham, H. A., Rosenfeld, M. G., Finney, M., Ruvkun, G., and Horvitz, H. R. (1988) The POU domain: a large conserved region in the mammalian Pit-1, Oct-1, Oct-2, and *Caenorhabditis elegans* unc-86 gene products. *Genes Dev.*, **2**, 1513.
6. Scholer, H. R., Dressler, G. R., Balling, R., Rohdewohld, H., and Gruss, P. (1990) Oct-4: a germline-specific transcription factor mapping to the mouse t-complex. *EMBO J.*, **9**, 2185.
7. Sturm, R. A., Das, G., and Herr, W. (1988) The ubiquitous octamer-binding protein Oct-1 contains a POU domain with a homeo box subdomain. *Genes Dev.*, **2**, 1582.
8. Kemler, I. and Schaffner, W. (1990) Octamer transcription factors and the cell type-specificity of immunoglobulin gene expression. *FASEB J.*, **4**, 1444.
9. Corcoran, L., Karvelas, M., Nossal, G. J., Ye, Z. S., Jacks, T., and Baltimore, D. (1993) Oct-2, although not required for early B-cell development, is critical for later B-cell maturation. *Genes Dev.*, **7**, 570.
10. Tanaka, M., Grossniklaus, U., Herr, W., and Hernandez, N. (1988) Activation of the U2 snRNA promoter by the octamer motif defines a new class of RNA polymerase II enhancer elements. *Genes Dev.*, **2**, 1764.
11. Tanaka, M. and Herr, W. (1990) Differential transcriptional activation by Oct-1 and Oct-2: interdependent activation domains induce Oct-2 phosphorylation. *Cell*, **60**, 375.

12. Pierani, A., Heguy, A., Fujii, H., and Roeder, R. G. (1990) Activation of octamer-containing promoters by either octamer-binding transcription factor 1 (OTF-1) or OTF-2 and requirement of an additional B-cell-specific component for optimal transcription of immunoglobulin promoters. *Mol. Cell. Biol.*, **10**, 6204.
13. Luo, Y., Fujii, H., Gerster, T., and Roeder, R. G. (1992) A novel B-cell-derived coactivator potentiates the activation of immunoglobulin promoters by octamer-binding transcription factors. *Cell*, **71**, 231.
14. Gstaiger, M., Knoepfel, L., Georgiev, O., Schaffner, W., and Hovens, C. M. (1995) A B-cell coactivator of octamer-binding transcription factors. *Nature*, **373**, 360.
15. Strubin, M., Newell, J. W., and Matthias, P. (1995) OBF-1, a novel B-cell-specific coactivator that stimulates immunoglobulin promoter activity through association with octamer-binding proteins. *Cell*, **80**, 497.
16. LaBella, F., Sieve, H. L., Roeder, R. G., and Heintz, N. (1988) Cell-cycle regulation of a human histone H2B gene is mediated by the H2B sub-type-specific consensus element. *Genes Dev.*, **2**, 32.
17. Murphy, S., Pierani, A., Scheidereit, C., Melli, M., and Roeder, R. G. (1989) Purified octamer binding transcription factors stimulate RNA polymerase III-mediated transcription of the 7SK RNA gene. *Cell*, **59**, 1071.
18. Fletcher, C., Heintz, N., and Roeder, R. G. (1987) Purification and characterisation of OTF-1, a transcription factor regulating cell cyle expression of a human histone H2b gene. *Cell*, **51**, 773.
19. Segil, N., Roberts, S. B., and Heintz, N. (1991) Mitotic phosphorylation of the Oct-1 homeodomain and regulation of Oct-1 DNA binding activity. *Science*, **254**, 1814.
20. Voss, J. W., Wilson, L., and Rosenfeld, M. G. (1991) POU-domain proteins Pit-1 and Oct-1 interact to form a heteromeric complex and can cooperate to induce expression of the prolactin promoter. *Genes Dev.*, **5**, 1309.
21. Pfeuffer, I., Klein-Hessling, S., Heinfling, A., Chuvpilo, S., Brabletz, T., Hentsch, B., Schwarzenbach, H., Matthias, P., and Serfling, E. (1994) Octamer factors exert a dual effect on the IL-2 and IL-4 promoters. *J. Immunol.*, **153**, 5572.
22. Poellinger, L. and Roeder, R. G. (1989) Octamer transcription factors 1 and 2 each bind to two different functional elements in the immunoglobulin heavy-chain promoter. *Mol. Cell. Biol.*, **9**, 747.
23. Poellinger, L., Yoza, B. K., and Roeder, R. G. (1989) Functional cooperativity between protein molecules bound at two distinct sequence elements of the immunoglobulin heavy-chain promoter. *Nature*, **337**, 573.
24. LeBowitz, J. H., Clerc, R. G., Brenowitz, M., and Sharp, P. A. (1989) The Oct-2 protein binds cooperatively to adjacent octamer sites. *Genes Dev.*, **3**, 1625.
25. Verrijzer, C. P., van Oosterhout, J. A. W. M., and van der Vliet, P. C. (1992) The Oct-1 POU domain mediates interactions between Oct-1 and other POU proteins. *Mol. Cell. Biol.*, **12**, 542.
26. Verrijzer, C. P. and van der Vliet, P. C. (1993) POU domain transcription factors. *Biophys. Biochem. Acta*, **1173**, 1.
27. O'Hare, P. (1993) The virion transactivator of herpes simplex virus. In *The alpha-herpesviruses* (ed. J. Davidson), p. 145. Saunders Scientific Publications, Philadelphia.
28. Kristie, T. M., LeBowitz, J. H., and Sharp, P. A. (1989) The octamer-binding proteins form multi-protein DNA complexes with the HSV alpha TIF regulatory protein. *EMBO J.*, **8**, 4229.
29. Wilson, A. C., LaMarco, K., Peterson, M. G., and Herr, W. (1993) The VP16 accessory

protein HCF is a family of polypeptides processed from a large precursor protein. *Cell*, **74**, 115.

30. Verrijzer, C. P., Kal, A. J., and van der Vliet, P. C. (1990) The Oct-1 homeo domain contacts only part of the octamer sequence and full Oct-1 DNA-binding activity requires the POU-specific domain. *Genes Dev.*, **4**, 1964.
31. Verrijzer, C. P., Alkema, M. J., van Weperen, W. W., van Leeuwen, H. C., Strating, M. J. J., and van der Vliet, P. C. (1992) The DNA binding specificity of the bipartite POU domain and its subdomains. *EMBO J.*, **11**, 4993.
32. Aurora, R. and Herr, W. (1992) Segments of the POU domain influence one another's DNA binding specificity. *Mol. Cell. Biol.*, **12** ., 455.
33. Dekker, N., Cox, M., Boelens, R., Verrijzer, C. P., van der Vliet, P. C., and Kaptein, R. (1993) Solution structure of the POU-specific DNA-binding domain of Oct-1. *Nature*, **362**, 852.
34. Klemm, J. D., Rould, M. A., Aurora, R., Herr, W., and Pabo, C. O. (1994) Crystal structure of the Oct-1 POU domain bound to an octamer site: DNA recognition with tethered DNA-binding modules. *Cell*, **77**, 21.
35. Walker, S., Hayes, S., and O'Hare, P. (1994) Site-specific conformational alteration of the Oct-1 POU domain-DNA complex as the basis for differential recognition by Vmw65 (VP16). *Cell*, **79**, 841.
36. Stern, S., Tanaka, M., and Herr, W. (1989) The Oct-1 homeodomain directs formation of a multiprotein-DNA complex with the HSV transactivator VP16. *Nature*, **341**, 624.
37. Lai, J.-S., Cleary, M. A., and Herr, W. (1992) A single amino acid exchange transfers VP16-induced positive control from the Oct-1 to the Oct-2 homeo domain. *Genes Dev.*, **6**, 2058.
38. Pomerantz, J. L., Kristie, T. M., and Sharp, P. A. (1992) Recognition of the surface of a homeo domain protein. *Genes Dev.*, **6**, 2047.
39. Scholer, H. R., Ciesiolka, T., and Gruss, P. (1991) A nexus between Oct-4 and E1a: implications for gene regulation in embryonic stem cells. *Cell*, **66**, 291.
40. Imperiale, M. J., Kao, H. T., Feldman, L., Nevins, J., and Strickland, S. (1984) Common control of the heat shock gene and early adenovirus genes: evidence for a cellular E1A-like activity. *Mol. Cell. Biol.*, **4**, 867.
41. Herskowitz, I., Rine, J., and Strathern, J. N. (1992) Mating-type determination and mating-type interconversion in *Saccharomyces cerevisiae*. In *The molecular biology and cellular biology of the yeast Saccharomyces: gene expression* (ed. E. Jones, D. Pringle, and J. A. Broach), p. 583. Cold Spring Harbor Laboratory Press, Cold Spring Harbor, NY.
42. Johnson, A. D. and Herskowitz, I. (1985) A repressor (MATα2 product) and its operator control expression of a set of cell type specific genes in yeast. *Cell*, **42**, 237.
43. Bender, A. D. and Sprague, G. F., jun. (1987) MATα1 protein, a yeast transcription activator binds synergistically with a second protein to a set of cell-type-specific genes. *Cell*, **50**, 681.
44. Kronstad, J. W., Holly, J. S., and MacKay, V. L. (1987) A yeast operator overlaps an upstream activation site. *Cell*, **50**, 369.
45. Jarvis, E. E., Hagen, D. C., and Sprague, G. F., jun. (1988) Identification of a DNA segment that is necessary and sufficient for α-specific gene control in *Saccharomyces cerevisiae*: implications for regulation of α-specific and **a**-specific genes. *Mol. Cell. Biol.*, **8**, 309.
46. Elbel, R. and Tye, B.-K. (1991) Both activation and repression of **a**-mating-type-specific genes in yeast require transcription factor Mcm 1. *Proc. Natl Acad. Sci. USA*, **88**, 10966.

47. Hwang-Shum, J.-J., Hagen, D. C., Jarvis, E. E., Westby, C. A., and Sprague, G. F., jun. (1991) Relative contributions of MCM1 and STE12 to transcriptional activation of **a**- and α-specific genes from *Saccharomyces cerevisiae*. *Mol. Gen. Genet.*, **227**, 197.
48. Primig, M., Winkler, H., and Ammerer, G. (1991) The DNA binding and oligomerization domain of MCM1 is sufficient for its interaction with other regulatory proteins. *EMBO J.*, **10**, 4209.
49. Fields, S. and Herskowitz, I. (1985) The yeast STE12 product is required for expression of two sets of cell-type-specific genes. *Cell*, **42**, 923.
50. Fields, S., Chaleff, D. T., and Sprague, G. F., jun. (1988) Yeast *STE7*, *STE11*, and *STE12* genes are required for expression of cell-type-specific genes. *Mol. Cell. Biol.*, **8**, 551.
51. Dolan, J. W. and Fields, S. (1990) Overproduction of the yeast STE12 protein leads to constitutive transcriptional induction. *Genes Dev.*, **4**, 492.
52. Song, O., Dolan, J. W., Yuan, Y. O., and Fields, S. (1991) Pheromone-dependent phosphorylation of the yeast STE12 protein correlates with transcriptional activation. *Genes Dev.*, **5**, 741.
53. Dolan, J. W., Kirkman, C., and Fields, S. (1989) The yeast STE12 protein binds to the DNA sequence mediating pheromone induction. *Proc. Natl Acad. Sci. USA*, **86**, 5703.
54. Errede, B. and Ammerer, G. (1989) STE12, a protein involved in cell-type-specific transcription and signal transduction in yeast, is part of protein-DNA complexes. *Genes Dev.*, **3**, 1349.
55. Kirkman-Correia, C., Stroke, I. L., and Fields, S. (1993) Functional domains of the yeast STE12 protein, a pheromone-responsive transcriptional activator. *Mol. Cell. Biol.*, **13**, 3765.
56. Herskowitz, I. (1995) MAP kinase pathways in yeast: for mating and more. *Cell*, **80**, 187.
57. Keleher, C. A., Goutte, C., and Johnson, A. D. (1988) The yeast cell-type-specific repressor α2 acts cooperatively with a non-cell-type-specific protein. *Cell*, **53**, 927.
58. Keleher, C. A., Passmore, S., and Johnson, A. D. (1989) Yeast repressor α2 binds to its operator cooperatively with yeast protein Mcm 1. *Mol. Cell. Biol.*, **13**, 3765.
59. Sauer, R. T., Smith, D. L., and Johnson, A. D. (1988) Flexibility of the yeast α2 repressor enables it to occupy the ends of its operator leaving the center free. *Genes Dev.*, **2**, 807.
60. Hall, M. N. and Johnson, A. D. (1987) Homeo domain of the yeast repressor α2 is a sequence-specific DNA-binding domain but is not sufficient for repression. *Science*, **237**, 1001.
61. Vershon, A. K. and Johnson, A. D. (1993) A short disordered protein region mediates interactions between the homeodomains of the yeast α2 protein and the MCM1 protein. *Cell*, **72**, 105.
62. Ammerer, G. (1990) Identification, purification and cloning of a polypeptide (PRTF/GRM) that binds to mating-specific promoter elements in yeast. *Genes Dev.*, **4**, 299.
63. Flessel, M. C., Brake, A. J., and Thorner, J. (1989) The MFα1 gene of *Saccharomyces cerevisiae*: genetic mapping and mutational analysis of promoter elements. *Genetics*, **121**, 223.
64. Jarvis, E. E., Clark, K. L., and Sprague, G. F., jun. (1989) The yeast transcription activator PRTF, a homolog of the mammalian serum response factor, is encoded by the *MCM1* gene. *Genes Dev.*, **3**, 936.
65. Inokuchi, K., Nakayama, A., and Hishimuma, F. (1987) Identification of sequence elements that confer cell-type-specific control of MFα1 expression in *Saccharomyces cerevisiae*. *Mol. Cell. Biol.*, **7**, 3185.
66. Sengupta, P. and Cochran, B. H. (1990) The PRE and PQ box are functionally distinct yeast pheromone response elements. *Mol. Cell. Biol.*, **10**, 6809.

67. Hagen, D. C., Bruhn, L., Westby, C., and Sprague, G. F., jun. (1993) Transcription of α-specific genes in *Saccharomyces cerevisiae*: DNA sequence requirements for activity of the coregulator α1. *Mol. Cell. Biol.*, **13**, 6866.
68. Bruhn, L. and Sprague, G. F., jun. (1994) MCM1 point mutants deficient in expression of α-specific genes: residues important for interaction with α1. *Mol. Cell. Biol.*, **14**, 2534.
69. Tan, S., Ammerer, G., and Richmond, T. J. (1988) Interactions of purified transcription factors: binding of yeast MATα1 and PRTF to cell-type-specific, upstream activating sequences. *EMBO J.*, **7**, 4255.
70. Tan, S. and Richmond, T. J. (1990) DNA binding-induced conformational change of the yeast transcriptional activator PRTF. *Cell*, **62**, 367.
71. Yuan, Y. O., Stroke, I. L., and Fields, S. (1993) Coupling of cell identity to signal response in yeast: interaction between the α1 and the STE12 protein. *Genes Dev.*, **7**, 1584.
72. Lydall, D., Ammerer, G., and Nasmyth, K. (1991) A new role for MCM1 in yeast: cell cycle regulation of SWI5 transcription. *Genes Dev.*, **5**, 2405.
73. Hagen, D. C., McCaffrey, G., and Sprague, G. J., jun. (1991) Pheromone response elements are necessary and sufficient for basal and pheromone-induced transcription of the FUS1 gene of *Saccharomyces cerevisiae. Mol. Cell. Biol.*, **11**, 2952.
74. Yuan, Y. O. and Fields, S. (1991) Properties of the DNA-binding domain of the *S. cerevisiae* STE12 protein. *Mol. Cell. Biol.*, **11**, 5910.
75. Roberts, R. L. and Fink, G. R. (1994) Elements of a single MAP kinase cascade in *Saccharomyces cerevisiae* mediate two developmental programs in the same cell type: mating and invasive growth. *Genes Dev.*, **8**, 2974.
76. Treisman, R. (1992) The serum response element: structure and function. In *Nuclear processes and oncogenes* (ed. P. A. Sharp), p. 163. Academic Press, Inc. San Diego.
77. Norman, C., Runswick, M., Pollock, R., and Treisman, R. (1988) Isolation and properties of cDNA clones encoding SRF, a transcription factor that binds to the c-fos serum response element. *Cell*, **55**, 989.
78. Treisman, R. and Ammerer, G. (1992) The SRF and MCM1 transcription factors. *Curr. Opin. Genet. Dev.*, **2**, 221.
79. Treisman, R. (1994) Ternary complex factors: growth factor regulated transcriptional activators. *Curr. Opin. Genet. Dev.*, **4**, 96.
80. Herrera, R. E., Shaw, P. E., and Nordheim, A. (1989) Occupation of the *c-fos* serum response element in vivo by a multi-protein complex is unaltered by growth factor induction. *Nature*, **340**, 68.
81. Konig, H. (1991) Cell-type specific multiprotein complex formation over the c-fos serum response element *in vivo*: ternary complex formation is not required for the induction of c-fos. *Nucl. Acids Res.*, **19**, 3607.
82. Malik, R. K., Roe, M. W., and Blackshear, P. J. (1991) Epidermal growth factor and other mitogens induce binding of a protein complex to the c-fos serum response element in human astrocytoma and other cells. *J. Biol. Chem.*, **266**, 8576.
83. Treisman, R. (1986) Identification of a protein-binding site that mediates transcriptional response of the *c-fos* gene to serum factors. *Cell*, **46**, 567.
84. Gauthier-Rouviere, C., Basset, M., Blanchard, J.-M., Cavadore, J.-C., Fernandez, A., and Lamb, N. J. C. (1991) Casein kinase II induces c-fos expression via the serum response element pathway and p67/SRF phosphorylation in living fibroblasts. *EMBO J.*, **10**, 2921.
85. Graham, R. and Gilman, M. (1991) Distinct protein targets for signals acting at the *c-fos* serum response element. *Science*, **251**, 189.

86. Shaw, P. E., Schroter, H., and Nordheim, A. (1989) The ability of a ternary complex to form over the serum response element correlates with serum inducibility of the human *c-fos* promoter. *Cell*, **56**, 563.
87. Dalton, S. and Treisman, R. (1992) Characterisation of SAP-1, a protein recruited by Serum Response Factor to the c-fos serum response element. *Cell*, **68**, 597.
88. Rao, V. N., Huebner, K., Isobe, M., Ar-Rushdi, A., Croce, C. M., and Reddy, E. S. (1989) *elk*, tissue-specific *ets*-related genes on chromosomes X and 14 near translocation breakpoints. *Science*, **244**, 66.
89. Hipskind, R. A., Rao, V. N., Muller, C. G., Reddy, E. S., and Nordheim, A. (1991) Ets-related protein Elk-1 is homologous to the *c-fos* regulatory factor p62TCF. *Nature*, **354**, 531.
90. Marais, R., Wynne, J., and Treisman, R. (1993) The SRF accessory protein Elk-1 contains a growth factor-regulated transcriptional activation domain. *Cell*, **73**, 381.
91. Hill, C. S., Marais, R., John, S., Wynne, J., Dalton, S., and Treisman, R. (1993) Functional analysis of a growth factor-responsive transcription factor complex. *Cell*, **73**, 395.
92. Mohun, T., Garrett, N., and Treisman, R. (1987) *Xenopus* cytoskeletal actin and human *c-fos* gene promoters share a conserved protein-binding site. *EMBO J.*, **6**, 667.
93. Treisman, R. H. (1985) Transient accumulation of c-fos RNA following serum stimulation requires a conserved 5′ element and c-fos 3′ sequences. *Cell*, **42**, 889.
94. Treisman, R. (1987) Identification of a polypeptide that binds to the c-fos serum response element. *EMBO J.*, **6**, 2711.
95. Stewart, W. E. I. (1979) *The interferon system*. Springer-Verlag, NewYork.
96. Revel, M. and Chebath, J. (1986) Interferon activated genes. *Trends Biochem. Sci.*, **11**, 166.
97. Friedman, R. L., Manly, S. P., McMahon, M., Kerr, I. M., and Stark, G. R. (1984) Transcriptional and posttranscriptional regulation of interferon-induced gene expression in human cells. *Cell*, **38**, 745.
98. Levy, D. E., Kessler, D. S., Pine, R., Reich, N., and Darnell, J. E., jun. (1988) Interferon-induced nuclear factors that bind a shared promoter element correlate with positive and negative transcriptional control. *Genes Dev.*, **2**, 383.
99. Lew, D. J., Decker, T., Strehlow, I., and Darnell, J. E., jun. (1991) Overlapping elements in the guanylate-binding protein gene promoter mediate transcriptional induction by alpha and gamma interferons. *Mol. Cell. Biol.*, **11**, 182.
100. Decker, T., Lew, D. J., Cheng, Y. S., Levy, D. E., and Darnell, J. E., jun. (1989) Interactions of alpha- and gamma-interferon in the transcriptional regulation of the gene encoding a guanylate-binding protein. *EMBO J.*, **8**, 2009.
101. Dale, T. C., Imam, A. M. A., Kerr, I. M., and Stark, G. R. (1989) Rapid activation by interferon α of a latent DNA-binding protein present in the cytoplasm of untreated cells. *Proc. Natl Acad. Sci. USA*, **86**, 1203.
102. Levy, D. E., Kessler, D. S., Pine, R., and Darnell, J. E., jun. (1989) Cytoplasmic activation of ISGF3, the positive regulator of interferon α stimulated transcription, reconstituted in vitro. *Genes Dev.*, **3**, 1362.
103. Fu, X. Y., Kessler, D., Veals, S. A., Levy, D. E., and Darnell, J. E., jun. (1990) ISGF3, the transcriptional activator induced by IFNα, consists of multiple interacting polypeptide chains. *Proc. Natl Acad. Sci. USA*, **87**, 8555.
104. Fu, X. Y., Schindler, C., Improta, T., Aebersold, R., and Darnell, J. E., jun. (1992) The proteins of ISGF-3, the interferon α-induced transcriptional activator, define a gene family involved in signal transduction. *Proc. Natl Acad. Sci. USA*, **89**, 7840.
105. Schindler, C., Fu, X.-Y., Improta, T., Aebersold, R., and Darnell, J. E., jun. (1992) Pro-

teins of transcripton factor ISGF-3: one gene encodes the 91- and 84-kDa proteins that are activated by interferon α. *Proc. Natl Acad. Sci. USA*, **89**, 7836.

106. Schindler, C., Shuai, K., Prezioso, V. R., and Darnell, J. E. (1992) Interferon-dependent tyrosine phosphorylation of a latent cytoplasmic transcription factor. *Science*, **257**, 809.
107. Veals, S. A., Schindler, C., Leonard, D., Fu, X.-Y., Aebersold, R., Darnell, J. E., jun., and Levy, D. E. (1992) Subunit of an alpha-interferon-responsive transcription factor is related to IRF and Myb families of DNA-binding proteins. *Mol. Cell. Biol.*, **12**, 3315.
108. Pine, R., Decker, T., Kessler., D. S., Levy, D. E., and Darnell, J. E., jun. (1990) Purification and cloning of interferon-stimulated gene factor 2 (ISGF2); ISGF2 (IRF-1) can bind to the promoters of both beta-interferon and interferon-stimulated genes but is not a primary transcriptional activator of either. *Mol. Cell. Biol.*, **10**, 2448.
109. Decker, T., Lew, D. J., Mirkovitch, J., and Darnell, J. E., jun. (1991) Cytoplasmic activation of GAF, an IFN-gamma-regulated DNA-binding factor. *EMBO J.*, **10**, 927.
110. Shuai, K., Schindler, C., Prezioso, V. R., and Darnell, J. E., jun. (1992) Activation of transcription by IFN-γ: tyrosine phosphorylation of a 91-kDa DNA binding protein. *Science*, **258**, 1808.
111. Muller, M., Laxton, C., Briscoe, J., Schindler, C., Improta, T., Darnell, J. E., jun., Stark, G. E., and Kerr, I. M. (1993) Complementation of a mutant cell line: central role of the 91 kDa polypeptide of ISGF3 in the interferon-α and -γ signal transduction pathways. *EMBO J.*, **12**, 4221.
112. Velazquez, L., Fellous, M., Stark, G. R., and Pellegrini, S. (1992) A protein tyrosine kinase in the interferon α/β signaling pathway. *Cell*, **70**, 313.
113. Muller, M., Briscoe, J., Laxton, C., Guschin, D., Ziemiecki, A., Silvennoinen, O., Harpur, A. G., Barbieri, G., Witthuhn, B. A., Schindler, C., Pellegrini, S., Wilks, A. F., Ihle, J. N., Stark, G. R., and Kerr, I. M. (1993) The protein tyrosine kinase JAK1 complements defects in interferon-α/-β and -γ signal transduction. *Nature*, **366**, 129.
114. Watling, D., Guschin, D., Muller, M., Silvennoinen, O., Witthuhn, B. A., Quelle, F. W., Rogers, N. C., Schindler, C., Stark, G. R., Ihle, J. N., and Kerr, I. M. (1993) Complementation by the protein tyrosine kinase JAK2 of a mutant cell line defective in the interferon-γ signal transduction pathway. *Nature*, **366**, 166.
115. Shuai, K., Horvath, C. M., Tsai Huang, L. H., Qureshi, S. A., Cowburn, D., and Darnell, J. E., jun., (1994) Interferon activation of the transcription factor Stat91 involves dimerization through SH2-phoshotyrosyl peptide interactions. *Cell*, **76**, 821.
116. Darnell, J. E., Kerr, I. M., and Stark, G. R. (1994) JAK–STAT pathways and transcriptional activation in response to IFNs and other extracellular signalling proteins. *Science*, **264**, 1415.
117. Zhong, Z., Wen, Z. L., and Darnell, J. E. (1994) Stat3 - a stat family member activated by tyrosine phosphorylation in response to epidermal growth-factor and interleukin-6. *Science*, **264**, 95.
118. Pine, R., Canova, A., and Schindler, C. (1994) Tyrosine-phosphorylated p91 binds to a single-element in the ISGF2/IRF-1 promoter to mediate induction by IFN-alpha and IFN-gamma, and is likely to autoregulate the p91 gene. *EMBO J.*, **13**, 158.
119. Igarashi, K., David, M., Feldman, G. M., Larner, A. C., and Finbloom, D. S. (1993) In vitro activation by interferon-gamma of a DNA-binding factor that recognizes the gamma response region of the FCγRI promoter requires a membrane-associated tyrosine kinase and is mimicked by vanadate. *J. Immunol.*, **150**,
120. Harroch, S., Revel, M., and Chebath, J. (1994) Interleukin-6 signaling via four transcription factors binding palindromic enhancers of different genes. *J. Biol. Chem.*, **269**, 26191.

121. Reid, L. E., Brassnet, A. H., Gilbert, C. S., Porter, A. C., Gewert, D. R., Stark, G. R., and Kerr, I. M. (1989) A single DNA response element can confer inducibility by both α- and γ-interferons. *Proc. Natl Acad. Sci. USA*, **86**, 840.
122. Ihle, J., Witthuhn, B. A., Quelle, F. W., Yamamoto, K., Thierfelder, W. E., Kreider, B., and Silvennoinen, O. (1994) Signalling by the cytokine receptor superfamily: JAKs and STATs. *Trends Biochem. Sci.*, **19**, 222.
123. Briscoe, J., Guschin, D., and Muller, M. (1994) Just another signalling pathway. *Curr. Biol.*, **4**, 1033.
124. Ihle, J. N. and Kerr, I. M. (1995) Jaks and Stats in signaling by the cytokine receptor superfamily. *Trends Genet.*, **11**, 69.
125. Fu, X.-Y. and Zhang, J.-J. (1993) Transcription factor p91 interacts with the epidermal growth factor receptor and mediates activation of the c-*fos* gene promoter. *Cell*, **74**, 1135.
126. Shuai, K., Ziemiecki, A., Wilks, A. F., Harpur, A. G., Sadowski, H. B., Gilman, M. Z., and Darnell, jun., J. E. (1993) Polypeptide signalling to the nucleus through tyrosine phosphorylation of Jak and Stat proteins. *Nature*, **366**, 580.
127. Johnston, J. A., Kawamura, M., Kirken, R.A., Chen, Y.-Q, Blake, T.B., Shibuya, K., Ortaldo, J.R., McVicar, D.W., and O'Shea, J.J.,. (1994) Phosphorylation and activation of the JAK3 Janus kinase in response to interleukin-2. *Nature*, **370**, 151.
128. Wagner, B. J., Hayes, T. E., Hoban, C. J., and Cochran, B. H. (1990) The SIF binding element confers sis/PDGF inducibility onto the c-fos promoter. *EMBO J.*, **9**, 4477.
129. Sadowski, H. B., Shuai, K., Darnell, J. E., and Gilman, M. Z. (1993) A common nuclear signal-transduction pathway activated by growth-factor and cytokine receptors. *Science*, **261**, 1739.
130. Zhong, Z., Zilong, W., and Darnell, J. E. jun. (1994) Stat3: a new family member activated by tyrosine phosphorylation in response to epidermal growth factor and interleukin-6. *Science*, **264**, 95.
131. Silvennoinen, O., Schindler, C., Schlessinger, J., and Levy, D. E. (1993) Ras-independent growth-factor signaling by transcription factor tyrosine phosphorylation. *Science*, **261**, 1736.
132. DeMaeyer, E. and DeMaeyer-Guignard, J. (1988) *Interferons and other regulatory cytokines.* John Wiley and Sons, Chichester.
133. Maniatis, T. (1988) Mechanisms of human beta-interferon gene regulation. *Harvey Lect.*, **82**, 71.
134. Goodbourn, S. (1990) The regulation of β-interferon gene expression. *Sem. Cancer Biol.*, **1**, 89.
135. Fan, C.-M. and Maniatis, T. (1989) Two different virus-inducible elements are required for human β-interferon gene regulation. *EMBO J.*, **8**, 101.
136. Visvanathan, K. V. and Goodbourn, S. (1989) Double-stranded RNA activates binding of NF-κB to an inducible element in the human β-interferon promoter. *EMBO J.*, **8**, 1129.
137. Macdonald, N. J., Kuhl, D., Maguire, D., Naf, D., Gallant, P., Goswamy, A., Hug, H., Bueler, H., Chaturvedi, M., delaFuente, H., Ruffner, H., Meyer, F., and Weissmann, C. (1990) Different pathways mediate virus inducibility of the human IFN-α1 and IFN-β genes. *Cell*, **60**, 767.
138. Whiteside, S. T., Visvanathan, K. V., and Goodbourn, S. (1992) Identification of novel factors that bind to the PRD I region of the human β-interferon promoter. *Nucl. Acids Res.*, **20**, 1531.

139. Lacoste, J., D'Addario, M., Roulston, A., Wainberg, M. A., and Hiscott, J. (1990) Cell-specific differences in activation of NF-kappa B regulatory elements of human immunodeficiency virus and beta interferon promoters by tumor necrosis factor. *J. Virol.*, **64**, 4726.
140. Ringold, G. M., Dieckmann, B., Vannice, J. L., Trahey, M., and McCormick, F. (1984) Inhibition of protein synthesis stimulates the transcription of human beta-interferon genes in Chinese hamster ovary cells. *Proc. Natl Acad. Sci. USA*, **81**, 3964.
141. Enoch, T., Zinn, K., and Maniatis, T. (1986) Activation of the human β-interferon gene requires an interferon-inducible factor. *Mol. Cell. Biol.*, **6**, 801.
142. Zinn, K., DiMaio, D., and Maniatis, T. (1983) Identification of two distinct regulatory regions adjacent to the human β–interferon gene. *Cell*, **34**, 865.
143. Goodbourn, S., Burstein, H., and Maniatis, T. (1986) The human β-interferon gene enhancer is under negative control. *Cell*, 601.
144. Goodbourn, S. and Maniatis, T. (1988) Overlapping positive and negative regulatory domains of the human β-interferon gene. *Proc. Natl Acad. Sci. USA*, **85**, 1447.
145. Kuhl, D., de, l. F. J., Chaturvedi, M., Parimoo, S., Ryals, J., Meyer, F., and Weissmann, C. (1987) Reversible silencing of enhancers by sequences derived from the human IFN-alpha promoter. *Cell*, **50**, 1057.
146. King, P. and Goodbourn, S. (1994) The β-interferon promoter responds to priming through multiple independent regulatory elements. *J. Biol. Chem.*, **269**, 30609.
147. Jacobsen, H., Mestan, J., Mittnacht, S., and Dieffenbach, C. W. (1989) Beta interferon subtype 1 induction by tumor necrosis factor. *Mol. Cell. Biol.*, **9**, 3037.
148. Lenardo, M. J., Fan, C.-M., Maniatis, T., and Baltimore, D. (1989) The involvement of NF-κB in β-interferon gene regulation reveals its role as a widely inducible mediator of signal transduction. *Cell*, **57**, 287.
149. Fujita, T., Miyamoto, M., Kimura, Y., Hammer, J., and Taniguchi, T. (1989) Involvement of a cis-element that binds an H2TF-1/NF kappa B like factor(s) in the virus-induced interferon-beta gene expression. *Nucl. Acids Res.*, **17**, 3335.
150. Du, W. and Maniatis, T. (1992) An ATF/CREB binding site protein is required for virus induction of the human interferon β gene. *Proc. Natl Acad. Sci. USA*, **89**, 2150.
151. Du, W., Thanos, D., and Maniatis, T. (1993) Mechanisms of transcriptional synergism between distinct virus-inducible enhancer elements. *Cell*, **74.**, 887.
152. Miyamoto, M., Fujita, T., Kimura, Y., Maruyama, M., Harada, H., Sudo, Y., Miyata, T., and Taniguchi, T. (1988) Regulated expression of a gene encoding a nuclear factor, IRF-1, that specifically binds to IFN-β gene regulatory elements. *Cell*, **54**, 903.
153. Fujita, T., Kimura, Y., Miyamoto, M., Barsoumian, E. L., and Taniguchi, T. (1989) Induction of endogenous IFN-α and IFN-β genes by a regulatory transcription factor, IRF-1. *Nature*, **337**, 270.
154. Harada, H., Fujita, T., Miyamoto, M., Kimura, Y., Maruyama, M., Furia, A., Miyata, T., and Taniguchi, T. (1989) Structurally similar but functionally distinct factors, IRF-1 and IRF-2, bind to the same regulatory elements of IFN and IFN-inducible genes. *Cell*, **58**, 729.
155. Harada, H., Willison, K., Sakakibara, J., Miyamoto, M., Fujita, T., and Taniguchi, T. (1990) Absence of the type I IFN system in EC cells: transcriptional activator (IRF-1) and repressor (IRF-2) genes are developmentally regulated. *Cell*, **63**, 303.
156. Watanabe, N., Sakakibara, J., Hovanessian, A. G., Taniguchi, T., and Fujita, T. (1991) Activation of IFN-β element by IRF-1 requires a post-translational event in addition to IRF-1 synthesis. *Nucl. Acids Res.*, **19**, 4421.

157. Leblanc, J. F. and Hiscott, J. (1992) Differential response of human interferon-beta promoter elements to trans-activation by HSV VP16 and IRF-1. *Virology*, **186**, 760.
158. Matsuyama, T., Kimura, T., Kitagawa, M., Pfeffer, K., Kawakami, T., Watanabe, N., Kundig, T., Amakawa, R., Kishihara, K., Wakeham, A., Potter, J., Furlonger, C. L., Narendran, A., Suzuki, H., Ohashi, P. S., Paige, C. J., Taniguchi, T., and Mak, T. W. (1993) Targeted disruption of IRF-1 or IRF-2 results in abnormal type I IFN gene induction and aberrant lymphocyte development. *Cell*, **75**, 83.
159. Reis, L. F. L., Harada, H., Wolchok, J. D., Taniguchi, T., and Vilcek, J. (1992) Critical role of a common transcription factor, IRF-1, in the regulation of IFN-β and IFN-inducible genes. *EMBO J.*, **11**, 185.
160. Ruffner, H., Reis, L. F. L., Näf, D., and Weissmann, C. (1993) Induction of type I interferon genes and interferon-inducible genes in embryonal stem cells devoid of interferon regulatory factor 1. *Proc. Natl Acad. Sci. USA*, **90**, 11503.
161. Ellis, M. J. C. and Goodbourn, S. (1994) NF-κB-independent activation of ß-interferon expression in mouse F9 embryonal carcinoma cells. *Nucl. Acids Res.*, **22**, 4489.
162. Maran, A., Maitra, R. K., Kumar, A., Dong, B., Xiao, W., Li, G., Williams, B. R. G., Torrence, P. F., and Silverman, R. H. (1994) Blockage of NF-κB signaling by selective ablation of an mRNA target by 2-5A antisense chimeras. *Science*, **265**, 789.
163. Fujita, T., Reis, L. F., Watanabe, N., Kimura, Y., Taniguchi, T., and Vilcek, J. (1989) Induction of the transcription factor IRF-1 and interferon-beta mRNAs by cytokines and activators of second-messenger pathways. *Proc. Natl Acad. Sci. USA*, **86**, 9936.
164. Thanos, D. and Maniatis, T. (1995) NF-κB: a lesson in family values. *Cell*, **80**, 529.
165. Thanos, D. and Maniatis, T. (1992) The high mobility group protein HMGI(Y) is required for NF-κB-dependent virus induction of the human IFN-β gene. *Cell*, **71**, 777.
166. Ghosh, G., van Duyne, G., Ghosh, S., and Sigler, P. B. (1995) Structure of NF-κB p50 homodimers bound to a κB site. *Nature*, **373**, 303.
167. Muller, C. W., Rey, F. A., Sodeoka, M., Verdine, G. L., and Harrison, S. C. (1995) Structure of the NF-κB p50 homodimer bound to DNA. *Nature*, **373**, 311.
168. Baltimore, D. and Beg, A. A. (1995) A butterfly flutters by. *Nature*, **373**, 287.
169. Nolan, G. P. (1994) NF-AT–AP-1 and Rel-bZIP: hybrid vigour and binding under the influence. *Cell*, **77**, 795.
170. Lewis, H., Kaszubska, W., DeLamarter, J., and Whelan, J. (1994) Cooperativity between two NF-κB complexes, mediated by high-mobility-group protein I(Y), is essential for cytokine-induced expression of the E-selectin promoter. *Mol. Cell. Biol.*, **14**, 5701.
171. Kaszubska, W., Hooft van Huijsduijnen, R., Ghersa, P., DeRaemy-Schenk, A.-M., Chen, B. P. C., Hai, T., DeLamarter, J. F., and Whelan, J. (1993) Cyclic AMP-independent ATF family members interact with NF-κB and function in the activation of the E-selectin promoter in response to cytokines. *Mol. Cell. Biol.*, **13.**, 7180.
172. Whitley, M. Z., Thanos, D., Read, M. A., Maniatis, T., and Collins, T. (1994) A striking similarity in the organization of the E-selectin and beta-interferon gene promoters. *Mol. Cell. Biol.*, **14**, 6464.
173. Fashena, S. J., Reeves, R., and Ruddle, N. H. (1992) A poly(dA-dT) upstream activating sequence binds high-mobility group I protein and contributes to lymphotoxin (tumor necrosis factor-β) gene regulation. *Mol. Cell. Biol.*, **12**, 894.
174. John, S., Reeves, R. B., Lin, J.-X., Child, R., Leiden, J. M., Thompson, C. B., and Leonard, W. J. (1995) Regulation of cell-type-specific interleukin-2 receptor α chain gene expression: potential role of physical interactions between Elf-1, HMGI(Y), and NF-κB family proteins. *Mol. Cell. Biol.*, **15**, 1786.

175. Giese, K., Cox, J., and Grosschedl, R. (1992) The HMG domain of lymphoid enhancer factor 1 bends DNA and facilitates assembly of functional nucleoprotein structures. *Cell*, **69**, 185.
176. Carlsson, P., Waterman, M. L., and Jones, K. A. (1993) The hLEF/TCF-1a HMG protein contains a context-dependent transcriptional activation domain that induces the TCRα enhancer in T cells. *Genes Dev.*, **7**, 2418.
177. Giese, K. and Grosschedl, R. (1993) LEF-1 contains an activation domain that stimulates transcription only in a specific context of factor-binding sites. *EMBO J.*, **12**, 4667.
178. Grosschedl, R., Giese, K., and Pagel, J. (1994) HMG domain proteins: architectural elements in the assembly of nucleoprotein structures. *Trends Genet.*, **10**, 94.
179. Goodbourn, S. (1994) Transcriptional regulation in activated T cells. *Curr. Biol.*, **4**, 930.
180. Rao, A. (1994) NF-ATp: a transcription factor required for the co-ordinate induction of several cytokine genes. *Immunology Today*, **15**, 274.
181. Flanagan, W. M., Corthesy, B., Bram, R. J., and Crabtree, G. R. (1991) Nuclear association of a T cell transcription factor blocked by FK-506 and cyclosporin A. *Nature*, **352**, 803.
182. McKeon, F. (1991) When worlds collide: immunosuppressants meet protein phosphatases. *Cell*, **66**, 823.
183. O'Keefe, S. J., Tamura, J., Kincaid, R. L., Tocci, M. J., and O'Neill, E. A. (1992) FK-506 and CsA-sensitive activation of the interleukin-2 promoter by calcineurin. *Nature*, **357**, 692.
184. Clipstone, N. A. and Crabtree, G. R. (1992) Identification of calcineurin as a key signalling enzyme in T-lymphocyte activation. *Nature*, **357**, 692.
185. McCaffrey, P. G., Luo, C., Kerppola, T. K., Jain, J., Badalian, T. M., Ho, A. M., Burgeon, E., Lane, W. S., Lambert, J. N., Curran, T., Verdine, G. L., Rao, A., and Hogan, P. G. (1993) Isolation of the Cyclosporin-sensitive T cell transcription factor NFATp. *Science*, **262**, 750.
186. Northrop, J. P., Ho, S. N., Thomas, D. J., Chen, L., Timmerman, L. A., Nolan, G. P., Admon, A., and Crabtree, G. R. (1994) NF-AT components define a family of transcription factors targeted in T cell activation. *Nature*, **369**, 497.
187. Schmidt, A., Hennighausen, L., and Siebenlist, U. (1990) Inducible nuclear factor binding to the κB elements of the human immunodeficiency virus enhancer in T cells can be blocked by Cyclosporin A in a signal-dependent manner. *J. Virol.*, **64**, 4037.
188. Jain, J., McCaffrey, P. G., Valge-Archer, V. E., and Rao, A. (1992) Nuclear factor of activated T cells contains Fos and Jun. *Nature*, **356**, 801.
189. McCaffrey, P.G., Goldfeld, A.E., and Rao, A. (1994) The role of NFATp in cyclosporin A-sensitive tumor necrosis factor-α gene transcription. *J. Biol. Chem.* **269**, 30445.

6 The function of homeodomain proteins in *Drosophila* development

CORNELIUS T. GROSS and WILLIAM J. McGINNIS

1. Homeodomain proteins in *Drosophila* development

Homeodomain proteins were initially discovered as the products of the homeotic selector genes in the fruit fly *Drosophila melanogaster* (1, 2). These genes were known to be responsible for determining the identity of specific anterior–posterior regions of the developing fly and had been mapped to two closely linked regions, the Antennapedia-complex (ANT-C) and Bithorax-complex (BX-C), collectively known as the Homeotic-complex or HOM-C (3, 4, reviewed in 5). It was discovered that DNA from the coding region of the *Antennapedia* gene hybridized to a number of genomic fragments from neighbouring HOM-C regions (1). Upon closer analysis, each of the HOM-C genes was shown to share a highly homologous region of approximately 180 base pairs, coined the homeobox. Subsequently, related homeotic complexes have been found in many higher organisms, including the round worm *Caenorhabditis elegans* (reviewed in 5).

Since their initial discovery in the HOM-C, a great number of homeobox genes have been identified in *Drosophila* (6). Many of these have been found amongst genes previously identified in genetic screens for developmental mutants, including an extensive screen for patterning mutants of the larval cuticle (7). New homeobox genes have also been cloned by homology to known homeobox genes of both *Drosophila* and vertebrates.

In order to provide an impression of the level of the present understanding of homeodomain protein function in *Drosophila*, this chapter will focus on three types of homeodomain proteins that have been extensively studied at the molecular level. The first protein, bicoid, is part of one of the best characterized, simple regulatory hierarchies in *Drosophila* development. It is expressed in a concentration gradient along the anterior–posterior axis of the developing embryo, and is responsible for activating numerous genes in distinct anterior–posterior domains. The bicoid response elements of one of these target genes, *hunchback*, have been analysed in some detail, and represent one of the simplest homeodomain regulatory targets described so far. The second protein, fushi tarazu, functions primarily in the segmentation of the early embryo and has been shown to positively regulate expression of its own transcription unit. The fushi tarazu autoregulatory element is one of the best understood complex homeodomain targets. The third class of proteins to be discussed are the products of the homeotic selector genes. These proteins are responsible for distinct segmental identities in the developing fly, and are encoded by a highly related set of genes found in the Antennapedia and Bithorax complexes. Because the homeodomains of these proteins are so similar, they offer a particularly convenient experimental system for examining the basis of homeodomain regulatory specificity.

The challenge faced in studying the function of homeodomain proteins lies in trying to understand how such a highly conserved protein domain has nevertheless developed the ability to function as part of diverse developmental hierarchies. How has each protein acquired a distinct functional specificity? What sorts of interactions are mediated by the homeodomain and how do they provide this specificity?

2. Homeodomain proteins are transcription factors

Soon after the homeobox was discovered, it was recognized that the homeodomain shows faint similarity to the DNA-binding domains of the yeast transcription factors MAT**a**1 and MATα2, as well as to the DNA-binding domains of a family of prokaryotic transcription factors, including the λcI, 434, and trp repressors (8, 9). These proteins had been extensively studied and shown to regulate transcription by binding to specific sequences in the promoters of their targets, suggesting that the homeodomain might be a sequence-specific DNA-binding domain.

Indeed, this hypothesis was quickly confirmed when the engrailed homeodomain was shown to bind to specific DNA sequences using an immunoprecipitation assay (10). Quantitative binding studies show that homeodomain proteins binds to specific DNA sites with equilibrium dissociation constants between 10^{-9} M and 10^{-11} M, and to non-specific DNA with approximately 100-fold lower affinity (11, 12). The affinity of these proteins for non-specific DNA is unusually high, and suggests that large regions of DNA may be bound by homeodomain proteins in the embryo. Evidence to support this idea comes from UV-cross-linking experiments that directly measure DNA binding of homeodomain proteins to sequences in living embryos (13). The authors find that, when compared with the non-homeodomain transcription factor zeste, which binds detectably only to known regulatory elements in the *Ultrabithorax* gene, both the ftz and even-skipped proteins appear to bind highly promiscuously even to DNA sequences not thought to be regulatory targets of these proteins. This may reflect the binding of these proteins either to non-specific DNA, or to very small, and thus abundant, specific DNA sites.

A number of immunoprecipitation and DNase footprinting experiments have shown that different homeodomain proteins can bind the same sequences, with each protein showing only a slight preference for certain sites over others (14, 15). These results appear to be incompatible with the model, derived from our understanding of prokaryotic transcription factors, that functional specificity *in vivo* is encoded in the DNA-binding specificity. Subsequently, however, it was discovered that proteins differing at amino acid 50 in the third helix of the homeodomain do indeed show significant differences in DNA-binding specificity. Changing this residue from glutamine to lysine or serine can switch the DNA-binding specificity of the ftz protein to that of bicoid or paired (16, 17) suggesting that DNA-binding specificity may contribute to the regulatory specificity of these highly diverged classes of homeodomains in the embryo. However, it is still unclear what role is played by the small DNA-binding preferences reported for proteins with the same amino acid at position 50 (the homeotic selector proteins, for example, all have glutamine at this position).

A direct comparison of the structures for the Antennapedia, engrailed, and MATα2 homeodomains complexed with DNA, reveals a shared three-helix architecture in which the third helix makes extensive contacts in the major groove of DNA (18–20). Additional contacts are made in the minor groove of DNA by amino acids at the N-terminus of the homeodomain that seem to form a long unstructured arm extending around the side of the DNA helix. In the engrailed–DNA structure amino acids 3 to 7 form a beta strand-like structure with the side chains of amino acids 3 and 5 reaching into the minor groove, and those of amino acids 4 and 6 pointing away from the DNA. It is difficult to assess the role of amino acids before position 3 because they are poorly defined in both the NMR and X-ray crystallographic structures. At the C-terminal end of the homeodomain the third alpha helix extends to position 58, beyond which, again, both structures are poorly resolved. Residues conserved in even the most highly diverged homeodomains generally contribute to the hydrophobic core at the centre of the homeodomain that allows the first and second alpha helices to buttress the third helix against DNA. More closely related homeodomains, however, conserve a large number of residues that are not implicated in binding to DNA.

Extensive evidence also exists that homeodomain proteins can act as transcriptional regulators, both in cell culture and in cell-free systems (21–25). A recent study by Johnson and Krasnow (26) demonstrates that Ultrabithorax (Ubx) protein can activate, and even-skipped (eve) protein can repress, basal transcription from a fragment of the *Adh* promoter in a cell-free transcription assay. Regulation by both proteins is dependent on the presence of a homeodomain-binding sequence, and both proteins are ineffective if added after a basal transcription complex has been allowed to form, suggesting that they act to enhance or inhibit the formation of a functional basal transcription complex (see Chapter 1). Studies with fractionated nuclear extracts are now being undertaken to identify which transcription factors are required for regulation by these proteins and to determine precisely where they act in the assembly of the basal transcription complex (R. Austin and M. Biggin, unpublished data).

Taken together, these experiments establish the paradigm that homeodomain proteins act as transcription factors during development. At present, however, it remains unclear precisely how the homeodomain achieves this function. What are the co-factors postulated to be required for the function of a number of homeodomain proteins? Which surfaces are required for contacting these co-factors? What DNA sequences are bound by homeodomain proteins in the embryo? What is required of a regulatory element to respond to a homeodomain protein in the developing fly? These questions have begun to be addressed for a number of homeodomain proteins with sometimes surprising results.

3. bicoid protein

3.1 bicoid response elements

The bicoid protein is translated from messenger RNA trapped at the anterior tip of the early *Drosophila* embryo, and a concentration gradient of the protein is formed as it diffuses into nuclei located in more posterior regions of the still unicellular embryo. The gradient of bicoid protein is a principal determinant of the differential cellular identities formed in the anterior half of the embryo. There are presumably many genes that are transcriptionally activated in anterior nuclei in a manner dependent on bicoid protein. One such bicoid-regulated transcription unit, required for the identity of a large block of cells in the head and thorax, corresponds to the gap gene *hunchback*.

The *hunchback* gene is transcribed from both maternal and zygotic promoters. An early domain of zygotic expression in the anterior half of the blastoderm-stage embryo is controlled by the zygotic promoter (27). In embryos derived from *bicoid* mutant mothers, the anterior domain of hunchback transcription is missing, and the extent of the expression domain along the embryonic anterior–posterior axis varies in a manner that is directly correlated with the maternal *bicoid* gene and protein dosage. Important control sequences that are sufficient for the anterior hunchback expression domain were initially localized to a region spanning –300 to +400 relative to the *hunchback* zygotic promoter (28). Further studies defined an apparent bicoid response element (bicoid RE) to a region mapping between –318 and –55. This bicoid RE can activate either the *bicoid* or the *hsp70* basal promoters when attached to *lacZ* reporter genes, can work in either orientation relative to the promoter, and can be moved as much as 1 kb further upstream of a test promoter without much reduction in activity.

Using an immunoprecipitation assay, Driever and Nüsslein-Volhard (29) screened a large region of *hunchback* upstream and coding DNA for restriction fragments with affinity for *Escherichia coli*-produced bicoid protein. The fragment with the highest affinity in the immunoprecipitation assay mapped just upstream of the *hunchback* zygotic promoter from –298 to –55 relative to the transcription start site, and contains three moderate to high affinity bicoid protein binding sites as defined by DNase footprinting. These sites, along with a few other binding sites that mapped to fragments in further upstream DNA, allowed the definition of a bicoid protein consensus binding sequence (TCTAATCCC, or GGGATTAGA on the strand with which the ftz binding site is usually reported). Hanes and Brent (16) and Treisman *et al.* (17) have shown that the preferential binding of bicoid protein to this site over other TAAT binding sites is determined by the specific bases to the 3′ side of the TAAT, and residue 50 (lysine) of helix 3 of the bicoid homeodomain, which contacts these bases.

Many lines of evidence support the idea that both the number and affinity of bicoid binding sites within *hunchback* upstream *cis*-regulatory DNA are crucial for the *hunchback* anterior expression domain, and for the function of the apparent bicoid response element (bicoid RE) within the –318 to –55 region. The successive deletion of this bicoid RE, tested with *lacZ* reporter constructs in transgenic embryos, resulted in lower overall levels of expression and expression domains that were restricted to anterior embryonic regions that contained the highest levels of bicoid protein. A regulatory element that contained only a single high-affinity bicoid binding site (located in a test element of 200 base pairs that spanned the *hunchback* zygotic transcription start) was capable of reproducing a weak version of the *hunchback* anterior domain. However, the ability of a single or multimerized bicoid protein binding site to act as a bicoid RE appears to be dependent on the sequence context in which it is found. Struhl *et al.* (30) found no detectable anterior expression provided by another fragment of the bicoid RE spanning sequences from –298 to –216 that included one high affinity and two lower affinity bicoid binding sites. Another piece of evidence supporting the context dependence of bicoid binding sites is that the substitution of bicoid binding sites for fushi tarazu protein binding sites in a *fushi tarazu*-dependent enhancer does not generate an element that is regulated by wild-type bicoid protein in embryos (31).

This context dependence of bicoid binding sites, presumably due to the influence of other unknown transcription factors binding nearby, can apparently be partly overcome by simple multimerization of bicoid protein binding sites. Four copies of a 25-base pair sequence containing a single high-affinity bicoid binding site are capable of directing a *hunchback*-like anterior expression domain that is dependent on the endogenous function of bicoid protein. Recent work by Hanes *et al.* (32) has provided formal proof that this multimerized regulatory element is directly responding to bicoid protein. Hanes and co-workers changed the sequence of the homeodomain binding sites from bicoid high-affinity sites (TCTAATCCC) to Antp-type high-affinity sites (TCTAATTAG), which abolished the ability of the mutant element to respond to wild-type bicoid protein in embryos. The function of the mutant element could be restored by the introduction of a *bicoid* gene producing a mutant version of the bicoid protein in which amino acid 50 in helix 3 of the homeodomain was switched from lysine to an Antp-type glutamine residue.

These studies have also provided evidence that the context dependence of bicoid protein binding sites in *Drosophila* embryos may be due more to the proper spacing of the bicoid sites than to the influence of specific nearby sequences. Multimers of 11-bp sequences containing the high-affinity bicoid binding sites provided no detectable anterior domain of expression in early embryos, while multimers of 25-bp sequences were functional whether the sequence adjacent to the bicoid binding sites was the same as in a natural bicoid target element from the *hunchback* promoter, or whether the sequences adjacent to the bicoid binding sites were randomized. The exact spacing between sites also does not seem to be crucial, as a trimer of bicoid binding sites, with centre to centre distances of 11 and 14, was capable of providing an anterior cap of expression in a region of the embryo where the concentration of bicoid protein is high (33).

The size of the anterior expression domain that can be activated by bicoid protein—how far down the bicoid concentration gradient the protein still works to activate *hunchback* regulatory sequences—appears to be largely controlled by the affinity between bicoid protein and its binding sites. Regulatory elements that contain numerous low-affinity bicoid protein binding sites can direct abundant amounts of reporter gene expression, but only in regions at the anterior end of the embryo where high concentrations of bicoid protein exist (34). In addition, the threshold level of the bicoid response on *hunchback* regulatory elements apparently involves as yet poorly characterized co-operative interactions, as the on–off boundary for *hunchback* transcription (and for bicoid response elements with multiple bicoid binding sites) is much sharper than the gradient of bicoid protein concentration. Whether this is due to co-operative DNA-binding interactions of bicoid protein monomers, co-operative binding with other co-factors interacting with the bicoid RE, co-operative interactions with the basal transcription apparatus, or any combination of these, is as yet unknown.

3.2 bicoid protein

Sequence analysis of *bicoid* cDNAs indicates that it encodes two homeodomain-containing protein isoforms of 489 and 494 amino acids that differ only by the inclusion of five extra codons just upstream of the homedomain-encoding exon (35). Notable structural features of the proteins include a homeodomain with a lysine residue at the critical position 50 in helix 3, which has a profound effect on its sequence-specific binding (16, 17). The bicoid proteins also contain histidine-proline repeats near the N-terminus, a serine-threonine-rich region just downstream of the homeodomain, and a glutamine-rich region (also known as opa or M-repeats) in the C-terminal half of the protein farther downstream of the homeodomain.

Struhl *et al.* (30) explored the function of these domains and other regions of the bicoid protein by expressing wild-type and mutant versions of bicoid protein attached to lexA DNA-binding domains in yeast cells. The function of the bicoid–lexA fusion proteins was assayed by testing their ability to activate transcription through the –318 to –55 module of the bicoid RE of the *hunchback* gene. In yeast, the N-terminal region, including the His-Pro repeat, appeared to be dispensable for either DNA binding or transcriptional activation. Three different mutations in the homeodomain region (two mis-sense and one in-frame deletion) appear to abolish DNA-binding activity, as they fail to activate through the bicoid RE, but have some minimal activity on LEXA response elements in the same yeast cells. The homeodomain mutations also appear to diminish the transcriptional activation function of the proteins in yeast. Finally, mutations that progressively eliminate more of the protein from the carboxyl end, including the glutamine-rich and serine-threonine-rich regions, have progressively less ability to function as transcriptional activators in yeast.

Analogous studies on deleted versions of the bicoid protein by Driever *et al.* (36), in this case testing for function in both yeast and *Drosophila* embryos, also indicate that the N-terminal half of the protein including the homeodomain is required for DNA binding, while the C-terminal half is required for transcriptional activation. Driever and co-workers also showed that the C-terminal half could be functionally substituted by different heterologous acidic activation domains. Amazingly, in some cases these heterologous activation domains, when attached to amino acids 1–246 of bicoid protein, could completely rescue the morphological defects and lethality of *bicoid* mutant embryos.

The *hunchback* anterior expression domain begins as a cap covering the entire anterior half of the embryo, but at slighly later stages in still early embryos *hunchback* transcription is repressed in the most anterior regions of the embryo. Ronchi and co-workers (33) provided evidence indicating that this effect is directly controlled by the phosphorylation of bicoid protein in the most anterior regions of the embryo, rendering it no longer able to bind or activate bicoid REs. They showed that an anterior retraction of expression was also observed from a regulatory element consisting only of three bicoid protein binding sites with a few adjoining base pairs. This regulatory element did not exhibit anterior repression, nor did the bicoid protein exhibit its usual, more phosphorylated, forms in mutant embryos where the *torso* tyrosine kinase signalling cascade was not activated.

In summary, the bicoid protein appears capable of activating very simple regulatory elements in *Drosophila* embryos, consisting of little more than multiple bicoid binding sites with some constraints on the spacing of the sites. In the natural *bicoid*-dependent enhancer elements that exist in upstream *hunchback* sequences, multiple bicoid protein binding sites are found, and the affinities of those sites are important in setting the absolute boundaries of activation on the embryonic anterior–posterior axis in response to the bicoid protein gradient, but are likely to require as yet unidentified co-activators to achieve full activity of the bicoid REs. However, in contrast to some *Drosophila* homeodomain proteins that are expressed at later stages of development, no other binding sites or other proteins seem to be necessary to determine that the bicoid RE be regulated by bicoid homeodomain proteins and no others. In this respect, bicoid may be different from other *Drosophila* homeodomain proteins. It seems possible that the simplicity of the synthetic bicoid REs that do provide a close mimickry of the *hunchback* anterior expression domain may be due to the relatively small number of similar pattern-controlling transcription factors that are present and active in the very early *Drosophila* embryo. At this stage of development a simple cluster of homeodomain binding sites with a preference for bicoid protein would not have to discriminate between the scores or hundreds of other homeodomain proteins at various concentrations that will only later be expressed in embryonic cells.

4. fushi tarazu protein

4.1 fushi tarazu response element

At present, one of the best described embryonic DNA targets for a *Drosophila* HOM-C type homeodomain protein is an autoregulatory element that maps in *fushi tarazu* (*ftz*) upstream DNA. Although *ftz* primarily serves a pair-rule segmentation function, not a homeotic function, its close structural affinity with the HOM-C proteins suggests that any lessons learned about ftz protein-targeting specificity might usefully apply to the HOM-C class in general. The *ftz* gene is transcribed in seven stripes of cells in blastoderm embryos. The initiation of this striped pattern of transcription is under the control of the primary pair-rule genes such as *hairy*, *runt*, and *even-skipped*, but the amplification and refinement of the pattern is modulated by an autoregulatory circuit that involves the ftz protein acting on an upstream enhancer of its own transcription unit. A *ftz* autoregulatory element (AE) was originally localized to a 2.6-kb fragment that mapped between 6.1 and 3.5 kb upstream of the ftz transcription start (37). The most important sequences for the function of this element map in the proximal region of this fragment (38), and a 430-bp sequence in this region is sufficient to direct reporter gene transcription in *ftz*-dependent stripes (31).

Schier and Gehring (31) have used second-site suppression experiments to demonstrate that a modified 430-bp *ftz* AE is directly bound by ftz protein in embryos. They engineered their element to have five ftz protein binding sites as measured by *in vitro* footprints. When three of the five ftz binding sites in the modified AE were changed to bicoid homeodomain binding sites, reducing their individual affinities for ftz protein by ~40-fold and increasing their affinity for bicoid protein (12), the 430-bp AE gained the ability to respond to bicoid-like binding variants of the ftz protein, in which residue 50 in helix 3 of the homeodomain was changed from glutamine to lysine. Importantly, the AE with bicoid binding sites does not lose the ability to be activated by wild-type ftz protein in a manner that is partially dependent on the function of the substituted bicoid protein binding sites (31, 39). The substitution of the bicoid binding sites for ftz binding sites is not sufficient to place this element under the regulatory control of the wild-type bicoid protein in *Drosophila* embryos.

If all of the ftz binding sites are deleted or mutated, the embryonic function of the AE is abolished, so the element does require homeodomain binding sites for function. However, ftz binding sites are not sufficient. Schier and Gehring (40) did a systematic deletion analysis of the 430-bp AE and found that function is abolished in elements with a few ftz protein binding sites but with other crucial non-homeodomain binding sites deleted. None of the deletion mutants of the AE resulted in ectopic activation of the element in embryonic cells, so presumably the important non-homeodomain-binding sequence elements represent binding sites of co-factors that influence the specificity or activity of ftz protein on the AE. At least some of these presumed co-factor sites are not immediately adjacent to ftz protein binding sites in the AE. Some combination of the required co-activator protein activities is apparently restricted to a ftz-like pattern, as the ectopic expression of ftz protein under the control of a heatshock promoter is not capable of globally activating the function of the AE in embryonic cells.

The DNA sequence of the ftz 430-bp AE has been determined from two other *Drosophila* species that have been evolutionarily separated from *Drosophila melanogaster* for approximately 60 million years (40, 41). The location of important sequences as defined by the deletion mutagenesis of the *D. melanogaster* AE exhibits a rough correlation with the location of evolutionarily conserved blocks of regulatory sequence. Good matches to some of the evolutionarily conserved AE sequences that are likely to be binding sites for non-homeodomain co-factors, are found in intronic regulatory elements from the *Drosophila engrailed* gene. The *engrailed* intron is another likely target for ftz binding and regulation (42). Based on sequence comparisons, Schier and Gehring suggest that some of the evolutionarily conserved sequences might represent binding sites for the FTZF1 or tramtrack DNA-binding proteins (40).

Han *et al.* (43) have also studied the function of the same 430-bp AE, identifying sequence elements that bind to proteins in *Drosophila* embryonic nuclear extracts, and correlating those elements with sequences that are required for AE function as defined by deletion mutagenesis. One region that they found to be particularly important in the context of partially deleted elements with attenuated function includes a sequence called O6. The O6 sequence maps in a region found to be absolutely required by Schier and Gehring in similar assays (40). A 19-bp O6 oligonucleotide from the region forms gel-shift complexes that are likely to contain the FTZF1 and tramtrack proteins, since antiserum against either protein abolishes the presence of specific mobility-shift complexes. Besides those containing FTZF1 and tramtrack, other regions of the *ftz* AE generated at least eight more electrophoretically distinct complexes, suggesting that the regulatory inputs necessary for normal *ftz* AE function may be numerous. In these experiments, the embryonic nuclear extract was derived from 0–12-hour embryos, a considerable developmental period considering that the *ftz* AE is only active from about 2.5 to 3.5 hours; so, some or many of the mobility complexes might be meaningless for the regulation of *ftz* AE. Oddly, the only protein that is genetically proven to interact with the *ftz* AE, the ftz protein, has not yet been unambiguously identified as a contributor to any of the AE mobility-shift complexes when embryonic nuclear extract is used as the source of DNA-binding protein.

Results of the *ftz* AE research support the idea that in a functional ftz target element, multiple ftz binding sites exist that are important for fine tuning the specificity of the enhancer, but virtually any ATTA-containing homeodomain binding site is sufficient to support a ftz-specific response, which requires as yet poorly characterized co-factors. Presumably, differential protein–protein interactions between these co-factors and ftz protein play a crucial role in allowing ftz protein to select *ftz* AE as a target.

4.2 ftz protein

The *in vitro* binding specificity of the ftz homeodomain has been systematically studied by Florence *et al.* (44). They studied the binding properties of two different *E.coli*-produced ftz homeodomain-containing polypeptides (one including a few additional amino acid residues beyond the homeodomain (18 N-terminal and 11 C-terminal)). No significant differences were detected in the *in vitro* binding activity of the two polypeptides. They found that the ftz homeodomain exists as a monomer in solution as determined by gel filtration, and binds as a monomer to multimerized ftz binding sites arranged with a variety of spacings and orientations. No evidence for co-operative binding was observed. Using information from oligonucleotide affinity-selection experiments, and substitution of every position within an 11-bp region to every possible base, a consensus optimal affinity site was derived, AAG-CAATTAAG. Changes in any position of this optimal binding site reduced the *in vitro* affinity of the site for the ftz homeodomain, though the effects of many single substitutions were very modest (less than two-fold reductions in affinity), particularly for substitutions outside the ATTA core sequence. The monomer equilibrium dissociation constant for ftz homeodomain binding to this sequence in the context of a 59-bp fragment was 2.5×10^{-11}. The equilibrium dissociation constant for non-specific binding sites under the same conditions was 3×10^{-9}. This 100-fold difference between specific and non-specific affinity is relatively small compared with some prokaryotic DNA-binding proteins. No perfect matches to the optimal ftz binding sequence are found within the 430-bp ftz AE, but one natural AE site has a GCAATTA core, which includes all the nucleotides thought to make direct contacts with the ftz homeodomain. It is not evident that this site is any more important than the others in determining the *ftz*-dependent response of the 430-bp AE. Amplifying this point, ftz protein binding sites in the *ftz* AE have been changed to generic homeodomain binding sites (BS2-type) without significantly influencing the strength of the element or the specificity of its response to ftz protein regulation (39).

The N-terminus of the ftz homeodomain is crucial to its binding activity, as a deletion of six amino acid residues, including both arginines 3 and 5 that make minor groove contacts in the engrailed homeodomain–DNA co-crystal (19), reduced the affinity of the ftz homeodomain for DNA by more than 100-fold (12).

The *in vivo* studies of Morrissey *et al.* (45) and Furukubo-Tokunaga *et al.* (46) indicate that the ftz homeodomain sequence can tolerate significant changes without losing either activity or a significant amount of its functional specificity. Morrissey and co-workers substituted the homeobox from the *paired* gene (encoding amino acid residues 3–61 of the homeodomain) for that of *ftz* in a heatshock promoter–*ftz* cDNA expression construct. Transgenic embryos expressing the hsftz(prdHD) chimeric protein had severe segmental defects similar to those caused by ectopic expression of wild-type ftz protein under heatshock promoter control, due at least in part to the ability of the chimeric protein to activate the expression of the endogenous *ftz* transciption unit in a stripe of cells where it is normally repressed. Presumably this is mediated by the *ftz* AE in the endogenous gene. Similarly the substitution of Sex combs reduced-like homeodomain amino acid residues in place of ftz residues in the central region of the homeodomain (including five amino acid changes in the C-terminal region of helix 1, helix 2, and the N-terminal region of helix 3) had little effect on the ability of the ftz protein to functionally rescue the defects in *ftz* mutant embryos (46). More extensive substitutions in this same region, caused by introducing 19 amino acid substitutions from the msh protein, abolished the embryonic rescuing ability of the mutagenized homeodomain.

Interestingly, the results of Fitzpatrick *et al.* (47) indicate that the ftz protein has some regulatory activity in *Drosophila* embryos that is independent of its DNA-binding function. These authors generated heatshock ftz constructs in which the central region of the homeodomain was deleted, and found that these mutant ftz proteins were still capable of ectopically activating the expression of the endogenous ftz transcription unit, perhaps via the *ftz* AE. These authors suggest that protein–protein interactions, perhaps mediated by the shreds of homeodomain sequence that remain in the deleted protein, are sufficient to activate the *ftz* AE in these cells. Such interactions might be sufficient to recruit the deleted version of ftz protein into an activation complex that functions on the AE, or might involve the deleted protein acting to exert a repressive effect on a negative regulator of the *ftz* AE in some cells. Deletions of the N- or C-terminal regions of the ftz protein abolished this regulatory effect, consistent with the idea that these regions contain transcriptional activation domains required for ftz protein function.

5. The homeotic selector proteins

The homeotic selector genes form a unique class of homeobox genes because they encode highly related homeodomains. Pairwise comparison reveals between 60 and 95% identity among these homeodomains, with Antennapedia and abdominal-A, for example, differing by only three amino acids. In addition, these proteins share a four-amino acid motif, YPWM, found 19 residues N-terminal of the homeodomain.

The homeotic selector proteins are also unique because they reside in two closely linked clusters on the third chromosome, the Antennapedia complex (ANT-C) and the Bithorax complex (BX-C), collectively known as the Homeotic complex, or HOM-C. In 1978 Lewis postulated that these genes diverged from a common ancestor in order to supply the segmental specialization required for the advent of more complex organisms (3). This model is supported by the presence of highly similar homeoboxes in these genes, and by the survival of various related forms of the HOM-C in organisms today. For example, the red flower beetle, *Tribolium castaneum*, has a single contiguous HOM-C locus instead of the divided ANT-C and BX-C of *Drosophila* (48). The round worm, *Caenorhabditis elegans*, has only four genes in its HOM-C, structurally related to the *Drosophila labial*, *Deformed*, *Antennapedia*, and *Abdominal-B* genes (49). Because of their close evolutionary relation, these proteins offer a useful system for deciphering how homeodomains acquired distinct regulatory activities.

5.1 Functional dissection of homeotic selector proteins

An important molecular understanding of the function of the HOM-C genes came from studies of dominant mutations of the *Antennapedia* (*Antp*) gene, which cause legs instead of antennae to grow from the head. Antibody staining for the Antp protein showed that improper expression of the protein in the developing head redirected the fate of the antennae towards the thoracic fate normally determined by Antp (50). This effect could be mimicked by ectopically expressing the Antp protein under control of the heatshock-inducible *hsp70* promoter during development, demonstrating that the Antp protein itself was able to redirect the fate of tissues (51). Since then, this ectopic expression approach has been widely used to examine the function and specificity of HOM-C proteins.

Further analysis of the Antp protein by this method has sought to identify the regions of the protein required for its function (52). The Antp protein is composed of a long N-terminal region with the homeodomain located 22 amino acids from the C-terminus. Deletion of most of the N-terminal region has only a minor effect on the function of the protein, at least as regards its ability to transform anterior epidermal cells towards more posterior (thoracic) fates. Although sequences very close to the N-terminus are required for function of the protein, the function of this region does not seem to be specific to Antp, as the analogous region from the Sex combs reduced (Scr) protein is partially able to supply this function. As expected, removal of sequences containing the homeodomain abolishes its function.

In order to identify regions of the Antp protein responsible for regulatory specificity, hybrid proteins were made by swapping regions of the Antp and Scr proteins (52). Replacing the Antp homeodomain and flanking sequences with analogous regions from the Scr protein results in a protein now able to effect transformations like the full-length parental Scr protein. Further experiments show that sequences C-terminal to the homeodomain are required for the full potency of this hybrid but are not required for its functional specificity, suggesting that the homeodomain and the 79 amino acids N-terminal to it are sufficient to impart Scr specificity on an otherwise Antp protein. These results have been extended more recently to show that changes in the homeodomain are themselves sufficient to switch from Antp to Scr function (53, 54). Similar studies with mouse homologues of the Antp and Scr proteins, Hoxb6 and Hoxa5, also suggest that homeodomain sequences are sufficient to specify Antp and Scr function in the embryo (55, 56). These results are striking because the Antp and Scr homeodomains differ by only four amino acids (see Fig. 1). Superposition of these residues on the engrailed–DNA crystal structure shows them to be part of the long N-terminal arm that makes contacts in the minor groove of DNA. It seems likely that at least some of these residues are not in a position to

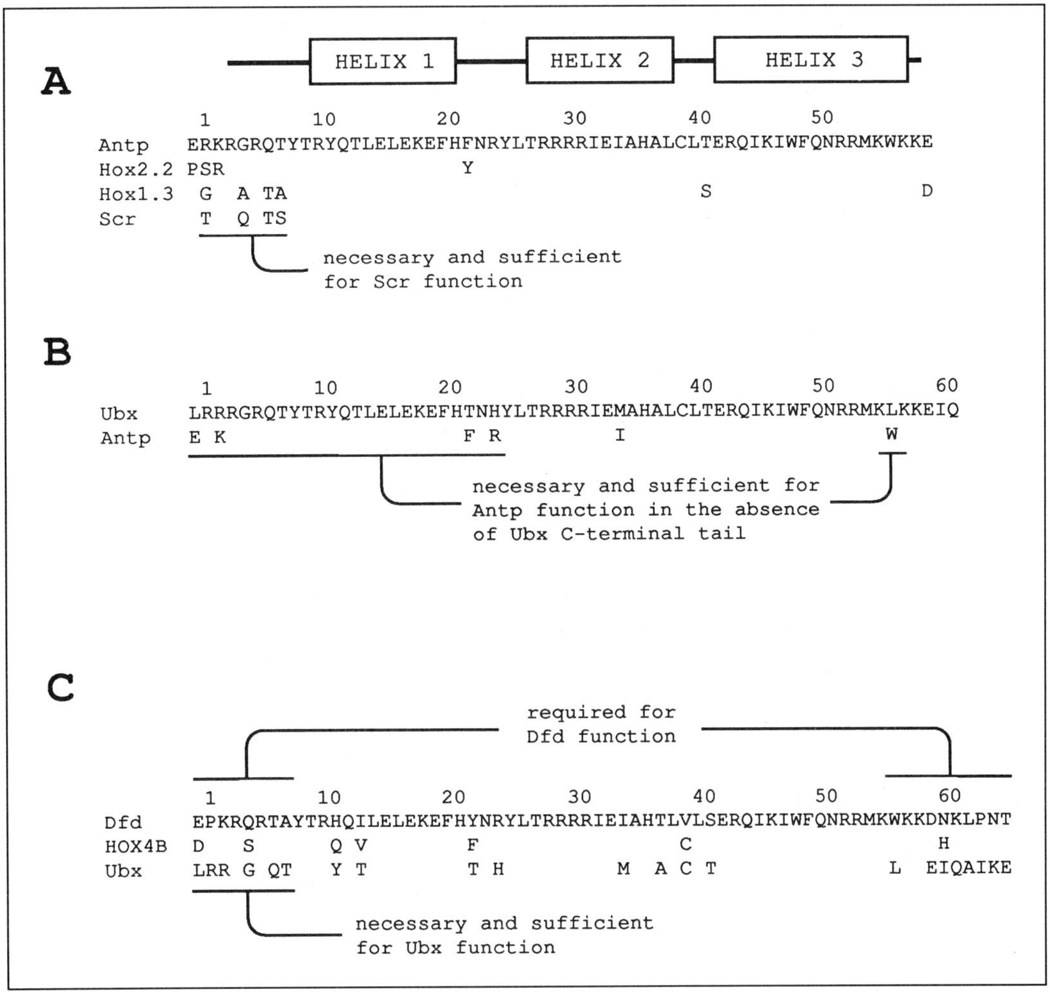

Fig. 1 Summary of homeodomain swap experiments between homeotic selector proteins. The function of these proteins was studied by an ectopic expression assay using the heatshock promoter. Structural motifs of the engrailed homeodomain are depicted above the sequences (18), and important regions identified in these studies are indicated with bars. (A) Antp and Scr swaps (47–49) and the Antp homologue Hoxb6 (50) and the mouse Scr homologue Hoxa5 (51). (B) Ubx and Antp swaps (52, 53). (C) Dfd and Ubx swaps (54, 55) and the human Dfd homologue HOX4B (53).

make contact with DNA, leaving open the possibility that they may be important for interactions of the homeodomain with as yet unidentified co-factors.

A similar series of swaps have been made between the Ubx and Antp proteins (57, 58). As in the above case, substituting the Antp homeodomain and C-terminal amino acids for corresponding regions of the Ubx protein successfully switches from Ubx to Antp specificity. However, swapping only the homeodomain from Ubx to Antp results in a protein that retains Ubx specificity, suggesting that residues C-terminal to the Ubx homeodomain are able to suppress the effect of swapping amino acids within the homeodomain. Removing these C-terminal residues successfully allows the Antp homeodomain to redirect the specificity of the Ubx protein. This C-terminal region may be acting as an independent domain of the Ubx protein because it continues to function even when distanced from the homeodomain by the insertion of five alanine residues. Chan and Mann (58) went on to show that all but two of the seven amino acid differences between the Ubx and Antp homeodomains are required for the specific function of these proteins (see Fig. 1). Three of these amino acids, threonine 23, histidine 25, and leucine 57, correspond to residues well defined in the crystal structure that may form a contiguous interaction surface on the back of the homeodomain.

In one other case it has also been possible to map regulatory specificity to particular homeodomain residues. Ectopic expression of the Deformed (Dfd) protein under control of the heatshock promoter causes head structures to form in the thorax of the *Drosophila* larva. A similar but weaker effect is obtained using a human homologue of Dfd, HOX4B, which is nearly identical within the homeodomain and virtually unrelated outside (59, see Fig. 1). This experiment points to the critical role of the homeodomain in the functional specificity of the Dfd protein. Precisely substituting the Ubx homeodomain into Dfd causes it to induce ectopic thoracic structures instead of head structures (60). Genetic tests and antibody staining show that the Dfd protein functions by activating its own transcription unit ectopically and that the Dfd protein with the Ubx homeodomain acts by ectopically activating the *Antp* gene, a normal target of the endogenous Ubx protein. Thus, the homeodomain substitution has successfully changed regulatory activity from one target to another. Further sub-swaps of homeodomain residues between the two proteins identified six amino acids in the N-terminus of the homeodomain that are both necessary and sufficient to switch from a Dfd to a Dfd/Ubx function (61, see Fig. 1). Four of these changes are found at positions identical to those sufficient to swap Antp and Scr activity (53). In fact, at three of these positions Dfd and Antp are identical (see Fig. 1), indicating that these proteins may present similar interaction surfaces here. Swaps at the C-terminus of the homeodomain show that both the N-terminus and the C-terminus are necessary for the Dfd protein, including tryptophan 57, which Chan and Mann (58) have shown to be required for Antp function. Because some of these critical residues lie outside the structurally well-defined regions of the homeodomain, it is difficult at this point to guess at their role in the function of these homeodomains. Presumably these amino acids are involved in contacting DNA or interacting with as yet unidentified co-factors.

Additional homeodomain swap experiments with the Dfd and rough proteins indicate that, despite the critical role of the homeodomain, regions outside the homeodomain are required for functional specificity (62). The rough protein is required for differentiation of photoreceptor cells in the *Drosophila* compound eye. For these swap experiments, it was not necessary to use an ectopic expression assay, because it was possible to completely rescue a mutation in the endogenous *rough* gene with a properly regulated copy of the gene using P-element-mediated transformation. Surprisingly, after substituting the entire Dfd homeodomain into the rough protein it was still possible to successfully rescue rough function, even though their homeodomains differ at 26 positions. Thus, in this situation at least, the Dfd homeodomain is not sufficient to activate Dfd targets.

5.2 DNA-binding specificity

A simple way to explain the different regulatory specificities of these proteins is to postulate differences in their abilities to bind regulatory sequences in their target genes. Early DNA-binding experiments with homeodomain proteins showed that proteins containing a glutamine at position 50 of the homeodomain bind to the same sequences with similar affinities, and, although small differences (2–10-fold depending on the binding site) in DNA binding between the Dfd and Ubx proteins have been well documented (63, 64), it is not clear to what extent these differences in affinity are exploited in the embryo. Binding studies with the Dfd/Ubx homeodomain swap proteins described by Lin and McGinnis (61) do show a correlation between their ability to regulate Dfd or Antp and their ability to bind optimized Dfd and Ubx binding sites, so there may be a DNA-binding component to the specificity of these proteins (64). However, here again the binding differences are relatively small. Beachy *et al.* (65) have recently presented evidence for co-operative binding of the Ubx protein *in vitro*, which they argue may be a general method used by homeodomain proteins to amplify small differences in DNA-binding affinity. However, because none of these binding sites have been shown to mediate regulation by these proteins in the embryo, it is difficult to assess the significance of these DNA-binding differences.

5.3 Homeotic selector response elements

In 1977 Garcia-Bellido postulated the existence of a set of genes downstream of the homeotic selector genes that would realize their different developmental programmes (66). He imagined that many of these genes would be common targets for the homeotic selector proteins, and that differential regulation of these genes would bring about their distinct segmental fates. Recently, a number of homeotic selector gene-responsive regulatory elements have been characterized that have made it possible to address the question of HOM-C protein regulatory specificity on targets of known function.

Vachon *et al.* (67) have shown that a small fragment from the upstream region of the *Distalless* gene is able to respond to the Ubx and abd-A proteins when reintroduced into the *Drosophila* germ-line. This 877-bp fragment drives expression of a reporter gene in three small patches of cells that are destined to form the limb structures of the first, second, and third thoracic segments. Larvae that harbour a deletion of the BX-C, and thus are missing the Ubx, abd-A, and Abd-B genes, show an expansion of patches into abdominal segments, demonstrating that these genes normally repress the *Distalless* regulatory element in these segments. DNA footprinting analysis of this element with purified Ubx and abd-A proteins reveals two sites bound by both proteins. Mutations in these sites eliminate footprinting by these proteins and, when elements containing these mutations are reintroduced into the fly germ-line, they drive expanded expression in abdominal segments just as the wild-type element does in the absence of the BX-C genes, arguing that these sites mediate this repression by serving as direct targets of homeodomain binding in the embryo. This result suggests that the Ubx and abd-A proteins can act through the same binding sites, a site, however, that does not seem to be regulated by the Antp protein. Studies with ectopically expressed chimeric proteins have demonstrated that the seven amino acid differences between Ubx and Antp are responsible for this specificity (58). It will be instructive to further dissect this regulatory element to identify in more detail the sequences required for repression, and to determine whether the Ubx binding site can act alone to mediate repression of heterologous regulatory elements.

For another homeotic selector target gene, *decapentaplegic* (*dpp*), compelling evidence now exists for direct regulation by the Ubx protein (68). In this experiment, patterned after similar experiments discussed above with ftz, mutations in a Ubx binding site were suppressed by compensatory mutations at position 50 in the third helix of the Ubx protein. These data argue that Ubx can bind directly to, and regulate transcription from, an *in vitro* defined binding site. It will now be possible to identify potential co-factors involved in the specificity of this Ubx response element.

A detailed dissection of a small regulatory element responsive to the Deformed protein has recently provided evidence for the presence of such co-factors (69). This 120-bp element contains three sequences required for strong expression in the maxillary segment: a single Deformed binding site, and two sequences not bound by Deformed protein. These additional sequences may be binding sites for co-factors that permit this element to be specifically regulated by Deformed. The existence of such sites supports an idea, suspected for some time, that the *Drosophila* homeotic selector proteins interact with other factors to achieve regulatory specificity (reviewed in 70).

6. Mechanism of homeodomain action

A considerable number of genes have been identified that respond directly to *Drosophila* homeodomain proteins in the embryo. In a number of cases, the regions of these proteins required for function have been described, and increasingly small regulatory elements exist that serve as faithful targets of homeodomain proteins in the embryo. Next, it will be necessary to inquire about the biochemical nature of regulation by these proteins. An understanding of homeodomain function at this level will require distinguishing between homeodomain residues that affect DNA binding and those critical for other interactions; and it will demand detailed analysis of the sequences required to construct a successful DNA response element and the identification of potential co-factors involved in this regulation. During this investigation, the question of regulatory specificity will continue to be an important theme, particularly since so many homeodomain proteins are able to bind similar DNA sequences *in vitro*. In preparation for this task, arguments can be made for a number of mechanistic models explaining the regulatory specificity of homeodomain proteins.

6.1 Role of DNA binding in regulatory specificity

The simplest model evokes DNA-binding differences to explain regulatory specificity. However, a large body of evidence now exists that argues that DNA binding is not sufficient to explain regulatory specificity. One of the most dramatic arguments comes from experiments with mutants of the ftz protein that have severely reduced affinity for ftz binding sites *in vitro*. In the course of their experiments to show direct binding of ftz to its autoregulatory element in the embryo, Schier and Gehring (39) showed that changing residue 50 of the ftz homeodomain from glutamine to lysine results in a protein that is still able to weakly autoregulate but now binds ~40-fold more weakly to the autoregulatory element *in vitro*. The DNA-binding affinity of the protein has been dramatically reduced but its specificity has been retained. A similar, but even more dramatic, effect was observed in the case of a glutamine to cysteine mutation at position 50 of the *C. elegans* MEC-3 homeodomain. This protein was able to completely rescue a deletion of the endogenous gene when reintroduced back into the organism (71). These results demonstrate that a reduction in affinity does not preclude regulation, and suggests that the small differences in affinity recorded between homeotic selector proteins may not be sufficient to confer specificity (64).

Another argument for the existence of mechanisms of regulatory specificity in addition to DNA binding is based on experiments in which homeodomain binding sites have been removed from their natural sequence context and then placed upstream of reporter genes and reintroduced into the fly germ-line. Vincent *et al.* (72) reported such an experiment using the 12-bp NP sequence isolated as a consensus of engrailed and ftz binding sites in the first *engrailed* intron (15). This sequence drives expression in a large number of glial cells in the embryonic peripheral nervous system, which bear no relation to the region of expression of any known homeodomain protein. However, it is likely that this element is nevertheless being regulated by a homeodomain protein because point mutations in the binding sequence that abolish ftz binding also abolish expression of the construct; and point mutations that have no effect on ftz binding also have no effect on the expression pattern (an argument supported by the ability of this sequence to selectively bind homeodomain proteins among the products of a *Drosophila* λgt11 expression library [73]). Surprisingly, they found that these sites function only when placed upstream of the *hsp70* basal promoter, but not the *ftz* or *engrailed* promoters, arguing for the requirement of sequences in addition to homeodomain binding sites to create a successful homeodomain response element.

Similar conclusions can be drawn from experiments with the *Deformed* autoregulatory element. It was found that the single Deformed binding site required for expression of this 120-bp element does not function faithfully when removed from its surrounding sequences (C. Zeng, unpublished data). Apparently, sequences lying outside the Deformed binding site are required for proper regulation by Deformed. This conclusion is confirmed by mutations in two flanking regions, which render the autoregulatory element inactive (69, see Section 5.3).

6.2 Homeodomain co-factors

Are there any candidates for homeodomain co-factors in *Drosophila*? An attractive protein for such a role is extradenticle (74, 75). Mutations in the *extradenticle* gene transform the identities of thoracic segments without affecting the expression patterns of the homeotic selector genes. These mutations have the same effect on the thoracic transformations caused by ectopic expression of Antp or Ubx, demonstrating that the extradenticle protein can function throughout the embryo to maintain proper regulation by Antp and Ubx. Flies that lack both the maternal as well as the zygotic component of *extradenticle* fail to form most specialized segmental structures, so it is likely that *extradenticle* has effects on other homeotic selector genes too. The *extradenticle* gene has recently been isolated and shown to encode a homeodomain protein significantly different from the HOM-C proteins, and homologous to human pbx proteins (75, 76).

Several examples of homeodomain–co-factor interactions from other organisms serve as models for such interactions in *Drosophila*. A number of these systems have been characterized quite extensively at this point and are useful in providing diverse examples of the strategies employed by homeodomains to achieve regulatory specificity.

Undoubtedly, the best understood homeodomains belong to the MAT**a**1 and MATα2 mating type proteins responsible for the expression of **a** and α haploid and **a**/α diploid cell-type specific genes in yeast (reviewed in 77 and 78; see Chapter 5). The MATα2 protein is able to repress **a**-specific genes in conjunction with the MCM1 protein, and repress haploid-specific genes in the presence of the MAT**a**1 protein. On the **a**-specific promoter MATα2 binds with low affinity to a site with dyad symmetry. However, it binds with similar affinity to either tandem or inverted repeat versions of this site with spacings ranging from 0 to 27 bp (79). Addition of the MCM1 protein increases its affinity specifically for an inverted repeat site with a 13-bp spacing like those found in promoters of **a**-specific genes. This shows how a second protein can contribute to the binding specificity of a homeodomain protein, and serves as one model for promoting the regulatory specificity of *Drosophila* homeodomain proteins that bind to similar sequences (see Section 5.2).

The interaction with MCM1 is mediated by a region of the MATα2 protein lying just N-terminal of the homeodomain that is not required for DNA binding in the absence of MCM1 (80). Transfering this domain to the homeodomain of the *Drosophila* engrailed protein allows this protein to co-operatively bind with MCM1 on the **a**-specific binding site, showing that this region can function as an independent protein–protein interaction domain. This domain lies just ahead of the N-terminal residues shown to be sufficient to swap specificity between certain *Drosophila* homeodomain proteins (see Section 5.1), and as a result it is unclear at this point to what extent this interaction will serve as a model for *Drosophila* proteins.

Interaction of MATα2 with the MAT**a**1 protein on haploid-specific genes, on the other hand, is mediated by amino acids C-terminal to the MATα2 homeodomain (81). This 14 amino acid, C-terminal tail permits MAT**a**1 to dimerize with MATα2 in solution and bind with high affinity to a response element that is essentially a deleted version of the MATα2/MCM1 site described above from which the central MCM1 binding region has been removed. This example shows how the removal of a co-factor binding site from a homeodomain response element can switch the ability of a homeodomain protein to interact with different co-factors. The C-terminal tail of MATα2 overlaps with the positions of amino acids necessary for Dfd function and those Ubx residues able to override Antp substitutions in the homeodomain (61, 58, Fig. 1, and see Section 5.1). Thus, in this case, a similar domain may have been adapted for use by both yeast and *Drosophila* proteins.

Further examples of homeodomain–co-factor interactions are provided by the POU domain proteins. Of these, the octamer-site binding proteins have been studied most extensively (see Chapter 5). The mammalian Oct-1 protein is able to activate simian virus 40 (SV40) intermediate early genes in conjunction with the viral VP16 protein. The highly related protein Oct-2, however, is unable to interact with the VP16 protein by virtue of seven amino acid differences within its homeodomain (reviewed in 82). Substitution of a single amino acid, from alanine to glutamic acid at position 22 in the end of helix 1 of the homeodomain, confers on Oct-2 the ability to interact successfully with VP16 (83, 84). In addition, in order to bind effectively, Oct-1 and VP16 require a complex of proteins, called HCF (85). Here we see how amino acids in the homeodomain can confer regulatory specificity via interaction with a multi-protein complex.

In the case of another mammalian POU domain protein, Pit-1, a single point mutation of arginine to tryptophan at position 58 of the homeodomain has been shown to cause combined pituitary hormone deficiency in humans (86). The mutant chromosome shows a partially dominant phenotype suggesting that the mutant protein may interfere with the activity of the wild-type protein. This idea is confirmed by cell culture transfection experiments in which the addition of mutant protein dramatically reduces the transcriptional activity of the wild-type protein. This result can be explained by a model in which the mutant protein interacts non-productively with a normal binding partner of the Pit-1 protein. Again, this mutation lies in the C-terminal homeodomain region found to be important for Dfd, Ubx, and Antp function (see Fig. 1).

In *Drosophila*, POU domain proteins have been isolated by homology with mammalian proteins. One of these proteins, I-POU, has been shown to bind to the Cf1a POU doman protein and prevent it from binding DNA (87). An intact homeodomain helix 1 and 2, but not helix 3, are required for this interaction. Alignment of the I-POU homeodomain with other POU homeodomains shows that it resembles other POU domain proteins except for a two-amino acid deletion in the N-terminal tail of the homeodomain. Indeed, reinserting two basic residues here disrupts its ability to bind Cf1a and confers upon it the ability to bind DNA. A careful study of the genomic sequences surrounding this part of the homeobox revealed the presence of an alternative intron encoding two basic residues at this position (88). Thus, two proteins are produced from the I-POU locus, one binds to the Cf1a protein and the other binds DNA. Here we see how residues in the N-terminal end of the homeodomain corresponding to amino acids critical for Antp, Scr, Ubx, and Dfd specificity (see Fig. 1) can be involved in both protein–protein and protein–DNA interactions.

Recently, evidence has accumulated for the co-operative action of the *Caenorhabditis elegans* homeodomain proteins UNC-86 and MEC-3 (71). The *unc-86* gene encodes a POU-type homeodomain protein responsible for establishing and maintaining touch-cell fate. The *mec-3* gene encodes a LIM homeodomain protein required for the proper differentiation of touch cells once they have been formed. Both of these proteins are required for the correct regulation of the *mec-3* gene itself. DNase footprinting of a *mec-3* autoregulatory element reveals three overlapping binding sites for the two proteins. Two of these sites are required for the function of the element. Gel retardation experiments with one of these sites reveal a novel complex seen only in the presence of both UNC-86 and MEC-3, reflecting heterodimerization between these proteins. Experiments with deletions of the MEC-3 protein show that the homeodomain and 16 amino acids C-terminal to it are sufficient for this interaction. A naturally occurring mutation in *mec-3*, which results in a single glutamine to cysteine change at position 63 of the homeodomain, severely reduces the ability of MEC-3 to heterodimerize with UNC-86 without compromising its ability to bind DNA. This system provides yet another example of C-terminal homeodomain amino acids involved in protein–protein interactions.

6.3 Models of homeodomain action

These examples allow one to propose a series of models to explain the mechanism of homeodomain function (Fig. 2). In the simplest case DNA-binding differences between homeodomains confer regulatory specificity (Fig. 2, model 1). Homeodomain protein A binds specifically and with high affinity to its target gene. Other homeodomain proteins fail to regulate this target because they have much lower affinities for this site. It is possible that this mechanism is used to discriminate between widely diverged homeodomains, but the evidence has been discussed above for why this model is unlikely to be sufficient to explain the regulatory specificities of similar homeodomain proteins.

The second model invokes a co-factor, B, that increases the affinity of homeodomain protein A for its site, to which it binds with low affinity in the absence of B (Fig. 2, model 2). In this case, residues outside the homeodomain might be necessary but not sufficient for this interaction, as is implied by swap experiments with the Deformed and rough proteins (61, and see Section 5.1).

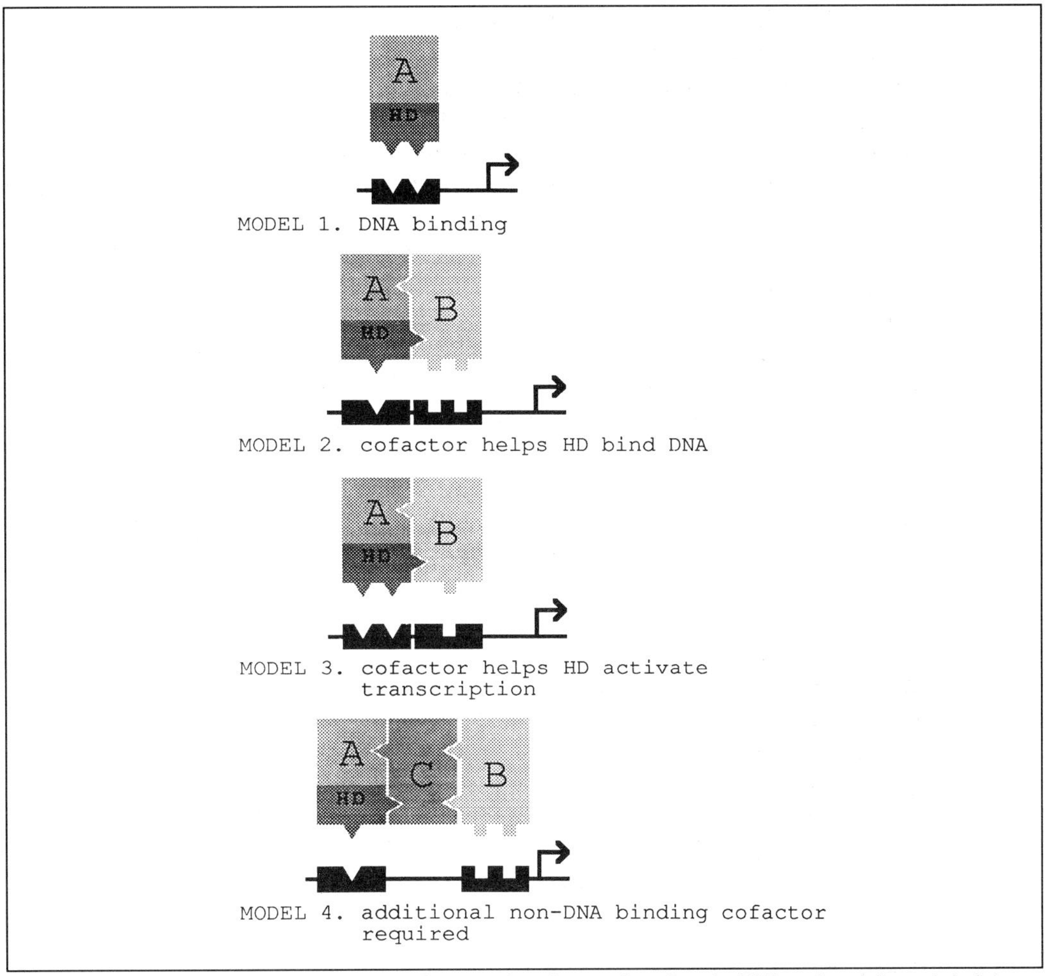

Fig. 2 Models for the functional specificity of homeodomain proteins. Protein–DNA and protein–protein interactions are depicted with notches: a single notch represents a low-affinity interaction, while two notches represent a high-affinity interaction. Model 1: homeodomain factor A binds with high affinity and specificity to its target site. Other homeodomain proteins cannot regulate this target because they fail to bind this site. Model 2: co-factor B helps homedomain protein A bind to its site by itself binding to a nearby site. Without co-factor B, A binds only weakly to its site. Specificity is provided by residues both inside and outside the homeodomain. Model 3: similar to model 2, except co-factor B is required for A to activate transcription from its site. Both A and other homeodomain proteins can bind with high affinity to this site. Model 4: a second co-factor mediates interaction between A and co-factor B. In this case, regulatory specificity can reside either in B or C even though C does not contact DNA itself.

The third model differs from the second model in that the co-factor B does not contribute significantly to the DNA-binding affinity of factor A for its site. Instead, it is required for factor A to become transcriptionally active once bound to its site (Fig. 2, model 3). Other homeodomain proteins can still bind to this site, but do so non-productively. This model allows for the binding of different homeodomain proteins to common sites as is suggested by DNA footprinting experiments and by the mapping of homeodomain protein–DNA complexes in the embryo by UV-cross-linking (13).

The fourth model is a variation on the second and third models in which a second co-factor, C, is required to mediate specific interaction between the first co-factor and the homeodomain protein (Fig. 2, model 4). In this case, specificity is contributed by a co-factor that does not contact DNA itself. The extradenticle protein may be involved in a complex of this sort because it binds DNA (75, 89) and is required for the regulatory specificity of at least three homeodomain proteins but may require additional factors to confer specificity among these proteins. Needless to say, it is likely that homeodomain proteins will use a combination of the models presented above. For example, the Oct-1–VP16–HCF complex appears to use aspects of both model 3 and model 4 (reviewed in 82). Oct-1–VP16 interaction requires the non-DNA-binding co-factor HCF; however, specificity seems to reside in the Oct-1–VP16 interaction itself.

How can such homeodomain co-factors best be isolated? Biochemical methods have been applied to identify such co-factors in a number of cases. Gel retardation experiments with embryonic protein extracts and the ftz autoregulatory element have presented evidence for the existence of numerous proteins that bind to this element, but as yet none of these proteins have been shown to contribute to ftz regulatory specificity (43, and see Section 4). Further biochemical studies will perhaps be more successful in this search. Alternatively, genetic selection for *Drosophila* factors that modulate the activity of homeodomain proteins may be possible in yeast. This approach may provide some distinct advantages over a strictly *in vitro* approach because, unlike DNA-binding techniques, this method has the potential to identify co-factors that affect all aspects of transcriptional regulation, and since it is a clonal method it may be able to detect proteins present in embryonic extracts at prohibitively low concentrations. This method has been used successfully to isolate a mammalian homeodomain protein that increases the DNA-binding avidity of the SRF protein (90).

Finally, the rich history of experimental genetics in *Drosophila* can be exploited to isolate genes affecting the regulatory specificity of homeodomain proteins (91, 92).

References

1. McGinnis, W., Levine, M., Hafen, E., Kuroiwa, A., and Gehring, W. J. (1984) A conserved DNA sequence found in homeotic genes of the *Drosophila* Antennapedia and Bithorax complexes. *Nature*, **308**, 428.
2. Scott, M. P. and Weiner, A. (1984) Structural relationships among genes that control development: sequence homology between the *Antennapedia*, *Ultrabithorax*, and *fushi tarazu* loci of *Drosophila*. *Proc. Natl Acad. Sci. USA*, **81**, 4115.
3. Lewis, E. B. (1978) A gene complex controlling segmentation in *Drosophila*. *Nature*, **276**, 565.
4. Kaufman, T. C., Lewis, R., and Wakimoto, B. (1980) Cytogenetic analysis of chromosome

3 in *Drosophila melanogaster*: the homeotic gene complex in polytene chromosome interval 84A-B. *Genetics*, **94**, 115.
5. McGinnis, W. and Krumlauf, R. (1992) Homeobox genes and axial patterning. *Cell*, **68**, 283.
6. Dessain, S. and McGinnis, W. (1993) *Drosophila* homeobox genes. In *Advances in developmental biochemistry* , Vol. 2, (ed. P. Wassarman), p. 1. JAI Press, Greenwich, CT.
7. Nusslein-Volhard, C. and Wieschaus, E. (1980) Mutations affecting segment number and polarity in *Drosophila. Nature*, **287**, 795.
8. Shepherd, J. C. W., McGinnis, W., Carrasco, A. E., DeRobertis, E. M., and Gehring, W. J. (1984) Fly and frog homeo domains show homologies with yeast mating type regulatory proteins. *Nature*, **310**, 70.
9. Laughon, A., Carroll, S. B., Storfer, E. A., Riley, P. D., and Scott, M. P. (1985) Common properties of proteins encoded by the Antennapedia complex proteins of *Drosophila melanogaster. Cold Spring Harbor Symp. Quant. Biol.*, **50**, 253.
10. Desplan, C., Theis, J., and O'Farrell, P. H. (1985) The *Drosophila* developmental gene, *engrailed*, encodes a sequence-specific DNA binding activity. *Nature*, **318**, 630.
11. Affolter, M., Percival-Smith, A., Müller, M., Leupin, W., and Gehring, W. J. (1990) DNA binding properties of the purified Antennapedia homeo domain. *Proc. Natl Acad. Sci. USA*, **87**, 4093.
12. Percival-Smith, A., Müller, M., Affolter, M., and Gehring, W. J. (1990) The interaction with DNA of wild-type and mutant *fushi tarazu* homeodomains. *EMBO J.*, **9**, 3967.
13. Walter, J., Dever, C. A., and Biggin, M. D. (1994) Two homeodomain proteins bind with similar specificity to a wide range of DNA sites in *Drosophila* embryos. *Genes Dev.*, **8**, 1678.
14. Hoey, T. and Levine, M. (1988) Divergent homeo box proteins recognize similar DNA sequences in *Drosophila. Nature*, **332**, 858.
15. Desplan, C., Theis, J., and O'Farrell, P. H. (1988) The sequence specificity of homeodomain-DNA interaction. *Cell*, **54**, 1081.
16. Hanes, S. D. and Brent, R. (1989) DNA specificity of the bicoid activator protein is determined by homeodomain recognition helix residue 9. *Cell*, **57**, 1275.
17. Treisman, J., Gönczy, P., Vashishtha, M., Harris, E., and Desplan, C. (1989) A single amino acid can determine the DNA binding specificity of homeodomain proteins. *Cell*, **59**, 553.
18. Otting, G., Qian, Y. Q., Billeter, M., Muller, M., Affolter, M., Gehring, W. J., and Wüthrich, K. (1990) Protein-DNA contacts in the structure of a homeodomain-DNA complex determined by nuclear magnetic resonance spectroscopy in solution. *EMBO J.*, **9**, 3085.
19. Kissinger, C. R., Liu, B., Martin-Blanco, E., Kornberg, T. B., and Pabo, C. O. (1990) Crystal structure of an engrailed homeodomain–DNA complex at 2.8 Å resolution: a framework for understanding homeodomain–DNA interactions. *Cell*, **63**, 579.
20. Wolberger, C., Vershon, A. K., Liu, B., Johnson, A. D., and Pabo, C. O. (1991) Crystal structure of a MATα2 homeodomain-operator complex suggests a general model for homeodomain-DNA interactions. *Cell*, **67**, 517.
21. Krasnow, M. A., Saffman, E. E., Kornfeld, K., and Hogness, D. S. (1989) Transcriptional activation and repression by *Ultrabithorax* proteins in cultured *Drosophila* cells. *Cell*, **57**, 1031.
22. Winslow, G. M., Hayashi, S., Krasnow, M., Hogness, D. S., and Scott, M. P. (1989) Transcriptional acitivation by the *Antennapedia* and *fushi tarazu* proteins in cultured *Drosophila* cells. *Cell*, **57**, 1017.

23. Jaynes, J. B. and O'Farrell, P. H. (1988) Activation and repression of transcription by homoeodomain-containing proteins that bind a common site. *Nature*, **336**, 744.
24. Biggin, M. D. and Tjian, R. (1989) A purified *Drosophila* homeodomain protein represses transcription *in vitro*. *Cell*, **58**, 433.
25. Ohkuma, Y., Horikoshi, M., Roeder, R. G., and Desplan, C. (1990) Binding site-dependent direct activation and repression of *in vitro* transcription by *Drosophila* homeodomain proteins. *Cell*, **61**, 475.
26. Johnson, F. B. and Krasnow, M.A. (1992) Differential regulation of transcription preinitiation complex assembly by activator and repressor homeodomain proteins. *Genes Dev.*, **6**, 2177.
27. Tautz, D. (1988) Regulation of the *Drosophila* segmentation gene *hunchback* by two maternal morphogenetic centers. *Nature*, **332**, 281.
28. Schröder, C., Tautz, D., Seifert, E., and Jäckle, H. (1988) Differential regulation of the two transcripts from the *Drosophila* gap segmentation gene *hunchback*. *EMBO J.*, **7**, 2881.
29. Driever, W. and Nüsslein-Volhard, C. (1989) The bicoid protein is a positive regulator of *hunchback* transcription in the early *Drosophila* embryo. *Nature*, **337**, 138.
30. Struhl, G., Struhl, K., and Macdonald, P. M. (1989) The gradient morphogen bicoid is a concentration-dependent transcriptional activator. *Cell*, **57**, 1259.
31. Schier, A. F. and Gehring, W. J. (1992) Direct homeodomain-DNA interaction in the autoregulation of the *fushi tarazu* gene. *Nature*, **356**, 804.
32. Hanes, S. D., Riddihough, G., Ish-Horowicz, D., and Brent, R. (1993) Specific DNA recognition and inter site spacing are critical for action of the bicoid morphogen in *Drosophila*. *Mol. Cell. Biol.*, **14**, 3364.
33. Ronchi, E., Treisman, J., Dostatni, N., Struhl, G., and Desplan, C. (1993) Down-regulation of the *Drosophila* morphogen bicoid by the torso receptor-mediated signal transduction cascade. *Cell*, **74**, 347.
34. Driever, W., Thoma, G., and Nüsslein-Volhard, C. (1989) Determination of spatial domains of zygotic gene expression in the *Drosophila* embryo by the affinity of binding sites for the bicoid morphogen. *Nature*, **340**, 363.
35. Berleth, T., Burri, M., Thoma, G., Bopp, D., Richstein, S., Frigerio, G., Noll, M., and Nüsslein-Volhard, C. (1988) The role of localization of *bicoid* RNA in organizing the anterior pattern of the *Drosophila* embryo. *EMBO J.*, **7**, 1749.
36. Driever, W., Ma, J., Nüsslein-Volhard, C., and Ptashne, M. (1989) Rescue of *bicoid* mutant *Drosophila* embryos by Bicoid fusion proteins containing heterologous activating sequences. *Nature*, **342**, 149.
37. Hiromi, Y., Kuroiwa, A., and Gehring, W. J. (1985) Control elements of the *Drosophila* segmentation gene *fushi tarazu*. *Cell*, **43**, 603.
38. Pick, L., Schier, A., Affolter, M., Schmidt-Glenewinkel, T., and Gehring, W. J. (1990) Analysis of the *ftz* upstream element: germ layer-specific enhancers are independently autoregulated. *Genes Dev.*, **4**, 1224.
39. Schier, A. F. and Gehring, W. J. (1993) Functional specificity of the homeodomain protein fushi tarazu: the role of DNA-binding specificity *in vivo*. *Proc. Natl Acad. Sci. USA*, **90**, 1450.
40. Schier, A. F. and Gehring, W. J. (1993) Analysis of a *fushi tarazu* autoregulatory element: multiple sequence elements contribute to enhancer activity. *EMBO J.*, **12**, 1111.
41. Maier, D., Preiss, A., and Powell, J. R. (1990) Regulation of the segmentation gene *fushi tarazu* has been functionally conserved in *Drosophila*. *EMBO J.*, **9**, 3957.
42. Kassis, J. A. (1990) Spatial and temporal control elements of the *Drosophila engrailed* gene. *Genes Dev.*, **4**, 433.

43. Han, W., Altan, N., and Pick, L. (1993) Multiple proteins interact with the *fushi tarazu* proximal enhancer. *Mol. Cell. Biol.*, **13**, 5549.
44. Florence, B., Handrow, R., and Laughon, A. (1991) DNA-binding specificity of the *fushi tarazu* homeodomain. *Mol. Cell. Biol.*, **11**, 3613.
45. Morrissey, D., Askew, D., Raj, L., and Weir, M. (1991) Functional dissection of the *paired* segmentation gene in *Drosophila* embryos. *Genes Dev.*, **5**, 1684.
46. Furukubo-Tokunaga, K., Müller, M., Affolter, M., Pick, L., Kloter, U., and Gehring, W. J. (1992) *In vivo* analysis of the helix-turn-helix motif of the fushi tarazu homeo domain of *Drosophila melanogaster*. *Genes Dev.*, **6**, 1082.
47. Fitzpatrick, V. D., Percival-Smith, A., Ingles, C. J., and Krause, H. M. (1992) Homeodomain-independent activity of the fushi tarazu polypeptide in *Drosophila* embryos. *Nature*, **356**, 610.
48. Stuart, J. J., Brown, S. J., Beeman, R. W., and Denell, R. E. (1991) A deficiency of the homeotic complex of the beetle *Tribolium*. *Nature*, **350**, 72.
49. Bürglin, T. R., Ruvkun, G., Coulson, A., Hawkins, N. C., McGhee, J. D., Schaller, D., Wittmann, C., Muller, F., and Waterston, R. H. (1991) Nematode homeobox cluster. *Nature*, **351**, 703.
50. Schneuwly, S., Klemenz, R., and Gehring, W. J. (1987) Redesigning the body plan of *Drosophila* by ectopic expression of the homoeotic gene *Antennapedia*. *Nature*, **325**, 816.
51. Gibson, G. and Gehring, W. J. (1988) Head and thoracic transformations caused by ectopic expression of *Antennapedia* during *Drosophila* development. *Development*, **102**, 657.
52. Gibson, G., Schier, A., LeMotte, P., and Gehring, W. J. (1990) The specificities of *Sex combs reduced* and *Antennapedia* are defined by a distinct portion of each protein that includes the homeodomain. *Cell*, **62**, 1087.
53. Zeng, W., Andrew, D. J., Mathies, L. D., Horner, M. A., and Scott, M. P. (1993) Ectopic expression and function of the *Antp* and *Scr* homeotic genes: the N terminus of the homeodomain is critical to functional specificity. *Development*, **118**, 339.
54. Furukubo-Tokunaga, K., Flister, S., and Gehring, W. (1993) Functional specificity of the *Antennapedia* homeodomain. *Proc. Natl Acad. Sci. USA*, **90**, 6360.
55. Malicki, J., Schughart, K., and McGinnis, W. (1990) Mouse *Hox-2.2* specifies thoracic segmental identity in *Drosophila* embryos and larvae. *Cell*, **63**, 961.
56. Zhao, J. J., Lazzarini, R. A., and Pick, L. (1993) The mouse *Hox-1.3* gene is functionally equivalent to the *Drosophila Sex combs reduced* gene. *Genes Dev.*, **7**, 343.
57. Mann, R. S. and Hogness, D. S. (1990) Functional dissection of Ultrabithorax proteins in *D. melanogaster*. *Cell*, **60**, 597.
58. Chan, S. and Mann, R. S. (1993) The segment identity functions of Ultrabithorax are contained within its homeo domain and carboxy-terminal sequences. *Genes Dev.*, **7**, 796.
59. McGinnis, N., Kuziora, M. A., and McGinnis, W. (1990) Human *Hox-4.2* and *Drosophila Deformed* encode similar regulatory specificities in *Drosophila* embryos and larvae. *Cell*, **63**, 969.
60. Kuziora, M. A. and McGinnis, W. (1989) A homeodomain substitution changes the regulatory specificity of the *Deformed* protein in *Drosophila* embryos. *Cell*, **59**, 563.
61. Lin, L. and McGinnis, W. (1992) Mapping functional specificity in the Dfd and Ubx homeodomains. *Genes Dev.*, **6**, 1071.
62. Lockett, T. J., Moretti, P. A. B., Lake, J., and Saint, R. (1993) *Rough* genes with *Deformed* homeobox substitutions exhibit *rough* regulatory specificity during *Drosophila* eye development. *Mech. Dev.*, **41**, 57.
63. Dessain, S., Gross, C. T., Kuziora, M., and McGinnis, W. (1992) Antp-type homeo-

domains have distinct DNA binding specificities that correlate with their target specificities in the embryo. *EMBO J.*, **11**, 991.

64. Ekker, S. C., von Kessler, D. P., and Beachy, P. A. (1992) Differential sequence recognition is a determinant of specificity in homeotic gene action. *EMBO J.*, **11**, 4059.
65. Beachy, P. A., Varkey, J., Young, K. E., von Kessler, D. P., Sun, B. I., and Ekker, S. C. (1993) Cooperative binding of an Ultrabithorax homeodomain protein to nearby and distant DNA sites. *Mol. Cell. Biol.*, **13**, 6941.
66. Garcia-Bellido, A. (1977) Homeotic and atavistic mutations in insects. *Amer. Zool.*, **17**, 613.
67. Vachon, G., Cohen, B., Pfeifle, C., McGuffin, M. E., Botas, J., and Cohen, S. M. (1992) Homeotic genes of the bithorax complex repress limb development in the abdomen of the *Drosophila* embryo through the target gene *Distal-less*. *Cell*, **71**, 437.
68. Capovilla, M., Brandt, M., and Botas, J. (1994) *decapentaplegic*, a TGF-β cognate, is a direct target of *Ultrabithorax* regulation required for a segmental identity function. *Cell*, **76**, 461.
69. Zeng, C., Pinsonneault, J., Gellon, G., McGinnis, N., and McGinnis, W. (1994) Deformed protein binding sites and cofactor binding sites are required for the function of a small, segment-specific regulatory element in *Drososphila* embryos. *EMBO J.*, **13**, 2362.
70. Hayashi, S. and Scott, M. P. (1990) What determines the specificity of action of *Drosophila* homeodomain proteins? *Cell*, **63**, 883.
71. Xue, D., Tu, Y., and Chalfie, M. (1992) Cooperative interactions between the *C. elegans* homeoproteins UNC-86 and MEC-3. *Science*, **261**, 1324.
72. Vincent, J. P., Kassis, J. A., and O'Farrell, P. H. (1990) A synthetic homeodomain binding site acts as a cell type specific, promoter specific enhancer in *Drosophila* embryos. *EMBO J.*, **9**, 2573.
73. Kalionis, B. and O'Farrell, P. H. (1993) A universal target sequence is bound *in vitro* by diverse homeodomains. *Mech. Dev.*, **43**, 57.
74. Peifer, M. and Weischaus, E. (1990) Mutations in the *Drosophila* gene *extradenticle* affect the way specific homeo domain proteins regulate segmental identity. *Genes Dev.*, **4**, 1209.
75. Rauskolb, C., Peifer, M., and Wieschaus, E. (1993) *extradenticle*, a regulator of homeotic gene activity, is a homolog of the homeobox-containing human proto-oncogene *pbx1*. *Cell*, **74**, 1101.
76. Flegel, W. A., Singson, A. W., Margolis, J. ., Bang, A. G., Posakony, J. W., and Murre, C. (1993) *Dpbx*, a new homeobox gene closely related to the human proto-oncogene *pbx1*: molecular structure and developmental expression. *Mech. Dev.*, **41**, 155.
77. Herskowitz, I. (1989) A regulatory hierarchy for cell specialization in yeast. *Nature*, **342**, 749.
78. Johnson, A. D. (1992) A combinatorial regulatory circuit in budding yeast. In *Transcriptional regulation*, Vol. 2, (ed. S. L. McKnight and K. R. Yamamoto), p. 975. CSH Laboratory Press, Cold Spring Harbor.
79. Smith, D. L. and Johnson, A. D. (1992) A molecular mechanism for combinatorial control in yeast: MCM1 protein sets the spacing and orientation of the homeodomains of an α2 dimer. *Cell*, **68**, 133.
80. Vershon, D. and Johnson, A. D. (1993) A short disordered protein region mediates interactions between the homeodomain of α2 protein and the MCM1 protein. *Cell*, **72**, 105.
81. Mak, A. and Johnson, A. D. (1993) The carboxy-terminal tail of the homeo domain protein α2 is required for function with a second homeo domain protein. *Genes Dev.*, **7**, 1862.
82. Herr, W. (1992) Oct-1 and Oct-2: differential transcriptional regulation by proteins that

bind to the same DNA sequence. In *Transcriptional regulation*, Vol. 2, (ed. S. L. McKnight and K. R. Yamamoto), p. 1103. CSH Laboratory Press, Cold Spring Harbor.

83. Lai, J.-S., Cleary, M. A., and Herr, W. (1992) A single amino acid exchange transfers VP16-induced positive control from the Oct-1 to the Oct-2 homeo domain. *Genes Dev.*, **6**, 2058.
84. Pomerantz, J. L., Kristie, T. M., and Sharp, P. A. (1992) Recognition of the surface of a homeo domain protein. *Genes Dev.*, **6**, 2047.
85. Wilson, A. C., LaMarco, K., Peterson, M. G., and Herr, W. (1993) The VP16 accessory protein HCF is a family of polypeptides processed from a large precursor protein. *Cell*, **74**, 115.
86. Radovik, R., Nations, M., Du, Y., Berg, L. A., Weintraub, B. D., and Wondisford, F. E. (1992) A mutation in the POU-homeodomain of Pit-1 responsible for combined pituitary hormone deficiency. *Science*, **257**, 1115.
87. Treacy, M. N., He, X., and Rosenfeld, M. G. (1992) I-POU: a POU domain protein that inhibits neuron-specific gene activation. *Nature*, **350**, 577.
88. Treacy, M. N., Neilson, L. I., Turner, E. E., He, X., and Rosenfeld, M. G. (1992) Twin of I-POU: a two amino acid difference in the I-POU homeodomain distinguishes an activator from an inhibitor of transcription. *Cell*, **68**, 491.
89. Van Dijk, M. A., Voorhoeve, M., and Murre, C. (1993) Pbx1 is converted into a transcriptional activator upon acquiring the N-terminal region of E2A in pre-B-cell acute lymphoblastoid leukemia. *Proc. Natl Acad. Sci. USA*, **90**, 6061.
90. Grueneberg, D. A., Natesan, S., Alexandre, C., and Gilman, M. Z. (1992) Human and *Drosophila* homeodomain proteins that enhance the DNA-binding activity of serum response factor. *Science*, **257**, 1089.
91. Kennison, J. A. and Tamkun, J. W. (1988) Dosage-dependent modifiers of *Polycomb* and *Antennapedia* mutations in *Drososphila*. *Proc. Natl. Acad. Sci. USA*, **85**, 8136.
92. Harding, K. W., Gellon, G., McGinnis, N., and McGinnis, W. (1995) A screen for modifiers of *Deformed* Function in *Drosophila*. *Genetics*, **140**, 1339.

7 | Transcriptional regulation of skeletal myogenesis

CRAIG NEVILLE and NADIA ROSENTHAL

1. Introduction

The recent excitement in the field of vertebrate skeletal myogenesis can be traced to the discovery that a muscle phenotype can be induced in cell culture by the controlled expression of specific regulatory genes. When the function of these genes is perturbed in the embryo, muscle cell development is profoundly affected. This confluence of molecular and genetic findings has established skeletal muscle as a model system for studying transcriptional regulation of developmental processes. The intricate patterns of contractile tissue formation in the adult animal depend on early inductive signals that specify particular myogenic fates of primordial precursor cells in the embryonic somites. These cells then migrate to predictable positions in the vertebrate body plan and undergo terminal differentiation at precisely defined developmental stages. Subsequent modulation of the myogenic programme by neuronal, hormonal, and mechanical stimuli results in more subtle structural distinctions, which reflect the diversity of functions in the adult skeletal musculature. The regulatory pathways underlying the generation of this diversity in skeletal muscle structure and function are a current focus of intense research activity.

Why has this tissue proven particularly amenable to molecular analysis? From an experimental standpoint, myogenic differentiation can be manipulated in culture by a relatively short exposure to serum-deficient medium, which results in fusion of monocytic myoblasts into multi-nucleate syncytia. Accompanying these morphological changes, the rapid activation of a battery of genes, including those that encode the components of the muscle contractile apparatus, is exquisitely co-ordinated between nuclei in the appropriate stoichiometry to produce a functional contractile tissue. Cell culture has enabled the characterization of multiple, muscle-specific DNA control elements associated with differentiation-specific gene expression. In many cases the DNA sequence motifs that are important for tissue-specific function constitute binding sites for regulatory proteins. Studies to define the genetic determinants of muscle stem cells, which give rise to committed myoblasts, have resulted in the identification of a family of transcription factors whose expression in non-muscle cells induces the myogenic phenotype. Through transgenic mouse technology, these factors have been shown to be critical components of the genetic pathways leading to muscle determination and differentiation in the embryo.

This chapter will summarize the recent progress in skeletal muscle molecular biology, emphasizing the central role of transcriptional regulation in each step of myogenesis, from the activation of contractile protein gene expression in post-mitotic muscle cultures, back to the initial selection of myogenic precursor cell populations from the embryonic mesoderm. Although skeletal muscle is presently one of the most well-characterized tissue systems in terms of molecular analysis, many of the players in this regulatory cascade have yet to be identified. Current research is poised to exploit a diversity of experimental systems to dissect the complex mechanisms underlying myogenic cell commitment and differentiation.

2. Transcriptional regulation of muscle-specific genes

Characterization of the primary genetic components responsible for the regulation of genes transcribed in a tissue- and stage-specific manner has been a necessary prerequisite for building models of the molecular pathways leading to muscle differentiation. Multiple *cis*-acting DNA elements responsible for muscle-specific transcription have been identified by the introduction of mutant or chimeric genes into cultured cells, and into the germ-line of transgenic mice. Promoter-associated sequences have been identified in the genes encoding muscle actins, myosin heavy and light chains, troponin T, and acetylcholine receptor subunits (1), which drive linked gene expression exclusively in muscle cell cultures.

In addition to promoter-associated regulatory sequences, several muscle genes are also regulated by enhancer elements, which are defined by their ability to increase expression of a linked gene, irrespective of their position and orientation relative to the start site of transcription. Tissue-specific enhancer elements lying upstream, downstream, or within the gene body have been defined for genes encoding muscle creatine kinase (MCK), myosin light chain (MLC)1/3, acetylcholine receptor α-subunit, and troponin I (1). Many of these enhancers also function appropriately in transgenic mice as potent, tissue-specific transcriptional regulatory elements (2, 3). Both muscle-specific promoters and enhancers share sequence motifs that comprise target sites for common transcription factors (Fig. 1). In general, transcriptional regulation of muscle genes appears to reflect the paradigms established for constitutive transcriptional regulatory elements, in which multiple proteins bind to a complex and often overlapping array of recognition sequences (see Chapters 5 and 6).

2.1 Muscle-specific gene regulation

Although a single *cis*-acting regulatory element present in all muscle-specific control regions has not been found, several similar sequence motifs in evolutionarily

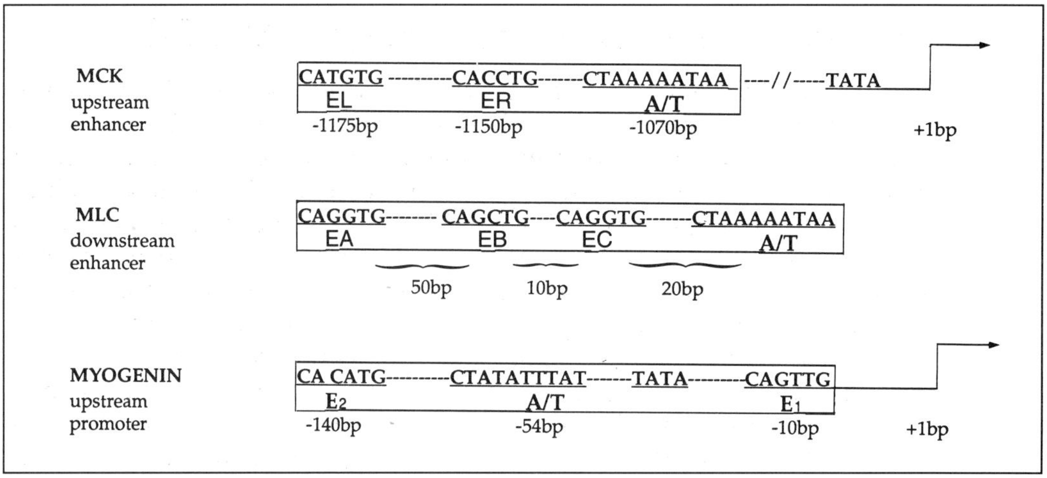

Fig. 1 Regulatory DNA elements of muscle-specific genes (1). Common sequence motifs are found in muscle-specific promoters and enhancers, arranged in different configurations. These motifs form binding sites for myogenic factors to augment transcription from the associated promoter. The enhancer upstream of the muscle creatine kinase gene (top) contains two bHLH binding sites (E boxes) and a single MEF2 binding site (A/T-rich sequence). The enhancer downstream of the MLC1/3 locus (middle) contains three E boxes and a MEF2 site, whereas the promoter of the myogenin gene (bottom) contains two E boxes, one of which is downstream of the TATA box, and a MEF2 site.

unrelated muscle genes have been shown to function as binding sites for common nuclear protein complexes (1). In most cases, deletion or mutation of these motifs impairs the ability of the control region to activate linked gene expression in both cell culture and transgenic animals, underscoring the functional importance of each motif in the transcriptional regulatory network.

Two such sequence motifs have recently emerged as critical components of the regulatory circuitry in muscle gene expression: the E box (CANNTG), and an A/T-rich motif ($CTA(A/T)_4$). Together these two sequences appear to form an essential composite for the activation of muscle-specific gene expression. In muscle-specific regulatory elements such as the MCK and MLC1/3 enhancer, E boxes are generally arranged in multiple copies within a given regulatory element (4, 5), where they act co-operatively in transcriptional activation of linked genes (Fig. 1). E boxes are often situated in close proximity to A/T-rich sequences, suggesting a possible synergistic relationship in the functions of the proteins that bind to the two motifs (Fig. 1). E boxes represent binding sites for the myogenic determination factor (MDF) family (6–12), whereas A/T motifs represent binding sites for the myocyte enhancer factor (MEF2) family (13–26), which are structurally related to serum response factors (27; see Chapter 5). Increasing the complexity of the situation, both E boxes and A/T motifs are also important functional components in the control regions of MDF genes themselves (21, 22, 28–32), presumably acting as part of an autoregulatory loop. Dissection of the circuitry responsible for maintaining the muscle phenotype will require an understanding of how E and A/T-boxes function as targets for their requisite transcription factors.

3. Myogenic determination factors (MDFs)

The four MDFs, MyoD, myogenin, myf5, and MRF4, are related transcription factors expressed exclusively in skeletal muscle lineages. MDFs were first characterized as master regulators of myogenesis as a result of their potent activation of the muscle phenotype when over-expressed in transfected non-muscle cell cultures (11, 13, 33–37). The importance of the MDFs in myogenesis is also evident from their evolutionary conservation. In vertebrates, the four MDF genes have a common three-exon structure and belong to a superfamily of genes encoding a related group of transcription factors (Fig. 2), which share an 80 amino acid basic–helix–loop–helix (bHLH) domain conserved in many species (Fig. 3; 8–43). The basic domain mediates DNA binding (44–47) and the HLH domain enables protein dimerization between different members of the bHLH family (28, 48, 49). This is important because bHLH proteins function in pairs to activate transcription by binding to their target sites. In muscle cells, the MDFs form heterodimers with other ubiquitously expressed bHLH proteins, known as E proteins, and activate gene transcription by binding to the E boxes in muscle-specific regulatory elements, such as those illustrated in Fig. 1. Although these E boxes all share the CANNTG sequence motif, differences in the binding site preferences of various bHLH heterodimers for different versions of the CANNTG motif and its surrounding sequences may play an important role in subtly regulating the expression of different muscle genes (47, 50; Fig. 4).

The crystal structure of the MDF–DNA complex has recently been solved for

Human	Mouse/rat	Chick/quail	Toad
myf3	myoD	CMD1/qmf1	XMyoD
myf4	myogenin	myogenin/qmf2	XMyogenin
myf5	myf5	myf5/qmf3	XMyf5
myf6	herculin/MRF4	herculin	XMRF4

Electric ray	Sea urchin	Fruit fly	Nematode
MyoD	SUM1	nautilus	hlh1
myogenin			
myf5			

Fig. 2 The MyoD gene family. The four myogenic factor genes are conserved amongst vertebrates. Only a single related gene has been found in the genome of each invertebrate examined, of which only the bHLH region has been conserved. Genes have been isolated from human (178), mouse and rat (33, 34, 36, 37, 86), chick and quail (74, 162, 179), toad (180–182), electric ray (183), sea urchin (41), fruit fly (184, 185), and nematode (186).

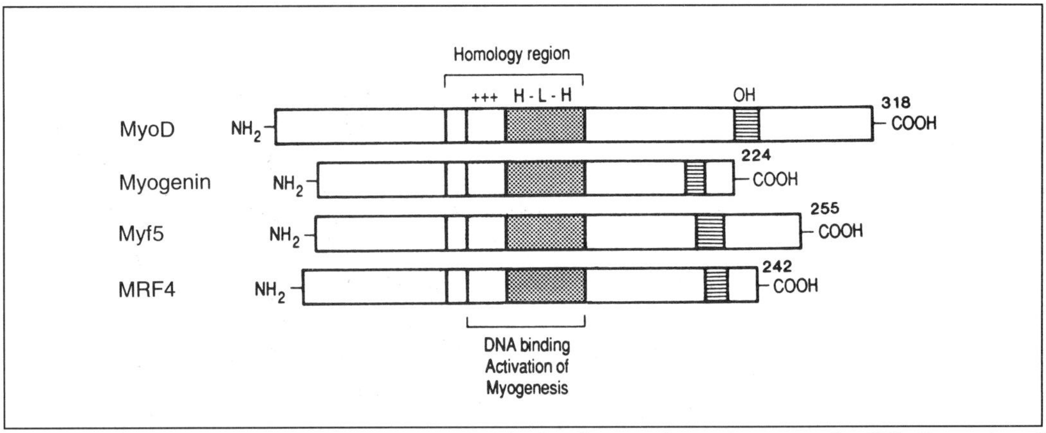

Fig. 3 Structural comparison of the mammalian bHLH family of myogenic regulatory proteins (38). The region of homology that encompasses the basic (+++) and HLH domains is shaded. The region that is sufficient for myogenesis is indicated at the bottom. A serine/threonine-rich region of homology (OH) is indicated by stripes. The number of amino acids in each polypeptide is shown at the end of each box.

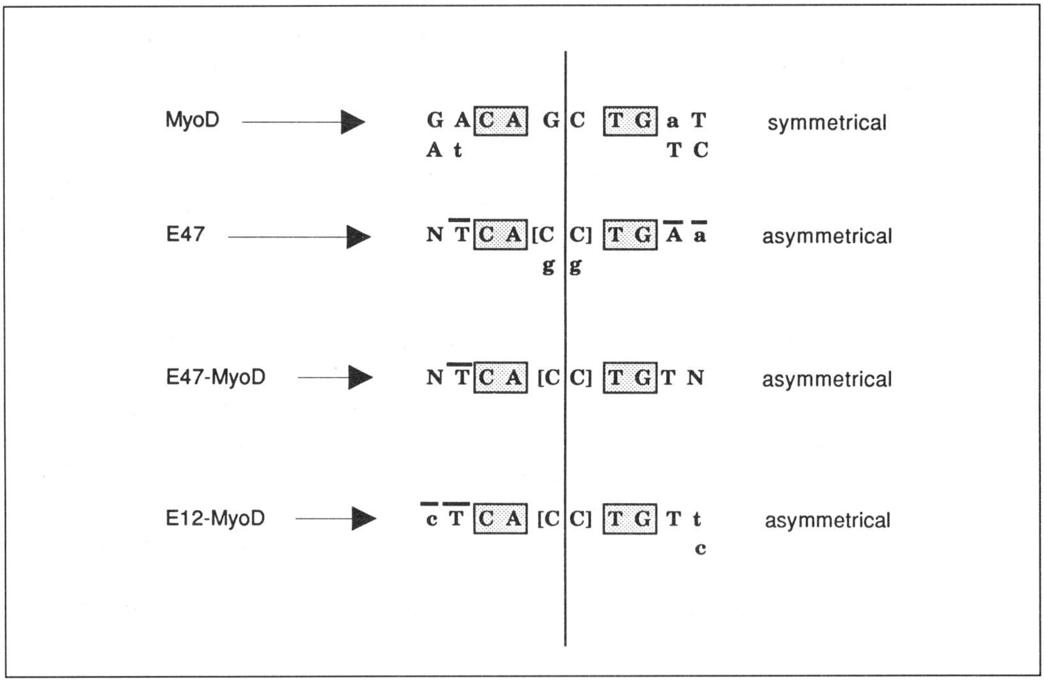

Fig. 4 Binding site consensus sites for MyoD and E12 or E47 protein complexes, as revealed by binding site selection (50). Preferences that are absolute or nearly so are indicated with upper case letters and incomplete preferences are indicated with lower case letters. A line drawn over an upper case letter indicates a base that is not present in that position. A line drawn over a lower case letter similarly indicates a decrease in use of that base. Pairs of nucleotides that are coupled are bracketed.

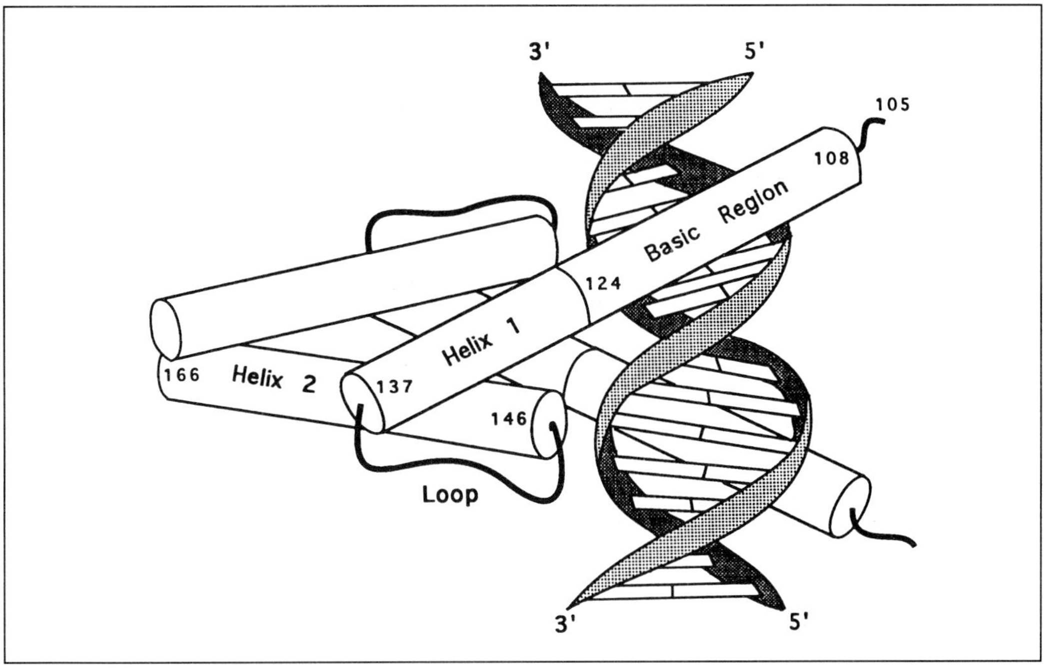

Fig. 5 Model of the MyoD–DNA complex (51). The alpha helices are represented as cylinders, and a few residue numbers are given for reference. The sugar–phosphate backbones of the DNA are represented by stippled ribbons.

MyoD (51). A MyoD protein homodimer makes direct contact with the DNA at the E box consensus site where two key amino acids (Ala-114 and Thr-115) in the basic domain, which are required for muscle-specific regulation by the MDF family, lie buried in the protein–DNA interface (Fig. 5). These two residues do not make direct contact with DNA but are probably important in positioning a third residue (Arg-111) in the major groove for transcriptional activation. The structural similarity between the DNA-binding domains of the MDFs and other bHLH proteins implies that heterodimers between different bHLH members may assume similar configurations when bound to DNA. This would suggest that the cell specificity of MDF action may lie in protein–protein interactions. It is presently unclear how these residues could directly participate in protein–protein interactions with other proteins, unless they impart subtle changes on overall protein conformation. Future studies in this area will undoubtedly focus on detailed structural analysis of mutations in these key residues, and their effects on interactions with other transcription factors involved in myogenic gene regulation.

3.1 Specificity of MDF function in muscle cell culture

All four MDFs can induce myogenesis when over-expressed in transfected non-muscle cell cultures, and they bind E box consensus sites *in vitro* with equal affinity.

However, they are not equally potent in *trans*-activating muscle gene expression (52–54), suggesting that they may possess distinct activities. The tissue specificity of MDF function is apparently modulated by a number of parameters, including the sequences within and surrounding the E box target (46, 47, 53), and a conserved pair of alanine and threonine residues in the MDF basic DNA-binding domain (44, 45). Perhaps further specificity is conferred by the formation of heterodimers between MDFs and different E proteins (28, 49, 55), some of which have yet to be characterized. Alternatively, the unique amino and carboxyl termini of each MDF may permit specific interactions with different subsets of transcription factors, which may bind other muscle control elements. This possibility is supported by domain-swapping studies, in which rearrangements of MDF-unique regions confer altered affinities for their target sites (46, 48, 53, 54). It is therefore likely that MDF target specificity may lie, at least partially, in the interactions of each MDF with different combinations of other transcription factors.

Over-expression of MDFs in non-muscle cell culture also induces transcription of endogenous MDF genes (56, 57), suggesting that continuous, autoregulatory activation of MDF gene expression may be an important component of the commitment to a myogenic phenotype. The fact that E boxes have been found in the promoters of the MDF genes themselves (29, 30; Fig. 1), suggests that the MDF genes are targets for an autoactivation pathway that may help stabilize or reinforce the cell's commitment to the myogenic pathway.

3.2 Regulation of MDF function in cell culture

The details of the interplay between the four individual MDFs in muscle cells remain to be determined, but some insight can be gained from their respective expression patterns in muscle cultures (Fig. 6). MyoD and myf5 gene expression is characteristic of proliferating myoblasts (58) where their function is repressed. In contrast, expression of the myogenin and MRF4 genes is activated only when myoblasts exit the cell cycle upon growth factor withdrawal (58, 59). Thus, the mere presence of one myogenic factor is not sufficient to activate expression of all the others. In fact, expression of MyoD or myf5 is mutually exclusive in established muscle cell lines, and the myf5 expression is reduced in some cells when transfected with a MyoD expression vector (60). It is possible that these two particular MDFs may actually inhibit each others expression, either through direct interaction with the corresponding control regions, or through more indirect pathways involving other transcription factors.

General pathways controlling cell proliferation also play an important role in regulating MDF function. For example, the insulin-like growth factors (IGFs) promote myogenic differentiation in muscle cells, apparently through the stimulation of myogenin gene expression (61). In contrast, the action of myogenin is repressed by fibroblast growth factor (FGF), which stimulates protein kinase C to phosphorylate the conserved threonine in the basic region of the myogenin protein and thereby inhibit DNA binding (62). Growth factors also serve to maintain high

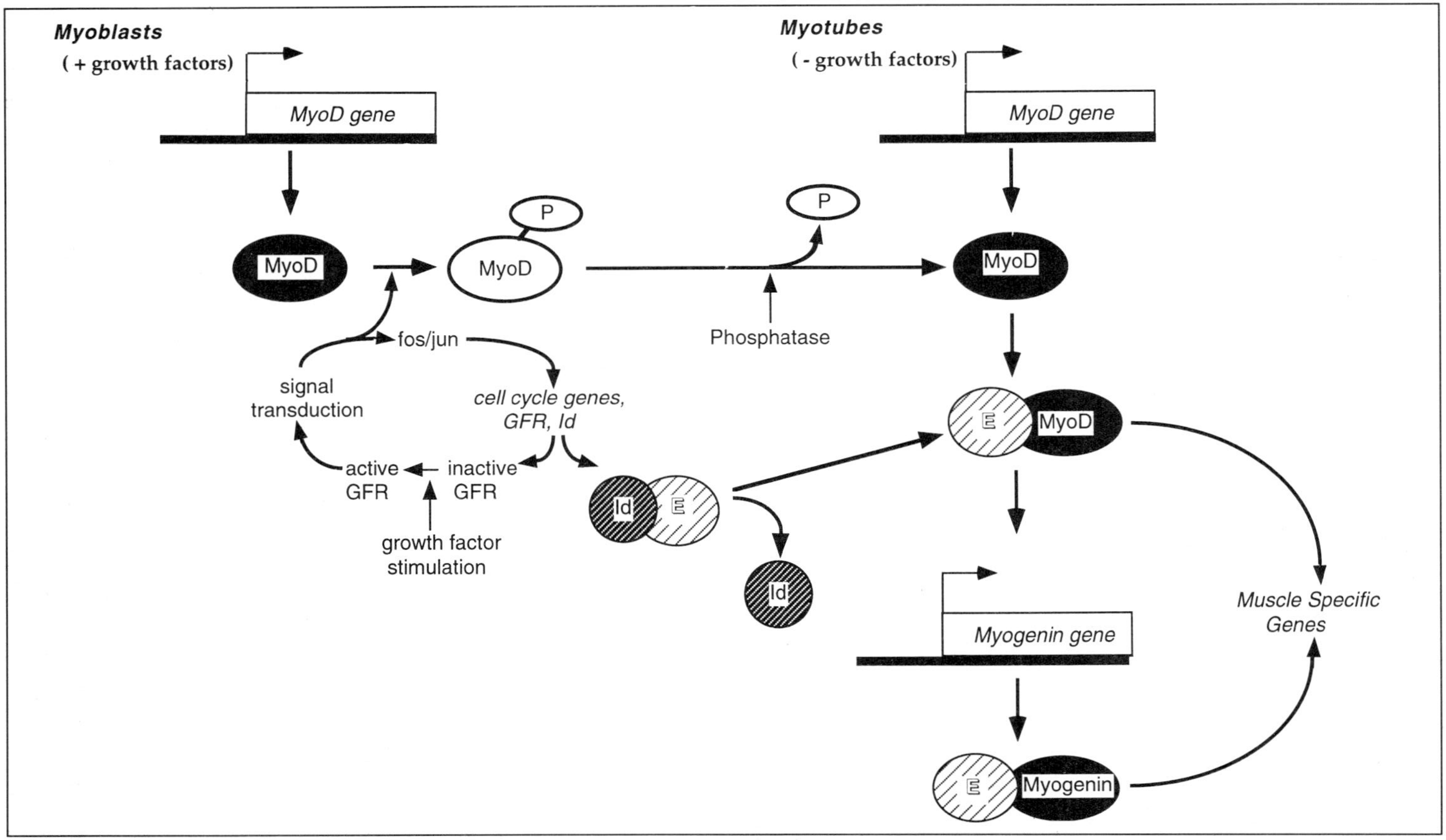

Fig. 6 Model for regulation of myogenic factors during differentiation (40). During myoblast proliferation (left), several inhibitory pathways prevent differentiation. First, MyoD is expressed but the protein is inactivated by phosphorylation. The signal transduction pathway leading to MyoD modification is stimulated by growth factors, which activate growth factor receptors (GFRs). Second, expression of fos/jun inhibits MyoD regulatory functions and maintains expression of cell-cycle genes,.as well as genes encoding growth factor receptors and Id. Third, E protein partners of MyoD are sequestered by binding to Id. When growth factors are removed (right), MyoD protein is activated by dephosphorylation and forms heterodimers with E proteins. These complexes participate in the activation of differentiation-dependent regulators, such as myogenin. Finally, expression of muscle-specific genes is intitiated and maintained by myogenic factors in active complexes.

expression of Id, an HLH protein that lacks a basic DNA-binding motif (63, 64). The function of Id *in vivo* may be related to cell proliferation (65), and in cell culture it is thought to repress MDF function by binding to the E2A gene products, sequestering them and thereby interfering with their interaction with MDFs (66). Other aspects of the cell-cycle regulatory circuitry, such as the expression of the complexed transcription factors fos/jun, also contribute to the repression of MyoD function (67, 68). Other regulatory factors implicated in cell-cycle control, such as the adenoviral E1A protein, inhibit myoblast differentiation and the *trans*-activation functions of MDFs (69, 70). Finally, a mechanism for the reciprocal relationship between withdrawal from the cell cycle and the induction of myogenesis is suggested by the interaction of the retinoblastoma (pRb) tumour supressor gene product with MyoD, which maintains pRb in its inactive state and leads to growth arrest (71). These studies, although still in the formative stage, suggest that the precise mechanisms by which growth factors and signal transduction pathways interact in the reciprocal control of the cell cycle and differentiation may be first worked out in the muscle system, and involves direct feedback on the MDF transcriptional regulatory family, about which much is already known.

4. MDF function in muscle development

The dominant role played by the MDFs in activating the myogenic programmes of cultured cells can be extended to muscle development *in vivo*. Vertebrate embryos derive their skeletal muscles from mesodermal tissues that condense into visceral arches in the head (the source of facial muscles), and into segmented structures called somites in the trunk, which mature in an anterior to posterior direction on either side of the neural tube (see Section 6.2 and Fig. 11 below). During the maturation of each somite, cells are further compartmentalized to form the myotome. Mononucleate myocytes in the myotome are the earliest site of MDF expression, and later give rise to multi-nucleate axial muscle tissues (72, 73). A subset of cells in the lateral part of the myotome is destined to become muscle as well, but does not express MDFs while in the somite. Rather, these cells migrate into the peripheral body and limbs as muscle precursors in which MDF expression is activated only upon arrival in the distal tissues, at which point terminal differentiation programmes are initiated (74–76). Axial and lateral muscle precursors are therefore spatially and temporally distinct in their MDF expression patterns.

During somitogenesis, expression of the four MDFs is activated sequentially in the developing myotome, preceding contractile protein gene activation. The order of MDF activation during development is species specific. In avian embryos, MyoD is activated first, followed by the activation of myf5 and myogenin (74), whereas in mammals, the order is myf5>myogenin>MyoD>MRF4 (75). Expression of myf5 is transient during mammalian embryogenesis, and expression of both myogenin and MyoD substantially decreases soon after birth, leaving MRF4 as the major MDF expressed in adult skeletal muscles. The order of MDF activation in mammalian embryogenesis is in apparent contradiction to the precocious expression of MyoD

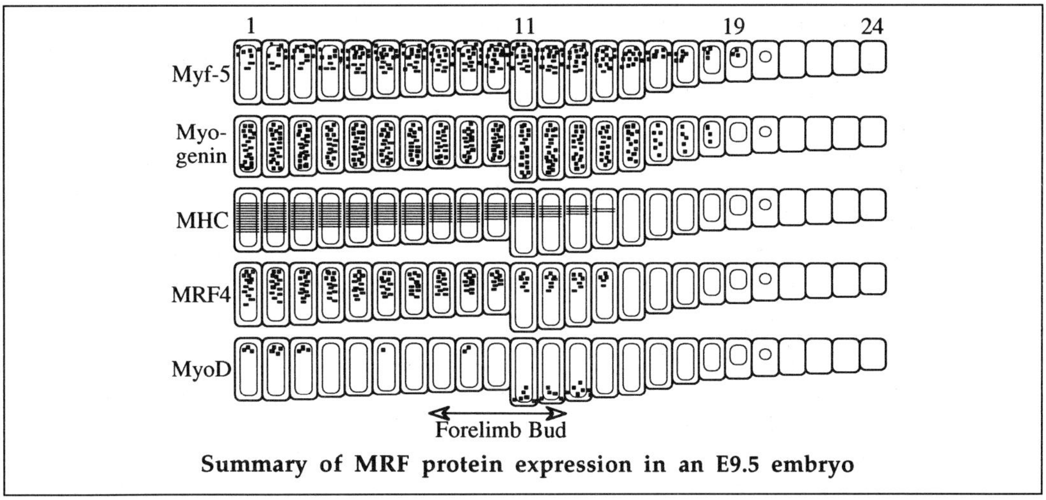

Fig. 7 Expression patterns of the four MDFs in the mouse embryo at 9.5 dpc (78). At this developmental stage, (i) myf5+ cells, though found in both dermatomes and mytomes, are restricted to the dorsal portion of the somites; (ii) MyoD+ cells are found in the ventral portion of the forelimb bud somites, but in the dorsal portion of neck somites; (iii) myogenin+ cells extend throughout the dorsal–ventral extent of the myotome; (iv) MRF4+ cells extend farther ventrally in the myotomes than myf5+ cells, but not as far as myogenin+ cells; and (v) myocytes that express myosin heavy chain (MyHC) are found in approximately the same areas as MRF4-expressing nuclei. Dots represent approximate distributions and numbers of nuclei, and are not actual data. (Figure courtesy of J. B. Miller.)

in myoblast cultures. The discrepancy between cell culture and embryos may be explained by a model in which two distinct lineages reside within the myotome: an early lineage that expresses myf5 gives rise to the first wave of mononucleate myotomal muscles, and a later lineage that generates multi-nucleate axial muscles (40, 77). In addition, subdomains within the myotome express different MDFs and combinations of MDFs at distinct developmental stages (Fig. 7), suggesting that the order of MDF activation may be disco-ordinate between different myotomal cells (78). This implies that different expression patterns of the four MDF proteins in the embryo may distinguish multiple myogenic cell lineages within the developing somites.

4.1 Perturbation of MDF expression during development

The potential roles of MDFs in myogenesis are further complicated by the results of transgenic mouse studies in which normal MDF expression has been perturbed. That MDFs function in a dominant fashion is graphically illustrated by the effects of over-expressing MyoD in heart muscle, which activates the ectopic expression of skeletal muscle-specific contractile protein genes (79). More widespread ectopic expression of MyoD in ectodermal and mesodermal lineages causes embryonic lethalities, although no ectopic myogenic conversion was seen (80). The unexpected phenotypes obtained by disrupting the function of endogenous MDF genes are

Genotype	**Phenotype**
MyoD –/–	• Viable and fertile • Normal muscle • 3.5-fold increase in myf5 expression
Myogenin –/–	• Peri-natal lethal • Normal number of myoblasts in limbs but little fusion • Expresses muscle-specfic differentiation markers • Significant axial, intercostal, and back muscle
Myf5 –/–	• Die at birth due to rib abnormalities • No muscle defects other than one day delay in development
MyoD +/–; Myf5 –/–	• Same as myf5 –/–
MyoD –/–; Myf5 +/–	• Peri-natal lethal • Fewer number of muscle fibres • Reduced mRNA levels of differentiation markers • Normal developmental pattern
MyoD –/–; Myf5 –/–	• Peri-natal lethal • No myoblasts nor muscle
MRF4 –/–	• ?

Fig. 8 Phenotypes of mice carrying null mutations in the myogenic factor genes (81–85).

equally difficult to interpret, and further highlight the complexity of their individual roles in transcriptional regulatory cascades (Fig. 8). Null mutations in the myf5 locus result in delayed myotome differentiation, but otherwise muscle development proceeds normally. Notably, rib formation is disrupted in myf5 knock-out lines, even though myf5 is never expressed in the rib precursors (81). This surprising abnormality is presumably due to the perturbation of local interactions in the somite between the developing myotome and the neighbouring sclerotome, which gives rise to ribs and axial cartilage.

Disruption of the myogenin gene locus produces a more predictable and dramatic phenotype in which terminal differentiation of myogenic precursors is largely blocked and muscle fails to form (82, 83). Embryos lacking myogenin die shortly after birth, presumably due to asphyxiation. Tissues where muscle should form are largely normal in morphology, yet contain few myofibres and do not express muscle-specific contractile protein genes. Nevertheless, cultured myoblasts from these myogenin knock-out animals are capable of differentiating upon mitogen depletion (83), indicating that myogenin is not necessary for terminal differentiation *per se*. This suggests that the role of myogenin *in vivo* may be to overcome a suppressive mechanism provided by the embryonic context, imposed by growth factors or cell–cell interactions. It is also possible that another MDF, MRF4, may be able to complement the function of myogenin in cell culture, by activating the same subset of downstream gene targets necessary for terminal differentiation.

Unexpectedly, mice carrying a null mutation in their MyoD loci are normal, underscoring the potential redundancy of MDF function *in vivo* (84). The muscles formed in the MyoD knock-out animals may be due to a compensatory increase in myf5 expression at later developmental stages, at which time the myf5 gene would normally be repressed. This possibility is supported by the extreme phenotype of mouse embryos carrying null mutations in both the myf5 and MyoD loci, generated by mating the two knock-out strains together (85). MyoD/myf5 knock-out animals lack all skeletal muscle, and present the skeletal defect in the ribs typical of myf5 null animals (81). Unlike the muscle beds of myogenin knock-out mice, which contain myoblasts that fail to differentiate, the presumptive muscle beds in the MyoD/myf5 double knock-out animals are totally devoid of myoblasts.

4.2 Interactions of the MDFs *in vivo*

A summary of the interplay of MDFs in generating the muscle phenotype is shown in Fig. 9. In this model, myf5 and MyoD have functionally redundant roles in activating the transcription of genes necessary for the generation of the myoblast phenotype during development. It is also evident that although both MyoD and myf5 are functionally inactive in cultured myoblasts (see Fig. 6), their expression is critical for earlier steps in myoblast formation in the developing embryo. Thus, both of these MDFs may have gene targets in myogenic cells that are distinct from the battery of structural genes activated during terminal differentiation. In contrast, myogenin appears to provide a non-overlapping function in the terminal phases of myogenic differentiation, which is likely to involve the relief of a transcriptional suppressive mechanism. MRF4 gene expression is activated at even later stages of differentiation after the formation of muscle fibres (86). A clearer picture of the relative roles of each MDF in muscle development will be gained from future analyses of mouse lines carrying different combinations of MDF gene disruptions.

5. MEF2 factors

It is clear from the complex functions of the MDFs described above that this family of transcription factors is part of a network of myogenic regulatory pathways. Although MDFs may control muscle structural gene expression directly, they may also activate expression of intermediate regulatory factors, or by co-operating with them control specific aspects of the myogenic phenotype. Current candidates include the myocyte enhancer factor 2 (MEF2) gene family of transcriptional regulators, which can act as intermediates in MDF action. MEF2 proteins are encoded by a group of four genes (MEF2-A, -B, -C, and -D) (Fig. 10), and are evolutionarily related to the MADS domain family of proteins, which include serum response factors

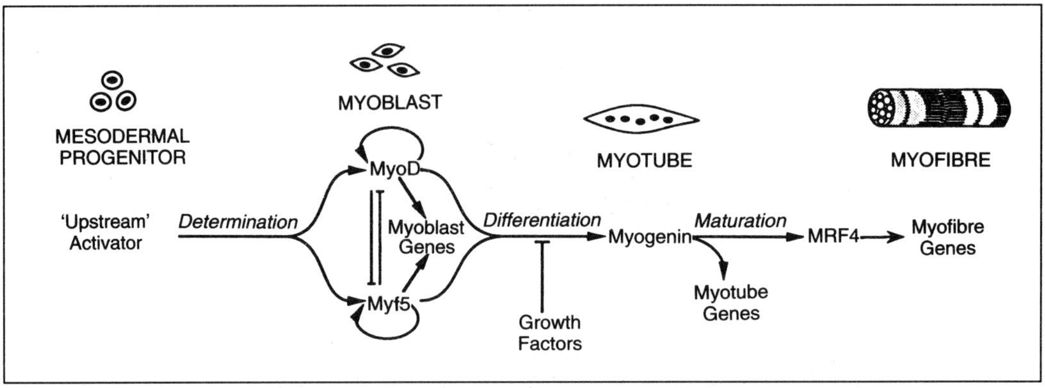

Fig. 9 Hypothetical regulatory pathway for muscle determination and differentiation during embryonic development (187). Either MyoD or myf5 are expressed as a consequence of myogenic lineage determination. MyoD and myf5 autoactivate their expression, but they may negatively regulate one another. Myoblast-specific genes are regulated in proliferating myoblasts by MyoD or myf5. A signal involving exogenous growth factors may activate differentiation *in vivo* as it does *in vitro*, whereupon MyoD and myf5 activate expression of myogenin, which activates myotube-specific genes. During myofibre maturation, myogenin expression declines and MRF4 is induced to control fibre-specific genes.

(26, 27; see Chapter 5). This regulatory family is characterized by a 56-amino acid domain, called the MADS box, that is essential for both DNA binding and dimerization. MEF2 factors contain an addtional motif adjacent to the MADS box, called the MEF2 domain (26, 27, 87–89) . MEF2 genes exhibit a complex pattern of activation during embryogenesis (31). The cell-type specificity of MEF2 gene expression is also modulated by alternative mRNA splicing and other post-transcriptional modification (26, 27, 31, 87–89). MEF2 proteins bind an A/T-rich motif present in the *cis*-regulatory regions of numerous muscle-specific genes, and are necessary for high levels of transcriptional activity (13–26). Increasing evidence points to this group of regulators as important modulators of myogenic gene expression.

5.1 MEF2 function in myogenesis

Several features of the MEF2 genes highlight their potential importance in myogenic gene regulation. First, MEF2 gene expression is induced in non-muscle cells in which MDFs are over-expressed (89), suggesting that they may be part of the myogenic regulatory circuit. Intriguingly, a MEF2 site in the myogenin promoter is essential for the activation of the myogenin gene (29, 30), underscoring an interdependency between the two groups of factors. Second, the MEF2 genes are activated in a wider range of cell types than are the MDFs, including cardiac and smooth, as well as striated, muscles (26, 27), and neuronal tissues (87–89), suggesting that they may participate in a number of cell-specific regulatory pathways. It is likely that certain MEF2 isoforms have distinct roles in the formation of certain skeletal muscle lineages, perhaps widening the scope of MDF function to include E box-independent gene activation, or maintaining the expression of the MDF genes themselves in a positive feedback loop. Finally, MEF2 gene expression can activate muscle gene

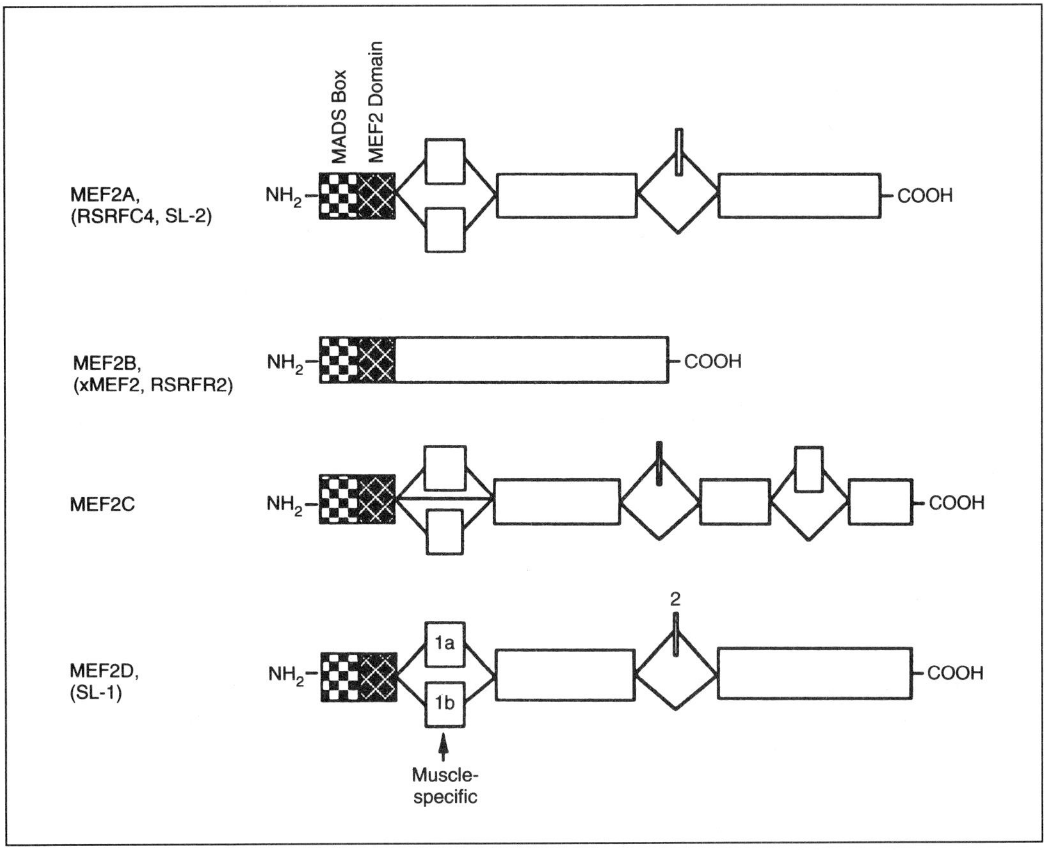

Fig. 10 Schematic respresentation of the MEF2 proteins described to date (31). The MADS (checkered) and MEF2 (cross-hatched) domains are indicated at the amino terminus of each protein. Alternative exons are shown.

expression in the absence of MDFs (17, 30), suggesting that it may participate in a parallel myogenic pathway. Future studies will explore the nature of the MDF–MEF2 interaction at the molecular level, and will assess the function of specific MEF2 isoforms in embryogenesis by targeted mutation. Given the broad spectrum of myogenic and neuronal cell types, it is likely that disruption of MEF2 function may have general consequences for development, although it may be necessary to knock out more than one of the MEF2 isoforms to observe these effects.

6. Regulation of muscle patterns during embryonic development

As seen from the studies described above, the formation of skeletal muscle in the vertebrate embryo can be attributed to the combined action of several classes of transcription factors expressed in myogenic lineages. Much less is currently known about the molecular cues that control the correct spatial and temporal expression of each of these factors. The complex, predictable organization of muscles in the vertebrate body plan must be generated early in development, guided by positional cues that co-ordinate their formation and morphology with other cells in the surrounding connective tissue, and with motor neurones that will eventually innervate each myotube in equally complex arrangements. The positional information that configures muscle pattern must therefore be integrated with regulatory programmes for muscle gene transcription. Recent studies have provided increasing evidence that the patterning of certain muscles in vertebrates and invertebrates may be directly determined by the same genetic pathways that control morphogenetic patterning in other cell types, such as connective tissue and the nervous system.

6.1 Homeobox genes as transcription factors controlling morphogenesis

An obvious candidate for the establishment of regional identity is the homeobox gene family, comprising an ancient, structurally related group of transcription factors (90; reviewed in Chapter 6, and see also Chapter 5). These genes are thought to regulate downstream gene expression in spatially and temporally restricted patterns to determine basic aspects of the animal body plan (91). A common structural motif, the homeobox, consists of a 183-bp stretch of nucleotides encoding the homeodomain that is capable of binding A/T-rich sequences in DNA (92–94). Homeobox genes can be divided into three distinct categories based on their functions in establishing tissue patterns. The first category includes the clustered HOM/Hox genes, which are conserved in both inverebrates (HOM) and vertebrates (Hox), expressed in spatially and temporally restricted patterns or codes to confer regional identities in tissues along embryonic axes. Perturbation of HOM/Hox expression results in characteristic homeotic transformations of cell identities. The second category includes many non-clustered homeobox-containing genes scattered throughout the genome, whose deletion results in the loss of cells in which they are normally expressed. Finally, a third category includes other non-clustered homeobox genes that exert indirect effects on morphogenetic processes by controlling the growth or differentiation of tissue-specific stem cells. Members from each of these three categories have been found to affect the pattern of cells committed to the myogenic phenotype.

6.2 HOM/Hox genes in invertebrate muscle patterning

The most direct evidence for the involvement of the HOM/Hox gene group in the control of muscle pattern comes from studies in Drosophila, demonstrating that the HOM genes belonging to the *Ultrabithorax-C* cluster control the regional identities of somatic abdominal muscles and their precursors. Ectopic expression of several genes from this category in the thoracic mesoderm confers an abdominal morphology on the resulting thoracic muscles, a classical example of homeotic transformation (95, 96). Notably, these homeotic genes may compete for the regulation of the common downstream gene *nautilis*, the Drosophila equivalent of the vertebrate MyoD gene (96). Positional fates of muscle precursors in Drosophila may therefore be established through direct regulation of the MDF gene family by clustered homeotic gene expression in developing mesoderm. It is likely that other targets are involved, since Nautilus is expressed ubiquitously in Drosophila muscle and therefore could not be expected to confer positional identity.

It remains to be seen whether these clustered homeobox genes play similar roles in the muscle patterning of higher organisms. A crucial but indirect role for the clustered Hox genes in muscle patterning has long been recognized in the vertebrate limb bud, where their expression in the surrounding mesenchyme determines the morphology of the muscle tissue that migrates into them (97; see Fig. 11). If Hox genes are directly involved in assigning muscle positional identity, Hox gene products must be present at some time in muscle development, either in committed myoblasts or in the mesodermal precursors that give rise to myogenic lineages. Hox gene expression is generally not detected in the mature myotome where the myogenic programme is already activated. However, transient expression of specific Hox genes has been documented throughout the early somite (98) at a time when positional identities in other somitic derivatives are established (99), and later in a regionally restricted group of developing myogenic cells at the most dorsal boundary of the dermamyotomal lip (100; see Fig. 11). These observations are intriguing, but their functional significance needs to be tested by perturbing the expression of specific Hox genes in the myogenic portion of the somite, and detecting a corresponding disruption in the resulting musculature.

6.3 Muscle-specific targets for Hox gene action

Conclusive evidence for the involvement of Hox genes in vertebrate muscle positional fates will probably require the identification of downstream targets in muscle cells or their precursors, since subtle changes in muscle morphology induced by ectopic Hox gene expression may be difficult to detect. Gene markers of muscle position have been revealed by studies in transgenic mice, in which several muscle-specific *cis*-regulatory regions driving reporter transgenes are responsive to positional cues in the embryo. One of these transgenes, a CAT reporter driven by the myosin light chain (MLC) 1 promoter and downstream enhancer (3), is expressed in a gradient in the developing somites, increasing in an anteroposterior direction and persisting in the segmented axial muscles of the adult (101–103). This expression pattern is an exclusive feature of the MLC-CAT transgene, since it is not reflected in the accumulation of the endogenous MLC1 transcript, and is probably attributable to the unique configuration of the MLC regulatory elements in the transgene construct. The signals responsible for the MLC-CAT transgene gradient in embryos appear to be cell autonomous, since dissociated myotomal cells maintain transgene expression levels characteristic of their anteroposterior position when induced to differentiate in culture (103). The clustered Hox genes are attractive candidates for

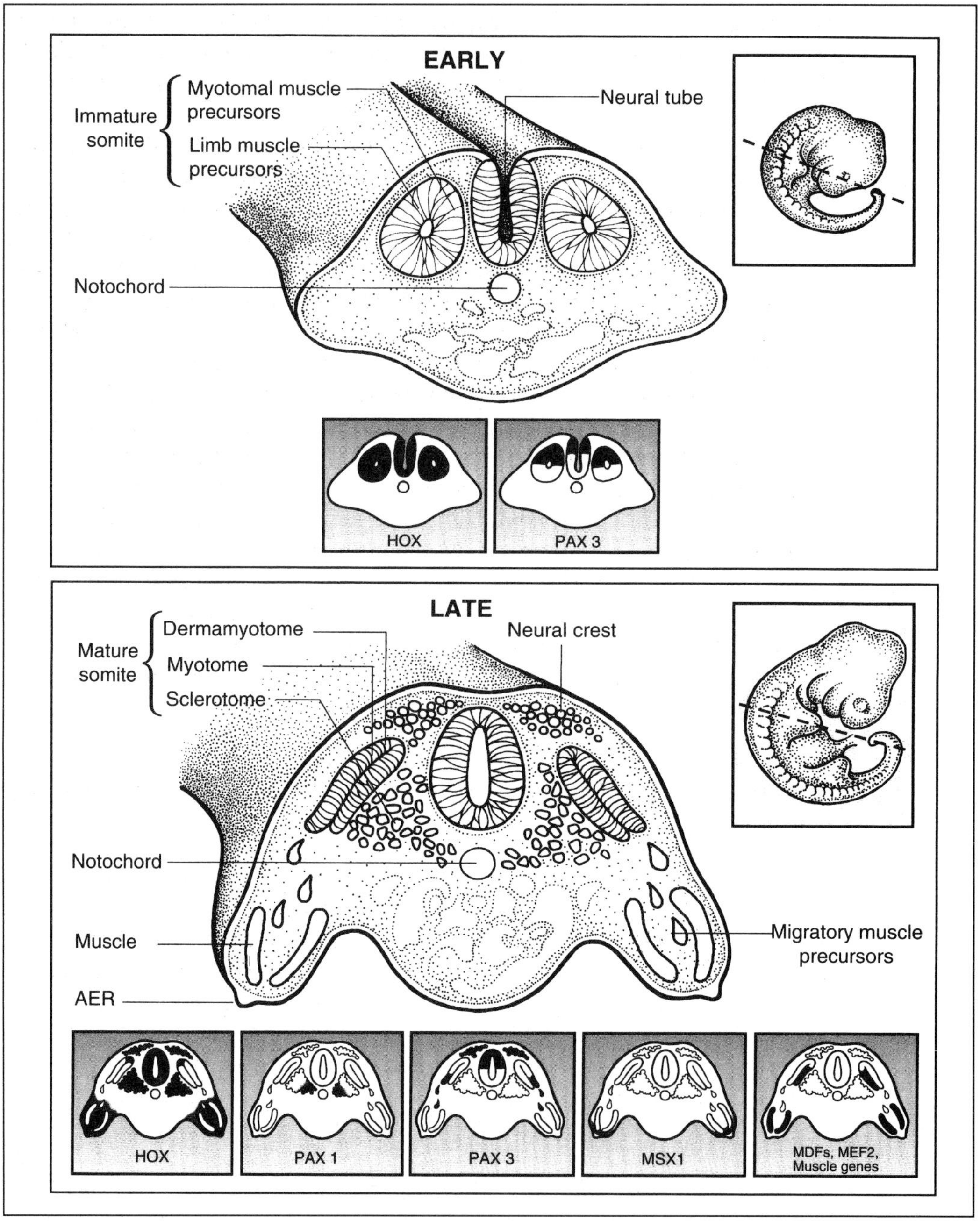

Fig. 11 Expression of homeobox genes in muscle cell precursors and their descendants in the vertebrate embryo (188). Top: diagram of early vertebrate embryo, transverse section. Cartoons in boxes show homeobox gene expression patterns (black) at this stage. Clustered Hox genes are expressed throughout the neural tube and paraxial mesoderm, including the immature somite. Pax-3 expression is restricted to the dorsal hemisphere of the uncompartmentalized somite. Bottom: diagram of later vertebrate embryo, transverse section. Left half of section shows the three cell lineages comprising the differentiated somite, and muscle precursor

establishing the MLC-CAT transgene gradient in the early somite, but it is not yet clear whether this is a direct result of Hox protein interaction with the MLC transgene regulatory elements, or whether intermediate transcription factors are involved.

Notably, muscle-specific promoters and enhancers often share sequence motifs, any of which could contribute to their responsiveness to positional cues in the early somite (see Fig. 1), either by direct or indirect interaction with Hox gene products. Of particular interest is a common A/T-rich site, which represents a binding site for the MADS box family of *trans*-activators, including the MEF2 family of myogenic factors (see Section 5). Functional interactions between homeodomain proteins and MADS factors have been demonstrated in yeast (104) as well as in animal cells (21, 105; see Chapter 5). A mutation in the myogenin promoter that abolishes the MEF2 binding site affects linked transgene expression in different subsets of myogenic precursors along the anteroposterior axis (29), suggesting that it may be a target for positional regulation in the embryo. In addition, the MCK enhancer contains an A/T-rich binding site, the gene product of the pHox/MHox gene, a non-clustered member of the homeobox gene superfamily, which is expressed in high levels in adult skeletal muscle. This site is essential for muscle-specific transcription and *trans*-activation by the MDF gene family (106). It is therefore feasible that Hox gene function in myogenic cells may require protein–protein interactions with selected muscle-specific factors, which in turn may influence the transcription of linked genes to assign positional identities in muscles according to a segmented body plan.

6.4 The role of other homeobox genes in regulating muscle identity

Several non-clustered homeobox genes have been implicated in the specification of muscle cells in both invertebrate and vertebrate systems. For example, the Drosophila homeobox gene *apterous* is expressed in a subset of mesodermal cells that generate six muscles in each abdominal hemisegment. These muscles are absent when the *apterous* gene is ablated, whereas ectopic Apterous expression produces additional muscles (107). Similarly, mutations in the homeobox-containing *tinman* gene, normally expressed in the Drosophila mesoderm, prevent the formation of specific somatic muscle founder cells (108). In vertebrates, the *Pax-3* gene, a member of the paired-type subclass of homeobox genes, is expressed in the mesodermal precursors that migrate from the somites to the developing limb buds to form muscle (109, 110; Fig. 11). These precursors do not express any known myo-

cells migrating from the lateral somite to form muscles in the developing limb bud. Boxes (left to right) show clustered Hox gene expression now limited to the neural crest, sclerotome, and the medial lip of the dermamyotome. Pax-1 is expressed in the sclerotomal cells nearest the notochord. Pax-3 is expressed in the lateral myotome and migrating limb muscle precursors, and in the upper half of the neural tube and neural crest. Msx-1 is expressed in the apical ectodermal ridge (AER) of the limb bud. Myogenic determination factors (MDFs and MEF2) and muscle-specific genes are expressed in the medial myotome and in fully differentiated limb muscles.

genic factors, nor do they appear to arise from cells that ever did (75). In mice lacking the *Pax-3* gene (a mutant called *Splotch*), limb muscle is absent (111), suggesting that the *Pax-3* gene product participates in the generation of a migratory muscle precursor cell population, by controlling either the formation in the somite, or the migration to the limb bud. It remains to be seen whether ectopic expression of the *Pax-3* gene product in non-muscle cells will induce extra muscle formation, as is the case for the *apterous* gene. However, the data at hand suggest that both these genes are directly involved in establishing the identity of specific muscle cells during development.

6.5 Homeobox genes as extrinsic effectors of muscle growth and differentiation

In the examples cited above, homeobox genes have been implicated in the developmental programmes intrinsic to the myogenic lineages. However, the spatially restricted expression patterns of several homeobox genes exert indirect effects on muscle formation in the vertebrate embryo, by inhibiting either the generation of myogenic precursors or the onset of myogenic differentiation. Two such examples are *Pax-1*, another paired-type gene, and *msx-1*, a non-clustered homeobox gene homologous to the *msh* gene in Drosophila (112). Both of these genes selectively inhibit myogenic differentiation, but at different stages of muscle development. *Pax-1* is normally expressed in the sclerotomal compartment of the mature somite, which gives rise to vertebral structures (113; Fig. 11). *Pax-1* may be a component in the dorsal–ventral patterning mechanism by which proliferation of cartilage is induced at the expense of myogenic tissue (114). At a later point in embryogenesis, expression of the *msx-1* gene is localized to the distal end of the growing limb bud. Although the muscle precursors are already in this region, terminal muscle differentiation is repressed, perhaps by a pathway involving *msx-1* (115; Fig. 11). Indeed, forced expression of *msx-1* blocks or delays muscle differentiation in cell culture, consistent with a role in the selective inhibition of myogenesis at the proliferative end of the developing limb bud.

The specific pathways by which homeobox genes indirectly control muscle gene transcription remain to be worked out, but the examples described above serve to underscore the importance of both inductive and inhibitory influences on appropriate patterning of muscle tissue. Understanding how positional information and cell identity conferred by homeobox genes affects cell-specific regulatory programmes is an important focus for future studies in this field, in which the muscle system is likely to play a prominent role.

7. Transcriptional control of skeletal muscle fibre type

Although skeletal muscle superficially appears to be a relatively homogeneous tissue, it is actually composed of a heterogeneous population of myofibres that, together, modulate the primary function of contractibility. The multiple types of fibres found in the skeletal muscle of adult vertebrates reflect an additional level of complexity in the regulation of muscle-specific gene transcription. The heterogeneous nature of muscle has been known for well over a century (116). Fibres were originally classified according to differences in evoked contraction rate, but biochemical, histochemical, and immunological assays have since been developed to allow typing of individual fibres within sections of muscle tissue. These techniques have revealed the molecular diversity amongst fibres, both at the cell surface, and in the energy-generating and contractile systems. Furthermore, changes in fibre composition represent a major component in muscle degeneration associated with both ageing and neuromuscular diseases. The differences can ultimately be traced to differences in gene expression, which contribute to the versatility in energy metabolism, fatigue resistance, and contraction rates that satisfy the functional requirements of individual muscles. Identification of the specific mediators that determine fibre type-specific gene expression is therefore a focus of current research.

7.1 Definition of muscle fibre types

The phenotypes of specific muscle fibres have been traditionally defined by the myosin heavy chain (MyHC) isoforms they express. The contraction rate of a muscle fibre is essentially regulated by its ATPase activity. This in turn is determined largely by the particular MyHC isoforms expressed in the fibre (Fig. 12). Each of the major fibre types contains a different MyHC that is expressed by a distinct gene. The MyHCs with the highest ATPase activity are expressed in those fibres that contract most rapidly, while MyHC isoforms with low rates of ATP hydrolysis are found in those muscle fibres that contract slowly (117, 118).

The diversity of MyHC genes expressed throughout the life span of an organism reflects the enormous versatility of skeletal muscle structure and function. At least ten genes that encode MyHCs expressed in skeletal muscle exist in mice, many of which have been shown to be clustered together in a single locus (119). It remains to be seen whether the chromosomal position of these genes within the MyHC locus governs the order of developmental expression in a manner similar to that of the homeobox and globin gene loci (120). Each MyHC isoform has a unique pattern of expression in development and is often expressed in several muscle cell types. These include isoforms that are transiently expressed in skeletal muscle fibres during development (embryonic, perinatal, and α-cardiac), two adult slow fibre-specific isoforms (Type I, or β-cardiac, and slow tonic), and five fast-specific isoforms (Types IIA, IIB, IIX, IIeom, and IIm). Types IIoem and IIm are highly specialized MyHCs found only in the superfast extraocular fibres and in certain jaw muscles of carnivores, respectively. Many adult fibres express more than one isoform at a time, and so are considered to be hybrid or mixed fibre types. This suggests that MyHC genes expressed together in certain fibres may contain common transcriptional regulatory sequences that are recognized by fibre-specific transcription factors, although these factors remain to be identified.

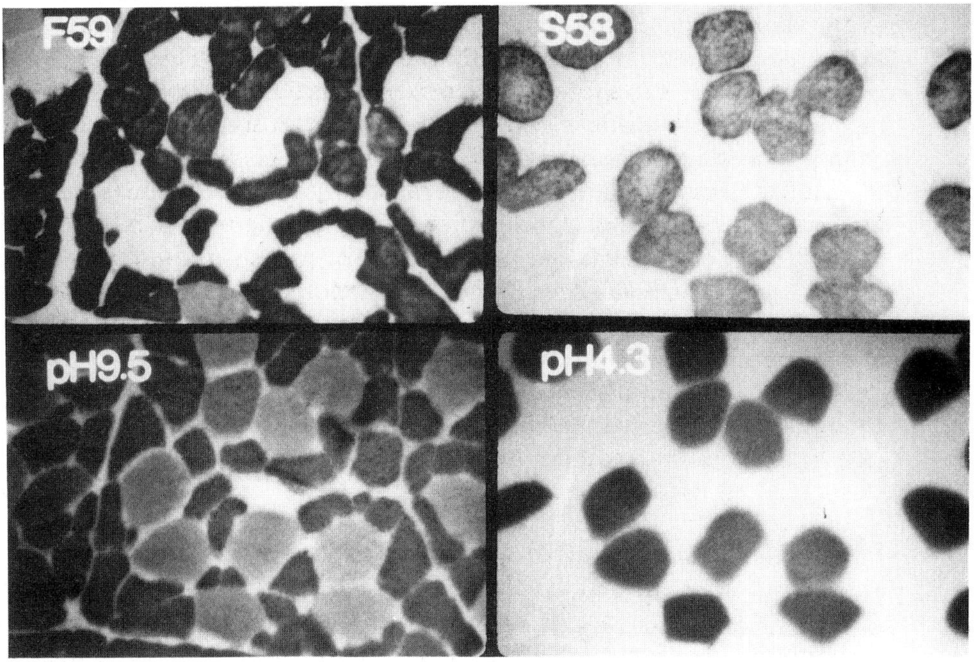

Fig. 12 The sartorius muscle of chick contains both fast and slow fibres. Two methods are in common use to determine fibre type in striated muscle. The four panels show parallel cross-sections of muscle, the top left is probed with the antibody F59 (189) that cross-reacts with all fast MyHC isoforms while the slow MyHC-specific antibody S58 is used in the top right. Antibody binding was visualized with a horseradish peroxidase-conjugated secondary antibody, so that the muscle fibres expressing particular MyHCs stain dark. The bottom two panels demonstrate differential histochemical staining of myofibral ATPase activity to determine fibre types. After pre-incubation of a serial cross- section at pH 9.5 (bottom left panel), the ATPase activity of Type II (fast) fibres is stable, while that of Type I is labile. Pre-incubation at pH 4.3 (bottom right) destroys Type II ATPase activity while preserving Type I activity. (Figure courtesy of J.B. Miller.)

In addition to the myosin heavy chains, many other muscle structural proteins are expressed in a fibre-specific manner. These include the troponin subunits, myosin light chains, α-tropomysin, and the sarcoplasmic reticulum Ca^{2+}–ATPase (120,121). There are separate isoforms expressed in slow and fast fibres for each of these proteins. Together with the MyHCs, these proteins are expressed in unique combinations that produce the necessary range of muscular function. The signals responsible for generating this diversity are complex and responsive to environmental factors but ultimately rely on gene expression patterns originally established in the developing embryo.

7.2 Emergence of fibre diversity in the embryo

The mechanisms by which specialized muscle fibres are generated from their less specialized embryonic precursors are of particular importance for understanding the molecular basis of fibre diversity. Myogenic precursor cells arise early in development in the dermamyotomal region of the somite (Fig. 11). These cells give rise to a population of embryonic myoblasts that are heterogeneous in character (77; Fig. 13). Upon differentiation, they express unique combinations of fibre-specific proteins. Current research is focused on the mechanisms that are responsible for generating fibre-specific gene expression programmes in the embryo.

There are currently several alternative models for the generation of fibre diversity, each invoking different pathways of transcriptional control. In the first model, myoblasts are homogeneous and capable of giving rise to all of the different types of fibres in skeletal muscle. Environmental influences, such as circulating hormones, growth factors, and innervation then play a crucial role in determining fibre-specific gene expression patterns. An environmental component is certainly evident in the adult, as illustrated by the imposition of fibre type by motor innervation. A classic experiment (122) demonstrated that the fibre type properties of an adult muscle can be transformed by cross-innervation with a foreign nerve. In these experiments a fast fibre gradually takes on the characteristics of a slow fibre upon reinnervation by a neurone with a slow pattern of electrical activity, with a corresponding change in gene expression. The effects of cross-innervation can be duplicated by using an appropriate frequency of electrical stimulation, directly demonstrating the role of electrical activity in modulating gene expression in the mature muscle fibre (123). From these results, it was inferred that the nerve most likely also instructed muscle fibre differentiation during development. However, this process is initiated prior to muscle innervation (124–128), and the initiation of myofibre diversity proceeds normally in embryonic chick and rat limbs that have been denervated *in vivo* by removal of the neural tube or treatment with the drug d-tubocurare (129). These results suggest that at least in early development, fibre-specific gene expression patterns must be determined by other mechanisms.

In the second model, muscle fibre type is dependent upon lineage directives that are inherited by differentiating cells from their progenitors. Myoblasts committed to the same fate must then fuse with each other, such that different fibre types would be derived from intrinsically distinct types or lineages of myoblasts. This model is supported by experiments involving the fusion of individual primary embryonic chick myoblasts propagated in culture. The resulting myotubes form three classes, those that express only fast isoforms, those that express only slow, with the remainder expressing both. A clonal population of cells is used to derive each set of myotubes and every cell within the set expresses the same MyHC isoforms. The initial myoblasts must therefore be intrinsically different in order to form myotubes with distinct patterns of transcription in the absence of innervation (130).

A third more complex model invokes an interplay between environmental signalling and intrinsic cellular programmes. In this multi-step model, the range of genes expressed in different lineages becomes increasingly restricted during development (120). Transcriptional regulatory pathways that determine a particular fibre type would result in specific patterns of cell surface markers, presented on the surface of muscle precursors destined to respond selectively to extracellular signals,

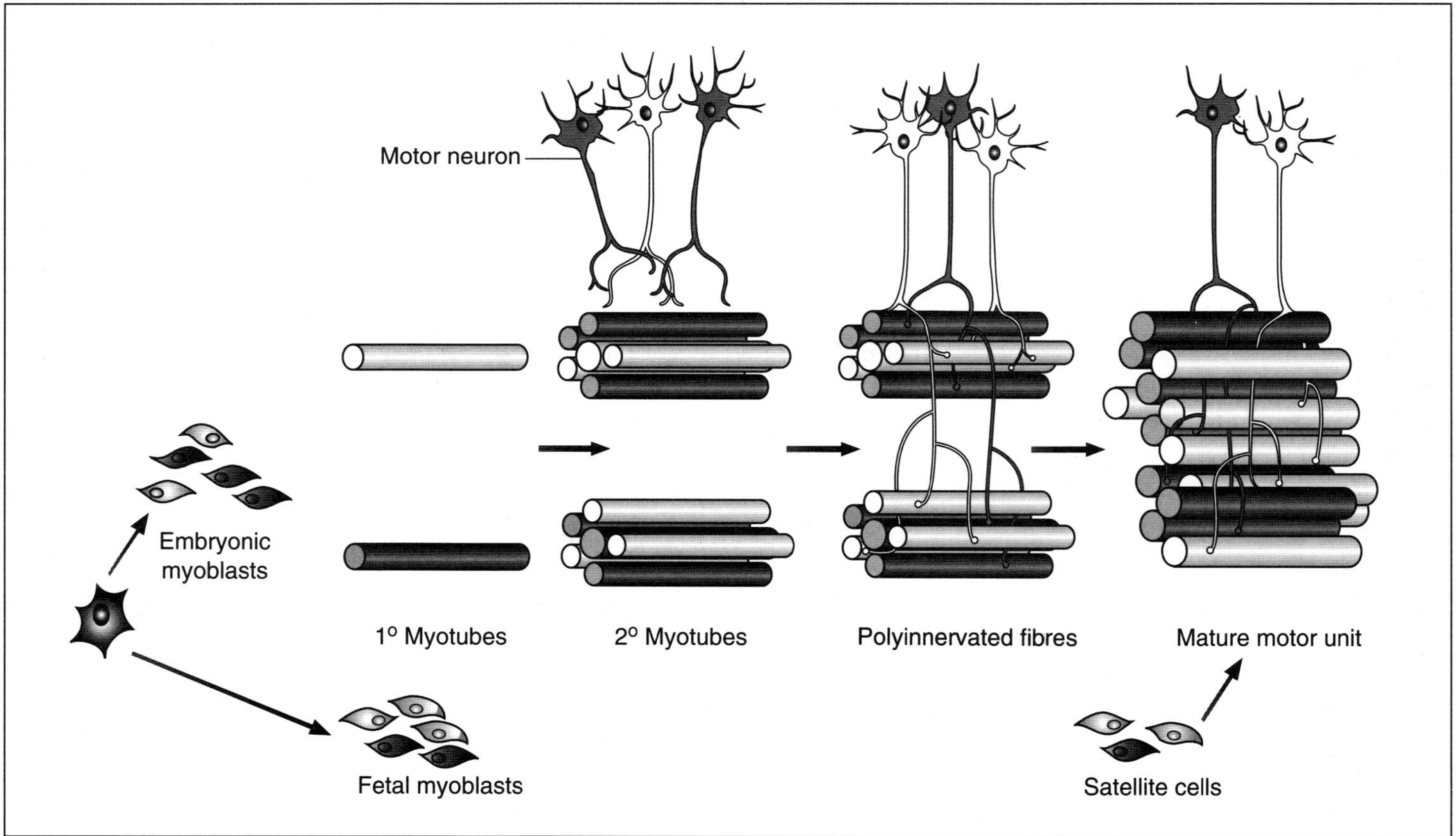

Fig. 13 Developmental stages in the generation of a motor unit. A population of myogenic precursor cells, under the influence of the neural tube, become determined embryonic and fetal myoblasts. At this period, myoblasts may already have intrinsic differences that contribute to the formation of fast and slow muscle fibres (dark and light cells). Embryonic myoblasts fuse to form large, multi-nucleated primary myotubes. A couple of days later, fetal myoblasts contribute to the formation of a rosette of smaller, secondary myotubes around the primary fibre. Both primary and secondary fibres then become innervated by multiple appropriate (fast or slow) motor neurones. During a protracted period after birth, most motor neurones die, leaving each muscle fibre with a single motor neurone contact. Satellite cells contribute to the repair of fibres damaged by injury or disease. (Figure adapted from 190.)

with innervation playing a role in maintaining selected fibre-specific gene expression in the mature myofibre. Although the molecular details of this hypothetical cascade are not yet known, the model intregrates current information and provides an attractive framework for further investigation.

7.3 Fibre-specific regulation of transgenes

The characterization of fibre-specific regulatory mechanisms in adult muscle has been largely carried out *in vivo* using transgenic mice, since muscle cell lines and primary muscle cultures can generate only limited fibre-specific phenotypes. Fibre-specific expression has thus far been demonstrated for several transgenes.

7.3.1 Troponin I gene regulation

To date the most extensive analysis of fast and slow fibre-specific gene regulation has been performed for the troponins. Fast and slow fibres have arisen by the time of birth in mammals; however, the various fast fibres types have not yet differentiated. This indicates the existence of two regulatory mechanisms contributing to fibre type-specific gene expression, one acting pre-natally, the other after birth when fibre subtypes are being established.

The expression pattern of the Troponin I (TnI_{fast}) transgene reflects this two-stage developmental process (131). Initially, the TnI_{fast} isoform is expressed exclusively in the early fast fibres, but at low and relatively homogeneous levels. The emergence of different fast fibre types happens in the initial two-week post-natal period, whereupon the expression of the TnI_{fast} gene is markedly increased in Type IIB fibres while remaining quite low in Types IIX and IIA. This developmental pattern is retained by a 2.5-kb fragment of DNA including the promoter for the quail TnI_{fast} gene, linked to a reporter transgene that is expressed exclusively in fast fibres, predominantly in Type IIB, and to a lesser degree in Types IIX and IIA in transgenic mice (132). As appears to be common with muscle-specific genes, the TnI_{fast} promoter has only weak activity by itself, and relies on an enhancer downstream of the RNA start site, in this case, within the first intron. These TnI_{fast} transcriptional control regions contain several *cis*-regulatory elements required for expression in cell culture, including a consensus binding site for the myogenic factors (133). Whether this site also confers fibre-specific expression is not yet known.

The *cis*-regulatory elements in the rat troponin I slow (TnI_{slow}) gene, conferring tissue specific expression as well as fibre specificity, have also been characterized in transgenic mice. TnI_{slow} is expressed exclusively in slow-twitch fibres. A transgene including the TnI_{slow} promoter spanning the first few hundred basepairs upstream of the transcription start point, and sequences contained within the first two introns, is preferentially expressed in the soleus (a muscle primarily composed of slow-twitch fibres) and is significantly lower in the extensor digitorum longus (EDL), gastrocnemius, and tibialis (all fast-twitch muscles). An analysis of the

TnI_{slow} transgene expression during development has not yet been performed, and a more precise identification of *cis*-regulatory elements awaits further investigation.

7.3.2 Myosin light chain gene regulation

Like the fast troponin I gene, the MLC1/3 locus is expressed at the highest levels in Type IIB fibres. The MLC1 isoform is expressed from a weak promoter that requires a downstream enhancer to direct efficient, muscle-specific expression in cell culture (134). A transgene carrying the rat MLC1 promoter and the downstream MLC enhancer is expressed preferentially in fast Type IIB fibres (135), suggesting that these elements are sufficient to respond to fibre-specific transcriptional regulatory signals. Further, the substitution of a heterologous promoter for the MLC1 promoter does not affect the expression patterns of the transgene construct, which narrows the fibre-specific *cis*-acting sequences to the downstream MLC enhancer (136). This enhancer contains binding sites for several myogenic transcription factors, including the MDFs and MEF2 family (Fig. 1), that may be involved in its fibre-specific activation pattern *in vivo*.

7.4 MDFs and regulation of fibre type

Due to the technical limitations in studying fibre-specific gene expression in cell culture, little is known about the *trans*-acting factors that regulate muscle gene transcription in a fibre-specific manner. The MDFs are not activated in fibre-specific patterns during development, although in adult muscle the relative expression levels of myogenin and MyoD have been shown to be higher in slow Type I and fast Type IIB fibres, respectively (137). A mechanism could be postulated where a threshhold amount of a particular myogenic factor is required to activate a programme of fibre-specific gene expression, by preferentially binding to specific E box targets (138). It is not clear, however, that all fibre specific transcriptional regulation could be mediated through E boxes alone. For example, an MyHC expressed specifically in Type IIB fibres does not require an E box for muscle-specific expression (139). Furthermore, it remains to be explained how the transcription factors MyoD and myogenin themselves are regulated at the transcriptional level in a fibre-specific manner.

8. Neuronal control of muscle gene transcription

Transcription in a fully differentiated muscle fibre is subject to extrinsic modulation, most dramatically by neuronal stimulation. Muscle cells, like neurones, can be electrically excited by signals emanating from the neuromuscular junction (NMJ), where synapses between motor neurones and muscle fibres are formed. The nuclei in the synaptic region of the multi-nucleate muscle fibre maintain characteristic patterns of transcription that are distinct from the nuclei in the rest of the fibre, preferentially expressing many synapse-specific proteins that contribute to neuromuscular signal transduction. The regulation of this selective gene expression is complicated, involving active repression of transcription in the extrajunctional nuclei, and very localized induction in those nuclei lying directly beneath the motor neurone. Recent advances in our understanding of the molecular mechanisms underlying neuromuscular stimulation have made skeletal muscle a valuable model system for studying how *trans*-synaptic signals can induce phenotypic changes by modulating gene expression.

The establishment of synapse-specific gene expression involves at least three different regulatory mechanisms of intracellular gene expression. The initial expression of synapse-specific genes accompanies myoblast fusion and is innervation independent. The induction of these genes is controlled by the myogenic factors, as part of the general programme of muscle differentiation. The second mechanism involves formation of a synapse with the muscle cell by the motor neurone, which controls both the distribution and number of neuromuscular receptors. Innervation causes a clustering of receptors beneath the nerve terminal, in part due to increased transcription in subjunctional nuclei. A third stage involves the repression of synapse-specific gene expression in extrasynaptic nuclei. This is achieved through the activation of a protein kinase by electrical activity, which in turn inactivates myogenic factors. Nerve-derived trophic factors appear to mediate at least a portion of the localized induction of transcription and may contribute to the formation of differences in the basal lamina of the synaptic and extrasynaptic regions responsible for a sustained extracellular synapse-specific signal. The molecular pathways governing these functions are outlined in more detail below.

8.1 Structure and function of the neuromuscular junction

In the 1930s, it was demonstrated that stimulating a motor neurone caused the release of acetylcholine (140), subsequently shown to be the neurotransmitter that stimulates a muscle to contract. The recipient of the signal on the post-synaptic side of the synapse, the nicotinic acetylcholine receptor (AChR) protein, is the dominant component of a mature neuromuscular synapse. The AChR is a transmembrane heteropentamer encoded by four homologous genes under co-ordinate transcriptional control.

The AChR is an agonist-gated cation channel that responds to acetylcholine by opening its channel to allow cations, principally sodium ions, to flow into the cell. The sodium depolarizes the cell, causing the voltage-sensitive sodium channels to open and propagate the depolarization (called the action potential or electrical activity) down the entire plasma membrane. This in turn causes the voltage-sensitive calcium channels to open. The sudden increase in cytosolic calcium then both elicits the contractile response of the muscle and activates the signal cascade responsible, maintaining appropriate muscle-specific gene transcription.

8.2 Initial expression of AChR is independent of neural input

As with other muscle-specific genes, transcription of the AChR subunit genes in the newly forming myotubes is independent of neuronal influence (141). The receptor genes are transcriptionally induced in myoblasts that have left the cell cycle and are competent to fuse. This induction can also be seen *in vitro*, both in muscle cell lines and cultured primary muscle cells.

Upon the formation of a mature NMJ, however, AChRs become restricted to the synapse. The density of the subjunctional AChRs directly beneath the motor nerve ending reaches 10 000 per square micrometre in a mature animal. This falls by more than 1000-fold only a few micrometres away from the NMJ. Other muscle proteins, such as acetylcholine esterase, sodium channel, dystrophin (142), 43 kDa protein (a protein that may play a role in AChR aggregation; 143), and several components of the cytoskeleton (144), also show localization to the synaptic region, suggesting a common mechanism for the co-ordinate regulation of these genes in subsynaptic nuclei.

8.3 Selective transcription in subsynaptic nuclei

The elimination of extrasynaptic receptors during development is mediated by restricting transcription to those nuclei lying directly beneath the motor neurone. This has been demonstrated with two separate procedures. *De novo* hybridization with intronic probes demonstrates that significant AChR transcription occurs only in subsynaptic nuclei (145, 146). In addition, transgenic mice carrying a reporter gene with a nuclear localization signal and the 5′ regulatory sequences of an AChR subunit gene contain reporter gene product localized to the synaptic region of the myofibre (145–147). Synapse-restricted gene expression is evident soon after the initial formation of synapses during development, indicating that the pathway regulating AChR transcription is also involved in the formation and maintenance of neuromuscular synapses (145, 146).

8.4 Electrical activity blocks extrasynaptic gene expression

Several lines of evidence have demonstrated the importance of electrical activity in repressing AChR subunit gene transcription of extrasynaptic nuclei. Denervation of skeletal muscle by severing the innervating motor neurone causes a dramatic increase in acetylcholine sensitivity in extrajunctional areas (148). The change is due to the reappearance of newly synthesized AChR on the entire muscle fibre surface (149). The normally quiescent AChR genes in the extrasynaptic nuclei become transcriptionally active (150), causing levels of AChR subunit mRNA to rise by up to 80-fold over a period of three days. If the muscle is allowed to reinnervate, extrasynaptic expression of AChR is again lost. The motor neurone therefore exerts a repressive effect upon extrasynaptic AChR gene expression, which could possibly be mediated either by a chemical signal from the nerve, or through electrical activity arising from synaptic transmission. In either case, synaptic signals must be selectively transduced to extrasynaptic nuclei in the muscle fibre.

Recent research in this area has focused on identifying the transcription factors that are responsive to innervation in both subsynaptic and extrasyanptic nuclei. All regulatory sequences of the AChR subunit genes contain E boxes shown to be essential for both stage- and tissue-specific expression when examined *in situ* (151–156). The E boxes are also present in the regulatory sequences conferring sensitivity to membrane electrical activity to transgenes in mice (136, 157–160). Recent work by Changeux and co-workers (161) demonstrates that of the two E boxes contributing to muscle-specific expression of the chick AChR α-subunit gene, one is required for electrical responsiveness, implicating the myogenic factors as possible mediators of the denervation signals. In fact, the myogenin message increases dramatically upon denervation in chick muscle (greater than 200-fold), with kinetics that are indistinguishable from increases in AChR α-subunit transcription (162). In contrast, the other myogenic factor genes have a minor denervation response (MyoD and myf5), or do not respond at all (MRF4). Moreover, only myogenin transcripts have been observed to decrease rapidly in response to inactivation by electrical activity (162), making myogenin a likely candidate for mediating innervation signals.

Regulation of myogenin transcription cannot account for the repression of synaptic proteins in extrasynaptic regions, however. Although the binding sites for myogenic factors have been shown to be essential for activating AChR subunit gene expression during denervation, expression of the myogenic factors, including myogenin, is not restricted to subsynaptic nuclei (146, 163). A mechanism to inactivate myogenin in the extrasynaptic nuclei of innervated myotubes, but allow for its activation upon removal of neural input, is likely to involve post-translational modification of myogenin protein itself.

Recent evidence for post-translational control of MDF activity supports a model, shown in Fig. 14, in which the phosphorylation status of the myogenin protein determines its transcriptional regulatory function. Myogenin is found in a phosphorylated state in electrically active myotubes. This is due to the influx of calcium upon myofibre depolarization, which activates protein kinase C (PKC; 151, 164–166). PKCα, the major isoform of the enzyme found in skeletal muscle, can inactivate myogenin by phosphorylation (167). This blocks autoactivation of the myogenin gene by its own product, and maintains low basal levels of synapse-specific genes, such as those encoding the AChR subunits. Blocking electrical activity or eliminating neural input causes dephosphorylation of the myogenin protein, a result also found after pharmacologically inhibiting PKC activity (168). In a dephosphorylated state, myogenin protein rapidly autoactivates its own promoter (22), stimulating its own transcription. When an adequate level of the myogenin protein is attained, transcription of the AChR subunit genes is induced. In this model, control of myogenin transcription is a focal point of neurone-regulated gene expression in the muscle fibre, but it is possible that other as yet unidentified regulatory genes may respond to the same signals. In addition, additional regulatory mechanisms must exist to restrict expression of specific proteins, such as the AChRs, to subsynaptic regions in the innervated muscle cell.

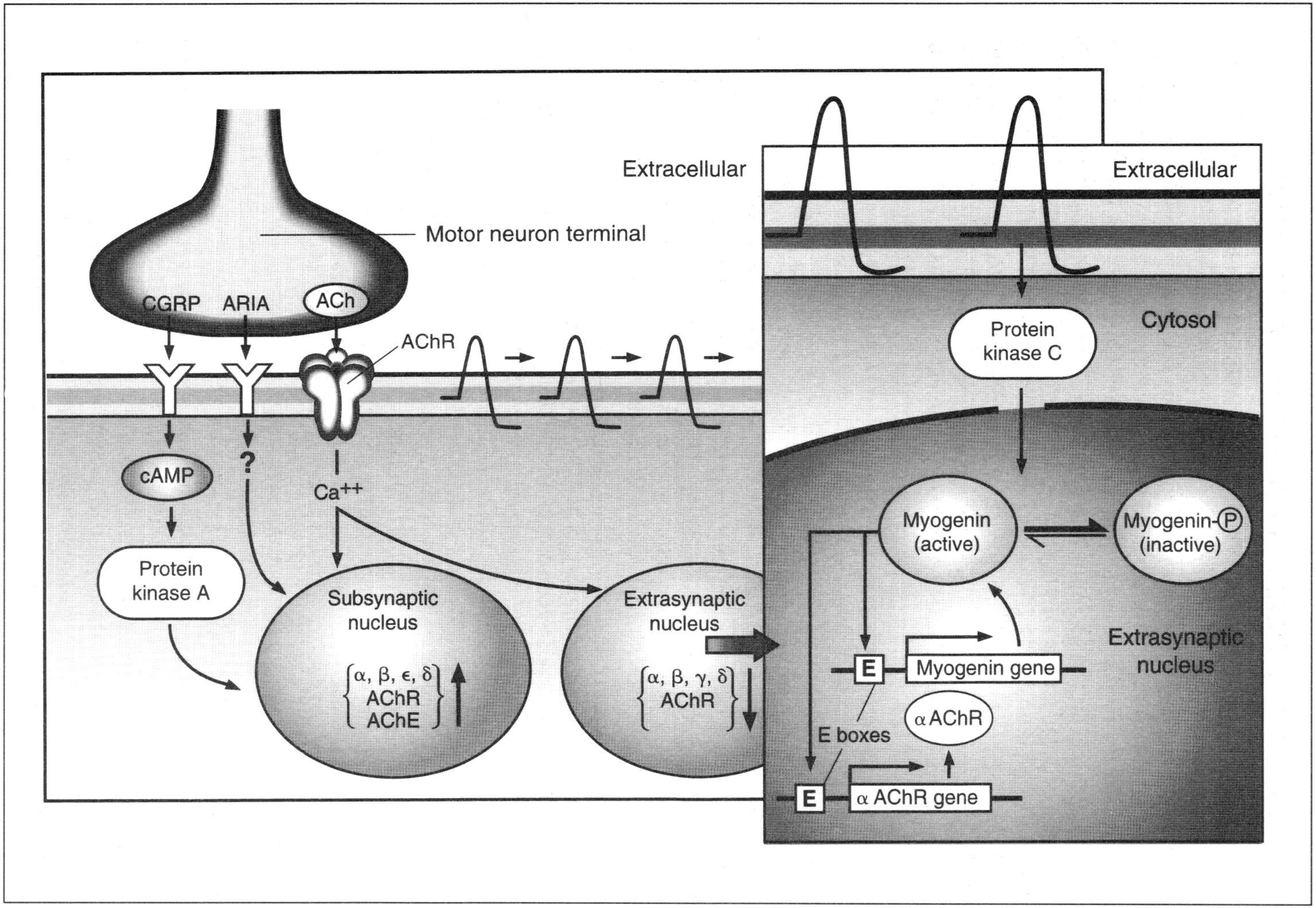

Fig. 14 Regulation of gene expression by the neuromuscular junction (insert: the extrajunctional region detailing the myogenin autocatalytic loop). A motor neurone forms a gap junction or synapse with a muscle fibre at the neuromuscular junction. Several factors produced by the innervating neurone, including the proteins ARIA and CGRP and the neurotransmitter acetylcholine, diffuse across the synapse to play an instructive role in the synthesis of AChR in the subsynaptic nuclei. AChR with bound acetylcholine allows ions to flow into the cell, initiating electical activity. The intracellular calcium activates subsynaptic AChR gene transcription via PKA and inhibits extrasynaptic expression via PKC. See text for details.

8.5 Trophic factors mediate synapse-specific transcription

Selective transcription of subjunctional nuclei is a highly regulated process, as the membrane beneath the nerve terminal occupies less than 0.1% of the muscle fibre surface. Soluble factors from the innervating motor neurone are believed to cause a local override of the global transcriptional repression caused by electrical activity, but have been particularly difficult to identify. The effects of the signal from the motor neurone are confined to a region within 50 micrometres of the synapse (146), indicating that the signalling pathway, including neuronally derived *trans*-synaptic substances and their post-synaptic receptors, as well as the signal cascade within the sarcoplasm, is confined to short distances. Candidates for synaptic signalling, particularly in establishing the initial synapse-specific transcription, include the calcitonin gene-related peptide (CGRP) and the acetylcholine receptor-inducing activity (ARIA).

8.5.1 Calcitonin gene-related peptide

The neuropeptide calcitonin gene-related peptide (CGRP) is found in the motor neurones of the spinal cord and in motor nerve endings (169). CGRP is released from the nerve terminal at the NMJ, and enhances muscle contraction and twitch potential. In addition, direct exposure of cultured primary chick myotubes to CGRP causes an increase of AChRs at the cell surface (170). Levels of precursor RNA encoding the receptor have also increased three-fold in these cells, indicating CGRP acts by stimulating transcription. CGRP also activates adenylate cyclase in muscle, which may induce AChR biosynthesis by a cAMP intermediate. Neither application of TTX, a blocker of electrical activity, nor the phorbol ester PMA, an inhibitor of the signal transduction pathway induced by electrical activity, have any effect on the response elicited by CGRP upon receptor synthesis. This indicates that CGRP and electrical activity regulate the transcription of the AChR genes via different intermediates or second messengers.

Binding sites for CGRP are present in large numbers in both cultured myotubes (170) and embryonic muscle (171). CGRP receptors are localized to the NMJ in adult muscle, suggesting CGRP acts together with other factors to control localized biosynthesis of NMJ structural components. However, the signal transduction pathway through which CGRP stimulates gene expression in subsynaptic nuclei is not yet known.

8.5.2 Acetylcholine receptor-inducing activity

Acetylcholine recector-inducing activity (ARIA) is another newly characterized factor, which may contribute to subsynaptic signal transduction. The ARIA glycoprotein was originally identified in chick brain (172), and increases the number of AChRs and sodium channels (173) both in cultured chick muscle and myoblasts. ARIA is first expressed in the spinal cord just prior to the onset of neuromuscular synapse formation, suggesting that it may induce AChR transcription at newly formed junctions. The ARIA gene has recently been identified (174) as an alternatively spliced form of the proto-oncogene *neu*, which encodes a growth factor first identified in central nervous system tumours (175)

The function of ARIA is still being investigated, but it is thought to bind to the basal lamina in the synaptic junction to prevent diffusion out of the synaptic cleft and so limit transcription induction to the immediate area near the nerve terminal (174). In addition, ARIA may be a marker for synaptic basal lamina, and act as a signal for NMJ nuclear location. When individual fibres within a denervated muscle in a mouse are destroyed, new fibres regenerate within the original basal lamina. If the fibres are prevented from becoming reinnervated, synapse-specific transcription still reinitiates in the precise locations that the original neuromuscular junction transcription had occurred (176, 177). This indicates that the synaptic basal lamina contains a long-lived extracellular signal for maintaining synapse-specific transcription. The signal might originate from either the myofibre or the neurone, with ARIA being a possible candidate.

9. Summary and future directions

The recent identification of multiple transcription factors important in the activation of skeletal muscle characteristics has led to a deeper appreciation of the complexity of transcriptional regulation, and provides a paradigm for the generation of other cell types. A compelling feature of research in this field has been the analysis of transcriptional control of the myogenic phenotype, both in cell culture and in the developing embryo, which has contibuted to the elaboration of a network of transcription factor interactions involved in the expression of muscle-specific genes. This approach will continue to be necessary for identifying upstream mesodermal regulators and factors from adjacent tissues that initiate myogenesis in the early embryo. Our current knowledge of muscle gene regulation also needs to be integrated with the influences of the environment of mature muscle tissue that provide neuronal and hormonal signals. The molecular biology of muscle development may prove to be a promising testing ground for new technologies. The conditional activation or suppression of transgenes in predetermined temporal and spatial patterns would provide a definitive test for the function of putative transcriptional regulators in specific muscle cell precursors. Finally, the study of gene regulation in myogenesis may contribute to our understanding of far ranging issues, such as the regulatory pathways integrating the cell cycle and the onset of differentiation. Studies of transcriptional control will undoubtedly be the focus of proposed research and promise to contribute important new principles in muscle cell, molecular, and developmental biology.

Acknowledgements

The authors wish to acknowledge the kind permission of Drs Charles Emerson, Jeffery Miller, Carl Pabo, and Eric Olson to reproduce their published and unpublished work. N. R. is an Established Investigator of the American Heart Association.

References

1. Rosenthal, N. (1989) Muscle cell differentiation. *Curr. Opin. Cell Biol.*, **1**, 1094.
2. Johnson, J. E., Wold, B. J., and Hauschka, S. D. (1989) Muscle creatine kinase sequence elements regulating skeletal and cardiac muscle expression in transgenic mice. *Mol. Cell.Biol.*, **9**, 3393.
3. Rosenthal, N., Kornhauser, J. M., Donoghue, M., Rosen, K. M., and Merlie, J. P. (1989) Myosin light chain enhancer activates muscle-specific, developmentally regulated gene expression in transgenic mice. *Proc. Natl Acad. Sci. USA*, **86**, 7780.
4. Weintraub, H., Davis, R., Lockshon, D., and Lassar, A. (1990) MyoD binds cooperatively to two sites in a target enhancer sequence: occupancy of two sites is required for activation.*Proc. Natl Acad. Sci. USA*, **87**, 5623.
5. Wentworth, B. M., Donoghue, M., Engert, J. C., Berglund, E. B., and Rosenthal, N. (1991)Paired MyoD-binding sites regulate myosin light chain gene expression. *Proc. Natl Acad. Sci. USA*, **88**, 1242.
6. Sternberg, E. A., Spizz, G., Perry, W. M., Vizard, D., Weil, T., and Olson, E. N. (1988) Identification of upstream and intragenic regulatory elements that confer cell-type-restricted and differentiation-specific expression on the muscle creatine kinase gene. *Mol. Cell. Biol.*, **8**, 2896.
7. Baldwin, T. J. and Burden, S. J. (1989) Muscle-specific gene expression controlled by a regulatory element lacking a MyoD1-binding site [published erratum appears in Nature (1990), **345**, 364]. *Nature*, **341**, 716.
8. Piette, J., Bessereau, J. L., Huchet, M., and Changeux, J. P. (1990) Two adjacent MyoD1-binding sites regulate expression of the acetylcholine receptor alpha-subunit gene. *Nature*, **345**, 353.
9. Sartorelli, V., Hong, N. A., Bishopric, N. H., and Kedes, L. (1992) Myocardial activation of the human cardiac alpha-actin promoter by helix-loop-helix proteins. *Proc. Natl Acad. Sci. USA.*, **89**, 4047.
10. Lassar, A. B., Buskin, J. N., Lockshon, D., Davis, R. L., Apone, S., Hauschka, S. D., and Weintraub, H. (1989) MyoD is a sequence-specific DNA binding protein requiring a region of myc homology to bind to the muscle creatine kinase enhancer. *Cell*, **58**, 823.
11. Braun, T., Bober, E., Winter, B., Rosenthal, N., and Arnold, H. H. (1990) Myf-6, a new member of the human gene family of myogenic determination factors: evidence for a gene cluster on chromosome 12. *EMBO J.*, **9**, 821.
12. Rosenthal, N., Berglund, E. B., Wentworth, B. M., Donoghue, M., Winter, B., Bober, E., Braun, T., and Arnold, H. H. (1990) A highly conserved enhancer downstream of the human MLC1/3 locus is a target for multiple myogenic determination factors. *Nucl. Acids Res.*, **18**, 6239.
13. Braun, T., Tannich, E., Buschhausen Denker, G., Arnold, H. H., Zhu, H., Garcia, A. V., Ross, R. S., Evans, S. M., and Chien, K. R. (1991) Promoter upstream elements of the chicken cardiac myosin light-chain 2-A gene interact with trans-acting regulatory factors for muscle-specific transcription: a conserved 28-base-pair element (HF-1) in the rat cardiac myosin light-chain-2 gene confers cardiac-specific and alpha-adrenergic-inducible expression in cultured neonatal rat myocardial cells. *Mol. Cell. Biol.*, **11**, 2273.
14. Molkentin, J. D. and Markham, B. E. (1993) Myocyte-specific enhancer-binding factor (MEF-2) regulates alpha-cardiac myosin heavy chain gene expression *in vitro* and *in vivo*. *J. Biol. Chem.*, **268**, 19512.

15. Hidaka, K., Yamamoto, I., Arai, Y., and Mukai, T. (1993) The MEF-3 motif is required for MEF-2-mediated skeletal muscle-specific induction of the rat aldolase A gene. *Mol. Cell. Biol.*, **13**, 6469.
16. Parmacek, M. S., Bengur, A. R., Vora, A. J., and Leiden, J. M. (1990) The structure and regulation of expression of the murine fast skeletal troponin C gene. Identification of a developmentally regulated, muscle-specific transcriptional enhancer. *J. Biol. Chem.*, **265**, 15970.
17. Gossett, L. A., Kelvin, D. J., Sternberg, E. A., and Olson, E. N. (1989) A new myocyte-specific enhancer-binding factor that recognizes a conserved element associated with multiple muscle-specific genes. *Mol. Cell. Biol.*, **9**, 5022.
18. Zhou, M. D., Goswami, S. K., Martin, M. E., and Siddiqui, M. A. (1993) A new serum-responsive, cardiac tissue-specific transcription factor that recognizes the MEF-2 site in the myosin light chain-2 promoter. *Mol. Cell. Biol.*, **13**, 1222.
19. Zhu, H., Garcia, A. V., Ross, R. S., Evans, S. M., and Chien, K. R. (1991) A conserved 28-base-pair element (HF-1) in the rat cardiac myosin light-chain-2 gene confers cardiac-specific and alpha-adrenergic-inducible expression in cultured neonatal rat myocardial cells. *Mol. Cell. Biol.*, **11**, 2273.
20. Muscat, G. E., Perry, S., Prentice, H., and Kedes, L. (1992) The human skeletal alpha-actin gene is regulated by a muscle-specific enhancer that binds three nuclear factors. *Gene Expr.*, **2,** 111.
21. Cserjesi, P. and Olson, E. N. (1991) Myogenin induces the myocyte-specific enhancer binding factor MEF-2 independently of other muscle-specific gene products. *Mol. Cell.Biol.*, **11**, 4854.
22. Edmondson, D. G., Cheng, T. C., Cserjesi, P., Chakraborty, T., and Olson, E. N. (1992) Analysis of the myogenin promoter reveals an indirect pathway for positive auto-regulation mediated by the muscle-specific enhancer factor MEF-2. *Mol. Cell. Biol.*, **12**, 3665.
23. Iannello, R. C., Mar, J. H., and Ordahl, C. P. (1991) Characterization of a promoter element required for transcription in myocardial cells. *J. Biol. Chem.*, **266**, 3309.
24. Li, Z. and Paulin, D. (1993) Different factors interact with myoblast-specific and myotube-specific enhancer regions of the human desmin gene. *J. Biol. Chem.*, **268**, 10403.
25. Nakatsuji, Y., Hidaka, K., Tsujino, S., Yamamoto, Y., Mukai, T., Yanagihara, T., Kishimoto, T., and Sakoda, S. (1992) A single MEF-2 site is a major positive regulatory element required for transcription of the muscle-specific subunit of the human phosphoglycerate mutase gene in skeletal and cardiac muscle cells. *Mol. Cell. Biol.*, **12**, 4384.
26. Yu, Y. T., Breitbart, R. E., Smoot, L. B., Lee, Y., Mahdavi, V., and Nadal Ginard, B. (1992) Human myocyte-specific enhancer factor 2 comprises a group of tissue-restricted MADS box transcription factors. *Genes Dev.*, **6**, 1783.
27. Pollock, R. and Treisman, R. (1991) Human SRF-related proteins: DNA-binding properties and potential regulatory targets. *Genes Dev.*, **5**, 2327.
28. Lassar, A. B., Davis, R. L., Wright, W. E., Kadesch, T., Murre, C., Voronova, A., Baltimore, D., and Weintraub, H. (1991) Functional activity of myogenic HLH proteins requires hetero-oligomerization with E12/E47-like proteins *in vivo*. *Cell*, **66**, 305.
29. Cheng, T. C., Wallace, M. C., Merlie, J. P., and Olson, E. N. (1993) Separable regulatory elements governing myogenin transcription in mouse embryogenesis. *Science*, **261**, 215.
30. Yee, S. P. and Rigby, P. W. (1993) The regulation of myogenin gene expression during the embryonic development of the mouse. *Genes Dev.*, **7**, 1277.
31. Martin, J. F., Miano, J. M., Hustad, C. M., Copeland, N. G., Jenkins, N. A., and Olson,

E.,N. (1994) A Mef2 gene that generates a muscle-specific isoform via alternative mRNA splicing. *Mol. Cell. Biol.*, **14,** 1647.

32. Leibham, D., Wong, M. W., Cheng, T. C., Schroeder, S., Weil, P. A., Olson, E. N., and Perry, M. (1994) Binding of TFIID and MEF2 to the TATA element activates transcription of the *Xenopus* MyoDa promoter. *Mol. Cell. Biol.*, **14**, 686.
33. Edmondson, D. G. and Olson, E. N. (1989) A gene with homology to the myc similarity region of MyoD1 is expressed during myogenesis and is sufficient to activate the muscle differentiation program [published erratum appears in *Genes Dev.* (1990) **4**, 1450]. *Genes Dev.*, **3**, 628.
34. Wright, W. E., Sassoon, D. A., and Lin, V. K. (1989) Myogenin, a factor regulating myogenesis, has a domain homologous to MyoD. *Cell*, **56**, 607.
35. Rhodes, S. J. and Konieczny, S. F. (1989) Identification of MRF4: a new member of the muscle regulatory factor gene family. *Genes Dev.*, **3**, 2050.
36. Miner, J. H. and Wold, B. (1990) Herculin, a fourth member of the MyoD family of myogenic regulatory genes. *Proc. Natl Acad. Sci. USA*, **87**, 1089.
37. Davis, R. L., Weintraub, H., and Lassar, A. B. (1987) Expression of a single transfected cDNA converts fibroblasts to myoblasts. *Cell*, **51**, 987.
38. Olson, E. N. (1990) MyoD family: a paradigm for development? *Genes Dev.*, **4**, 1454.
39. Weintraub, H., Davis, R., Tapscott, S., Thayer, M., Krause, M., Benezra, R., Blackwell, T. K., Turner, D., Rupp, R., and Hollenberg, S. (1991) The myoD gene family: nodal point during specification of the muscle cell lineage. *Science*, **251**, 761.
40. Emerson, C. P., jun. (1993) Skeletal myogenesis: genetics and embryology to the fore. *Curr. Opin. Genet. Develop.*, **3**, 265.
41. Venuti, J. M., Goldberg, L., Chakraborty, T., Olson, E. N., and Klein, W.,H. (1991) A myogenic factor from sea urchin embryos capable of programming muscle differentiation in mammalian cells. *Proc. Natl Acad. Sci. USA*, **88**, 6219.
42. Krause, M., Fire, A., White Harrison, S., Weintraub, H., and Tapscott, S. (1992) Functional conservation of nematode and vertebrate myogenic regulatory factors. *J. Cell Sci.* (Suppl.), **16**, 111.
43. Muneoka, K. and Sassoon, D. (1992) Molecular aspects of regeneration in developing vertebrate limbs. *Dev. Biol.*, **152**, 37.
44. Brennan, T. J., Chakraborty, T., and Olson, E. N. (1991) Mutagenesis of the myogenin basic region identifies an ancient protein motif critical for activation of myogenesis. *Proc. Natl Acad. Sci. USA*, **88**, 5675.
45. Davis, R. L. and Weintraub, H. (1992) Acquisition of myogenic specificity by replacement of three amino acid residues from MyoD into E12. *Science*, **256**, 1027.
46. Schwarz, J. J., Chakraborty, T., Martin, J., Zhou, J. M., and Olson, E. N. (1992) The basic region of myogenin cooperates with two transcription activation domains to induce muscle-specific transcription. *Mol. Cell. Biol.*, **12**, 266.
47. Yutzey, K. E. and Konieczny, S. F. (1992) Different E box regulatory sequences are functionally distinct when placed within the context of the troponin I enhancer. *Nucl. Acids Res.*, **20**, 5105.
48. Winter, B., Braun, T., and Arnold, H. H. (1992) Co-operativity of functional domains in the muscle-specific transcription factor Myf-5. *EMBO J.*, **11**, 1843.
49. Rashbass, J., Taylor, M. V., and Gurdon, J. B. (1992) The DNA-binding protein E12 co-operates with XMyoD in the activation of muscle-specific gene expression in *Xenopus* embryos. *EMBO J.*, **11**, 2981.
50. Blackwell, T. K. and Weintraub, H. (1990) Differences and similarities in DNA-binding

preferences of MyoD and E2A protein complexes revealed by binding site selection. *Science*, **250**, 1104.

51. Ma, P. C. M., Rould, M. A., Weintraub, H., and Pabo, C. O. (1994) Crystal structure of MyoD bHLH domain-DNA complex: perspectives on DNA recognition and implications for transcriptional activation. *Cell*, **77**, 451.
52. Fujisawa Sehara, A., Nabeshima, Y., Komiya, T., Uetsuki, T., and Asakura, A. (1992) Differential trans-activation of muscle-specific regulatory elements including the mysosin light chain box by chicken MyoD, myogenin, and MRF4. *J. Biol. Chem.*, **267**, 10031.
53. Chakraborty, T. and Olson, E. N. (1991) Domains outside of the DNA-binding domain impart target gene specificity to myogenin and MRF4. *Mol. Cell. Biol.*, **11**, 6103.
54. Mak, K. L., To, R. Q., Kong, Y., and Konieczny, S. F. (1992) The MRF4 activation domain is required to induce muscle-specific gene expression. *Mol. Cell. Biol.*, **12**, 4334.
55. Hu, J. S., Olson, E. N., and Kingston, R. E. (1992) HEB, a helix-loop-helix protein related to E2A and ITF2 that can modulate the DNA-binding ability of myogenic regulatory factors. *Mol. Cell. Biol.*, **12**, 1031.
56. Braun, T., Bober, E., Buschhausen Denker, G., Kohtz, S., Grzeschik, K. H., Arnold, H. H., and Kotz, S. (1989) Differential expression of myogenic determination genes in muscle cells: possible autoactivation by the Myf gene products [published erratum appears in *EMBO J.* (1989),**8**, 4358]. *EMBO J.*, **8**, 3617.
57. Thayer, M. J., Tapscott, S. J., Davis, R. L., Wright, W. E., Lassar, A. B., and Weintraub, H. (1989) Positive autoregulation of the myogenic determination gene MyoD1. *Cell*, **58**, 241.
58. Montarras, D., Chelly, J., Bober, E., Arnold, H., Ott, M. O., Gros, F., and Pinset, C. (1991) Developmental patterns in the expression of myf5, MyoD, myogenin, and MRF4 during myogenesis. *New Biol.*, **3**, 592.
59. Hinterberger, T. J., Mays, J. L., and Konieczny, S. F. (1992) Structure and myofiber-specific expression of the rat muscle regulatory gene MRF4. *Gene*, **117**, 201.
60. Peterson, C. A., Gordon, H., Hall, Z. W., Paterson, B. M., and Blau, H. M. (1990) Negative control of the helix-loop-helix family of myogenic regulators in the NFB mutant. *Cell*, **62**, 493.
61. Florini, J. R. and Ewton, D. Z. (1992) Induction of gene expression in muscle by the IGFs. *Growth Reg.*, **2**, 23.
62. Li, L., Zhou ,J., James, G., Heller Harrison, R., Czech, M. P., and Olson, E. N. (1992) FGF inactivates myogenic helix-loop-helix proteins through phosphorylation of a conserved protein kinase C site in their DNA-binding domains. *Cell*, **71**, 1181.
63. Benezra, R., Davis, R. L., Lassar, A., Tapscott, S., Thayer, M., Lockshon, D., and Weintraub, H. (1990) Id: a negative regulator of helix-loop-helix DNA binding proteins. Control of terminal myogenic differentiation. *Ann. N. Y. Acad. Sci.*, **599**, 1.
64. Benezra, R., Davis, R. L., Lockshon, D., Turner, D. L., and Weintraub, H. (1990) The protein Id: a negative regulator of helix-loop-helix DNA binding proteins. *Cell*, **61**, 49.
65. Wang, Y., Benezra, R., and Sassoon, D. A. (1992) Id expression during mouse development: a role in morphogenesis. *Dev. Dyn.*, **194**, 222.
66. Jen, Y., Weintraub, H., and Benezra, R. (1992) Overexpression of Id protein inhibits the muscle differentiation program: *in vivo* association of Id with E2A proteins. *Genes Dev.*, **6**, 1466.
67. Bengal, E., Ransone, L., Scharfmann, R., Dwarki, V. J., Tapscott, S. J., Weintraub, H., and Verma, I. M. (1992) Functional antagonism between c-Jun and MyoD proteins: a direct physical association. *Cell*, **68**, 507.

68. Li, L., Chambard, J. C., Karin, M., and Olson, E. N. (1992) Fos and Jun repress transcriptional activation by myogenin and MyoD: the amino terminus of Jun can mediate repression. *Genes Dev.*, **6**, 676.
69. Enkemann, S. A., Konieczny, S. F., and Taparowsky, E. J. (1990) Adenovirus 5 E1A represses muscle-specific enhancers and inhibits expression of the myogenic regulatory factor genes, MyoD1 and myogenin. *Cell Growth Diff.*, **1**, 375.
70. Braun, T., Bober, E., and Arnold, H. H. (1992) Inhibition of muscle differentiation by the adenovirus E1a protein: repression of the transcriptional activating function of the HLH protein Myf-5. *Genes Dev.*, **6**, 888.
71. Gu, W., Schneider, J. W., Condorelli, G., Kaushal, S., Mahdavi, V., and Nadal Ginard, B. (1993) Interaction of myogenic factors and the retinoblastoma protein mediates muscle cell commitment and differentiation. *Cell*, **72**, 309.
72. Lyons, G. E. and Buckingham, M. E. (1992) Developmental regulation of myogenesis in the mouse. *Semin. Dev. Biol.*, **3**, 243.
73. Pownall, M.,E. and Emerson, C.P., jun. (1992) Molecular and embryological studies of avian myogenesis. *Semin. Cell. Biol.*, **3**, 229.
74. Pownall, M. E. and Emerson, jun., C.P. (1992) Sequential activation of three myogenic regulatory genes during somite morphogenesis in quail embryos. *Develop. Biol.*, **151**, 67.
75. Ott, M. O., Bober, E., Lyons, G., Arnold, H., and Buckingham, M. (1991) Early expression of the myogenic regulatory gene, myf-5, in precursor cells of skeletal muscle in the mouse embryo. *Development*, **111**, 1097.
76. de la Brousse, F. C. and Emerson, C.P., jun. (1990) Localized expression of a myogenic regulatory gene, qmf1, in the somite dermatome of avian embryos. *Genes Dev.*, **4**, 567.
77. Stockdale, F. E. (1992) Myogenic cell lineages. *Dev. Biol.*, **154**, 284.
78. Smith, T. H., Kachinsky, A. M., and Miller, J. B. (1994) Somite subdomains, muscle cell origins, and the four muscle regulatory factor proteins. *J. Cell Biol.*, **127**, 95.
79. Miner, J. H., Miller, J. B., and Wold, B. J. (1992) Skeletal muscle phenotypes initiated by ectopic MyoD in transgenic mouse heart. *Development*, **114**, 853.
80. Faerman, A., Pearson White, S., Emerson, C., and Shani, M. (1993) Ectopic expression of MyoD1 in mice causes prenatal lethalities. *Dev. Dyn.*, **196**, 165.
81. Braun, T., Rudnicki, M. A., Arnold, H. H., and Jaenisch, R. (1992) Targeted inactivation of the muscle regulatory gene Myf-5 results in abnormal rib development and perinatal death. *Cell*, **71**, 369.
82. Hasty, P., Bradley, A., Morris, J. H., Edmondson, D. G., Venuti, J. M., Olson, E. N., and Klein, W. H. (1993) Muscle deficiency and neonatal death in mice with a targeted mutation in the myogenin gene. *Nature*, **364**, 501.
83. Nabeshima, Y., Hanaoka, K., Hayasaka, M., Esumi, E., Li, S., and Nonaka, I. (1993) Myogenin gene disruption results in perinatal lethality because of severe muscle defect. *Nature*, **364**, 532.
84. Rudnicki, M. A., Braun, T., Hinuma, S., and Jaenisch, R. (1992) Inactivation of MyoD in mice leads to up-regulation of the myogenic HLH gene Myf-5 and results in apparently normal muscle development. *Cell*, **71**, 383.
85. Rudnicki, M. A., Schnegelsberg, P. N. J., Stead, R. H., Braun, T., Arnold, H. H., and Jaenisch, R. (1993) MyoD or myf-5 is required for the formation of skeletal muscle. *Cell*, **75**, 1351.
86. Hinterberger, T. J., Sassoon, D. A., Rhodes, S. J., and Konieczny, S. F. (1991) Expression of the muscle regulatory factor MRF4 during somite and skeletal myofiber development. *Develop. Biol.*, **147**, 144.

87. McDermott, J. C., Cardoso, M. C., Yu, Y. T., Andres, V., Leifer, D., Krainc, D., Lipton, S. A., and Nadal Ginard, B. (1993) hMEF2C gene encodes skeletal muscle- and brain-specific transcription factors. *Mol. Cell. Biol.*, **13**, 2564.
88. Leifer, D., Krainc, D., Yu, Y. T., McDermott, J., Breitbart, R. E., Heng, J., Neve, R. L., Kosofsky, B., Nadal Ginard, B., and Lipton, S. A. (1993) MEF2C, a MADS/MEF2-family transcription factor expressed in a laminar distribution in cerebral cortex. *Proc. Natl Acad. Sci. USA*, **90**, 1546.
89. Martin, J. F., Schwarz, J. J., and Olson, E. N. (1993) Myocyte enhancer factor (MEF) 2C: a tissue-restricted member of the MEF-2 family of transcription factors. *Proc. Natl Acad. Sci. USA*, **90**, 5282.
90. Kappen, C. and Ruddle, F. (1993) Evolution of a regulatory gene family: *HOM/HOX* genes. *Curr. Opin. Genet. Devel.*, **3**, 931.
91. Botas, J. (1993) Control of morphogenisis and differentiation by *HOM/HOX* genes. *Curr. Opin. Cell Biol.*, **5**, 1015.
92. Gehring, W. J., Muller, M., Affolter, M., Percival-Smith, A., Billeter, M., Qian, Y. Q., Otting, G., and Wuthrich, K. (1990) The structure of the homeodomain and its functional implications. *Trends Genet.*, **6**, 323.
93. Hayashi, S. and Scott, M. P. (1990) What determines the specificity of action of *Drosophila* homeodomain proteins? *Cell*, **63**, 883.
94. Treisman, J., Harris, H., Wilson, D., and Despain, C. (1992) The homeodomain: a new face for the helix-turn-helix? *Bioessays*, **14**, 145.
95. Greig, S. and Akam, M. (1993) Homeotic genes autonomously specify one aspect of pattern in the *Drosophila* mesoderm. *Nature*, **362**, 630.
96. Michelson, A. M. (1994) Muscle pattern diversification in *Drosophila* is determined by the autonomous function of homeotic genes in the embryonic mesoderm. *Development*, **120**, 755.
97. Morgan, B. A., Izpisua-Belmonte, J. C., Deloule, D., and Tabin, C. J. (1992) Targeted misexpression of *Hox-4.6* in the avian limb bud causes apparent homeotic transformations. *Nature*, **358**, 236.
98. Behringer, R., Crotty, D. A., Tennyson, V. M., Brinster, R. L., Palmiter, R. D., and Wolgemuth, D. J. (1993) Sequences 5′ of the homeobox of the *Hox-1.4* gene direct tissue-specific expression of *lacZ* during mouse development. *Development.*, **117**, 823.
99. Kessel, M. and Gruss, P. (1990) Murine development control genes. *Science*, **249**, 374.
100. Levi, G., Corsetti, M. T., Boncinelli, E., and Corte, G. (1993) Changes in the prevalence of an homeobox gene product during muscle differentiation. *Mech. Dev.*, **43**, 111.
101. Donoghue, M. J., Merlie, J. P., Rosenthal, N., and Sanes, J. R. (1991) Rostrocaudal gradient of transgene expression in adult skeletal muscle. *Proc. Natl Acad. Sci. USA*, **88**, 5847.
102. Donoghue, M. J., Morris Valero, R., Johnson, Y. R., Merlie, J. P., and Sanes, J. R. (1992) Mammalian muscle cells bear a cell-autonomous, heritable memory of their rostrocaudal position. *Cell*, **69**, 67.
103. Grieshammer, U., Sassoon, D., and Rosenthal, N. (1992) A transgene target for positional regulators marks early rostrocaudal specification of myogenic lineages. *Cell*, **69**, 79.
104. Herskowitz, I. (1989) A regulatory hierarchy for cell specialization in yeast. *Nature*, **342**, 749.
105. Grueneberg, D. A., Natesan, S., Alexandre, C., and Gilman, M. Z. (1992) Human and *Drosophila* homeodomain proteins that enhance the DNA-binding activity of serum response factor. *Science*, **257**, 1089.

106. Cserjesi, P., Lilly, B., Bryson, L., Wang, Y., Sassoon, D. A., and Olson, E. N. (1992) MHox: a mesodermally restricted homeodomain protein that binds an essential site in the muscle creatine kinase enhancer. *Development*, **115**, 1087.
107. Bourgouin, C., Lundgren, S. E., and Thomas, J. B. (1992) Apterous is a *Drosophila* LIM domain gene required for the development of a subset of embryonic muscles. *Neuron*, **9**, 549.
108. Azpiazu, N. and Frasch, M. (1993) *tinman* and *bagman*: two homeo box genes that determine cell fates in the dorsal mesoderm in *Drosophila*. *Genes Dev.*, **7**, 1325.
109. Goulding, M., Lumsden, A., and Gruss, P. (1994) Regulation of *Pax-3* expression in the dermomyotome and its role in muscle development. *Development*, **120**, 957.
110. Williams, B. A. and Ordahl, C. P. (1994) *Pax-3* expression in segmental mesoderm marks early stages in myogenic cell specification. *Development*, **120**, 785.
111. Bober, E., Franz, T., Arnold, H. H., Gruss, P., and Tremblay, P. (1994) *Pax-3* is required for the development of limb muscles: a possible role for the migration of dermomyotomal muscle progenitor cells. *Development*, **120**, 603.
112. Robert, B., Sassoon, D., Jacq, B., Gehring, W., and Buckingham, M. (1989) Hox-7, a mouse homeobox gene with a novel pattern of expression during embryogenesis. *EMBO J.*, **8**, 91.
113. Balling, R., Deutsch, U., and Gruss, P. (1988) undulated, a mutation affecting the development of the mouse skeleton, has a point mutation in the paired box of Pax 1. *Cell*, **55**, 531.
114. Pourquie, O., Coltey, M., Teillet, M. A., Ordahl, C., and Le Douarin, N. M. (1993) Control of dorsoventral patterning of somitic derivitives by the notocord and floor plate. *Proc. Natl Acad. Sci. USA*, **90**, 5242.
115. Song, K., Wang, Y., and Sassoon, D. (1992) Expression of Hox-7.1 in myoblasts inhibits terminal differentiation and induces cell transformation. *Nature*, **360**, 477.
116. Ranvier, L. (1874) Some observations relating to the histology and physiology of striated muscle. *Arch. Physiol. Norm. Path.*, **1**, 5.
117. Barany, M. (1967) ATPase activity of myosin correlated with speed of muscle shortening. *J. Gen. Physiol.*, **50**, 197.
118. Guth, L. and Samaha, F. (1969) Qualitative differences between actomyosin ATPase of slow and fast mammalian muscle. *Exp. Neurol.*, **25**, 138.
119. Cox, R. D., Weydert, A., Barlow, D., and Buckingham, M. E. (1991) Three linked myosin heavy chain genes clustered within 370 kb of each other show independent transcriptional and post-transcriptional regulation during differentiation of a mouse mlc line. *Develop. Biol.*, **143**, 36.
120. Cox, R., Weydert, A., Barlow, D., and Buckingham, M. (1991) Three linked myosin heavy chain genes clustered within 370 kb of each other show independent transcriptional and post-transcriptional regulation during differentiation of a mouse muscle cell line. *Devel. Biol.*, **143**, 36.
121. Kaprielain, Z. and Fambrough, D. M. (1987) Expression of fast and slow isoforms of the Ca2+-ATPase in developing chick skeletal muscle. *Develop. Biol.*, **124**, 490.
122. Buller, A. J., Eccles, J. C., and Eccles, R. M. (1960) Differentiation of fast and slow muscles in the cat hind limb. *J. Physiol.*, **150**, 399.
123. Lomo, T. and Rosenthal, G. (1972) Control of ACh sensitivity by muscle activity in the rat. *J. Physiol.*, **221**, 493.
124. Crow, M. T. and Stockdale, F .E. (1986) Myosin expression and specialization among the earliest muscle fibers of the developing avian limb. *Develop. Biol.*, **113**, 238.

125. Butler, J., Cosmos, E., Brierly, J., and Hauschka, S. D. (1982) Differentiation of muscle fiber types in aneurogenic brachial arches of the chick embryo. *J. Exp. Zool.*, **224**, 65.
126. McLennan, I. S. (1983) Differentiation of muscle fiber type in the chicken hindlimb. *Develop. Biol.*, **97**, 222.
127. Phillips, W. D., Everett, A. W., and Bennett, M. R. (1986) The role of innervation in the establishment of the topographical distribution of primary myotubes during development. *J. Neurocyt.*, **15**, 397.
128. Condon, K., Silberstein, L., Blau, H. M., and Thompson, W. J. (1990) Differentiation of fiber types in aneural musculature of the prenatal rat hindlimb. *Develop. Biol.*, **138**, 275.
129. Fredette, B. J. and Landmesser, L. T. (1991) A re-evaluation of the role of innervation in primary and secondary myogenesis in developing chick muscle. *Develop. Biol.*, **143**, 19.
130. White, N. K., Bonner, P. H., Nelsen, D. R., and Hauschka, S. D. (1975) Clonal analysis of vertebrate myogenesis. IV. Medium-dependent classification of colony-forming cells. *Develop. Biol.*, **44**, 346.
131. Banerjee Basu, S. and Buonanno, A. (1993) *cis*-acting sequences of the rat troponin I slow gene confer tissue- and development-specific transcription in cultured muscle cells as well as fiber type specificity in transgenic mice. *Mol. Cell. Biol.*, **13**, 7019.
132. Hallauer, P. L., Bradshaw, H. L., and Hastings, K. E. (1993) Complex fiber-type-specific expression of fast skeletal muscle troponin I gene constructs in transgenic mice. *Development*, **119**, 691.
133. Lin, H., Yutzey, K. E., and Konieczny, S. F. (1991) Muscle-specific expression of the troponin I gene requires interactions between helix-loop-helix muscle regulatory factors and ubiquitous transcription factors. *Mol. Cell. Biol.*, **11**, 267.
134. Donoghue, M., Ernst, H., Wentworth, B., Nadal Ginard, B., and Rosenthal, N. (1988) A muscle-specific enhancer is located at the 3′ end of the myosin light-chain 1/3 gene locus. *Genes Dev.*, **2**, 1779.
135. Donoghue, M. J., Alvarez, J. D., Merlie, J. P., and Sanes, J. R. (1991) Fiber type- and position-dependent expression of a myosin light chain-CAT transgene detected with a novel histochemical stain for CAT. *J. Cell Biol.*, **115**, 423.
136. Merlie, J. P., Mudd, J., Cheng, T.-C., and Olson, E. N. (1994) Myogenin and acetylcholine receptor α gene promoters mediate transcriptional regulation in response to motor innervation. *J. Biol. Chem.*, **269**, 2461.
137. Hughes, S. M., Taylor, J. M., Tapscott, S. J., Gurley, C. M., Carter, W. J., and Peterson, C. A. (1993) Selective accumulation of MyoD and myogenin mRNAs in fast and slow adult skeletal muscle is controlled by innervation and hormones. *Development*, **118**, 1137.
138. Weintraub, H. (1993) The MyoD family and myogenesis: redundancy, networks, and thresholds. *Cell*, **75**, 1241.
139. Thompson, W. R., Nadal Ginard, B., and Mahdavi, V. (1991) A MyoD1-independent muscle-specific enhancer controls the expression of the beta-myosin heavy chain gene in skeletal and cardiac muscle cells. *J. Biol. Chem.*, **266**, 22678.
140. Dale, H. H., Feldberg, W., and Vogt, M. (1936) Release of acetylcholine at voluntary nerve endings. *J. Physiol.*, **86**, 353.
141. Cohen, S. A. and Fischbach, G. D. (1973) Regulation of acetylcholine sensitivity by muscle activity in cell culture. *Science*, **181**, 76.
142. Byers, T. J., Kunkel, L. M., and Watkins, S. C. (1991) The subcellular distribution of dystrophin in mouse skeletal, cardiac and smooth muscle. *J. Cell Biol.*, **115**, 411.
143. Frail, D. E., McLaughlin, L. L., Mudd, J., and Merlie, J. P. (1988) Identification of the

mouse muscle 43,000 dalton acetylcholine receptor-associated protein (RAPsyn) by cDNA cloning. *J. Biol. Chem.*, **263**, 15602.
144. Bloch, R. J. and Hall, Z. (1983) Cytoskeletal components of the vertebrate neuro-muscular junction: vinculin, alpha-actinin, and filamin. *J. Cell Biol.*, **97**, 217.
145. Sanes, J. R., Johnson, Y. R., Kotzbauer, P. T., Mudd, J., Hanley, T., Martinou, J. C., and Merlie, J. P. (1991) Selective expression of an acetylcholine receptor-lacZ transgene in synaptic nuclei of adult muscle fibers. *Development*, **113**, 1181.
146. Simon, A. M., Hoppe, P., and Burden, S. J. (1992) Spatial restriction of AChR gene expression to subsynaptic nuclei. *Development*, **114**, 545.
147. Klarsfeld, A., Bessereau, J. L., Salmon, A. M., Triller, A., Babiner, C., and Changeux, J. P. (1991) An acetylcholine receptor α-subunit promoter conferring proferential expression in muscle of transgenic mice. *EMBO J.*, **10**, 625.
148. Axelsson, J. and Thesleff, S. (1959) A study of supersensitivity in denervated mammalian skeletal muscle. *J. Physiol.*, **147**, 178.
149. Fambrough, D. M. (1970) Acetylcholine sensitivity at muscle fiber membrane: mechanism of regulation by motoneurons. *Science*, **168**, 372.
150. Tsay, H. J. and Schmidt, J. (1989) Skeletal muscle denervation activates acetylcholine receptor genes. *J. Cell Biol.*, **108**, 1523.
151. Klarsfeld, A., Daubas, P., Bourachot, B., and Changeux, J. P. (1987) A 5′ flanking region of the chicken acetylcholine receptor α-subunit gene confers tissue-specificity and developmental control of expression in transfected cells. *Mol. Cell. Biol.*, **7**, 951.
152. Wang, Y., Xu, H. P., Wang, X. M., Ballivet, M., and Schmidt, J. (1988) A cell type-specific enhancer drives expression of the chick muscle acetylcholine receptor alpha-subunit gene. *Neuron*, **1**, 527.
153. Wang, X. M., Tsay, H. J., and Schmidt, J. (1990) Expression of the acetylcholine receptor delta-subunit gene in differentiating chick muscle cells is activated by an element that contains two 16 bp copies of a segment of the alpha-subunit enhancer. *EMBO J.*, **9**, 783.
154. Gilmour, B. P., Fanger, G. R., Newton, C., Evans, S. M., and Gardner, P. D. (1991) Multiple binding sites for myogenic regulatory factors are required for expression of the acetylcholine receptor gamma-subunit gene. *J. Biol. Chem.*, **266**, 19871.
155. Simon, A. M. and Burden, S. J. (1993) An E box mediates activation and repression of the acetylcholine receptor delta-subunit gene during myogenesis. *Mol. Cell. Biol.*, **13**, 5133.
156. Numberger, M., Durr, I., Kues, W., Koenen, M., and Witzemann, V. (1991) Different mechanisms regulate muscle-specific AChR gamma- and epsilon-subunit gene expression. *EMBO J.*, **10**, 2957.
157. Merlie, J. P. and Kornhauser, J. M. (1989) Neural regulation of gene expression by an acetylcholine receptor promoter in muscle of transgenic mice. *Neuron*, **2**, 1295.
158. Chahine, K. G., Walke, W., and Goldman, D. (1992) A 102 base pair sequence of the nicotinic acetylcholine receptor delta-subunit gene confers regulation by muscle electrical activity. *Development*, **115**, 213.
159. Dutton, E. K., Simon, A. M., and Burden, S. J. (1993) Electrical activity-dependent regulation of the acetylcholine receptor delta-subunit gene, MyoD, and myogenin in primary myotubes. *Proc. Natl Acad. Sci. USA*, **90**, 2040.
160. Chahine, K. G., Baracchini, E., and Goldman, D. (1993) Coupling muscle electrical activity to gene expression via a cAMP-dependent second messenger system. *J. Biol. Chem.*, **268**, 2893.
161. Bessereau, J. L., Stratford-Perricaudet, L. D., Piette, J., Le Poupon, C., and Changeux, J.

P. (1994) *In vivo* and *in vitro* analysis of electrical activity-dependent expression of muscle acetylcholine receptor genes using adenovirus. *Proc. Natl Acad. Sci. USA*, **91**, 1304.
162. Neville, C. M., Schmidt, M., and Schmidt, J. (1992) Response of myogenic determination factors to cessation and resumption of electrical activity in skeletal muscle: a possible role for myogenin in denervation supersensitivity. *Cell. Mol. Neuro.*, **12**, 511.
163. Piette, J., Huchet, M., Duclert, A., Fujisawa Sehara, A., and Changeux, J. P. (1992) Localization of mRNAs coding for CMD1, myogenin and the alpha-subunit of the acetylcholine receptor during skeletal muscle development in the chicken. *Mech. Dev.*, **37**, 95.
164. Bursztajn, S., Schneider, L. W., Jong, Y. J., and Berman, S. A. (1988) *Biol. Cell*, **63**, 57.
165. Laufer, R., Klarsfeld, A., and Changeux, J. P. (1991) Phorbol esters inhibit the activity of the chicken acetylcholine receptor alpha-subunit gene promoter. Role of myogenic regulators. *Eur. J. Biochem.*, **202**, 813.
166. Huang, C. F., Neville, C. M., and Schmidt, J. (1993) Control of myogenic factor genes by the membrane depolarization/protein kinase C cascade in chick skeletal muscle. *FEBS Lett.*, **319**, 21.
167. Li, L., Heller Harrison, R., Czech, M., and Olson, E. N. (1992) Cyclic AMP-dependent protein kinase inhibits the activity of myogenic helix-loop-helix proteins. *Mol. Cell. Biol.*, **12**, 4478.
168. Nghiem, H.-O., Mendelzon, D., and Changeux, J. P. (1994) Phosphorylation of myogenin and regulation of the acetylcholine receptor genes expression by electrical activity and by protein kinase C. *J. Cell. Biochem.*, **18D**, 510.
169. Takamori, M. and Yoshikawa, H. (1989) Effect of calcitonin gene-related peptide on skeletal muscle via specific binding site and G protein. *J. Neurol. Sci.*, **90**, 99.
170. Jennings, C. G. and Mudge, A. W. (1989) Chick myotubes in culture express high-affinity receptors for calcitonin gene-related peptide. *Brain Res.*, **504**, 199.
171. Roa, M. and Changeux, J. P. (1991) Characterization and developmental evolution of a high-affinity binding site for calcitonin gene-related peptide on chick skeletal muscle membrane. *Neuroscience*, **41**, 563.
172. Harris, D. A., Falls, D. L., and Fischbach, G. D. (1989) Differential activation of myotube nuclei following exposure to an acetylcholine receptor-inducing factor. *Nature*, **337**, 173.
173. Corfas, G. and Fischbach, G. D. (1993) The number of Na+ channels in cultured chick muscle is increased by ARIA, an acetylcholine receptor-inducing activity. *J. Neurosci.*, **13**, 2118.
174. Falls, D. L., Rosen, K. M., Corfas, G., Lane, W. S., and Fischbach, G. D. (1993) ARIA, a protein that stimulates acetylcholine receptor synthesis, is a member of the neu ligand family. *Cell*, **72**, 801.
175. Padhy, L. C., Shih, C., Cowing, D., Finkelstein, R., and Weinberg, R. (1982) Identification of a phosphoprotein specifically induced by the transforming DNA of rat neuroblastomas. *Cell*, **28**, 865.
176. Jo, S. A. and Burden, S. J. (1992) Synaptic basal lamina contains a signal for synapse-specific transcription. *Development*, **115**, 673.
177. Brenner, H. L. R., Herczeg, A., and Slater, C. R. (1992) Synapse-specific expression of acetylcholine receptor genes and their products at original synaptic sites in rat soleus muscle fibres regenerating in the absence of innervation. *Development*, **116**, 41.
178. Braun, T. and Arnold, H. H. (1991) The four human muscle regulatory helix-loop-helix proteins Myf3-Myf6 exhibit similar hetero-dimerization and DNA binding properties. *Nucl. Acids Res.*, **19**, 5645.
179. Lin, Z. Y., Dechesne, C. A., Eldridge, J., and Paterson, B. M. (1989) An avian muscle

factor related to MyoD1 activates muscle-specific promoters in nonmuscle cells of different germ-layer origin and in BrdU-treated myoblasts. *Genes Dev.*, **3**, 986.

180. Hopwood, N. D., Pluck, A., and Gurdon, J. B. (1989) MyoD expression in the forming somites is an early response to mesoderm induction in *Xenopus* embryos. *EMBO J.*, **8**, 3409.
181. Jennings, C. G. (1992) Expression of the myogenic gene MRF4 during *Xenopus* development [corrected and republished article originally printed in *Develop. Biol.* (1992), **150**, 121]. *Develop. Biol.*, **151**, 319.
182. Hopwood, N. D., Pluck, A., and Gurdon, J. B. (1991) *Xenopus* Myf-5 marks early muscle cells and can activate muscle genes ectopically in early embryos. *Development*, **111**, 551.
183. Neville, C. M. and Schmidt, J. (1992) Expression of myogenic factors in skeletal muscle and electric organ of *Torpedo californica*. *FEBS Lett.*, **305**, 23.
184. Michelson, A. M., Abmayr, S. M., Bate, M., Arias, A. M., and Maniatis, T. (1990) Expression of a MyoD family member prefigures muscle pattern in Drosophila embryos. *Genes Dev.*, **4**, 2086.
185. Paterson, B. M., Walldorf, U., Eldridge, J., Dubendorfer, A., Frasch, M., and Gehring, W. J. (1991) The *Drosophila* homologue of vertebrate myogenic-determination genes encodes a transiently expressed nuclear protein marking primary myogenic cells. *Proc. Natl Acad. Sci. USA*, **88**, 3782.
186. Krause, M., Fire, A., Harrison, S. W., Priess, J., and Weintraub, H. (1990) CeMyoD accumulation defines the body wall muscle cell fate during *C. elegans* embryogenesis. *Cell*, **63**, 907.
187. Olson, E. N. and Klein, W. H. (1994) bHLH factors in muscle development: dead lines and commitments, what to leave in and what to leave out. *Genes Dev.*, **8**, 1.
188. Olson, E. N. and Rosenthal, N. (1994) Homeobox genes and muscle patterning. *Cell*, **79**, 9.
189. Page, S., Miller, J. B., DiMario, J. X., Hager, E. J., Moser, A., and Stockdale, F. E. (1992) Developmentally regulated expression of three slow isoforms of myosin heavy chain: diversity among the first fibers to form in avian muscle. *Develop. Biol.*, **154**, 118.
190. Miller, J. B. (1991) Myoblasts, myosins, MyoDs, and the diversification of muscle fibers. *Neuromusc. Disord.*, **1**, 7.

8 Transcriptional regulation during the eukaryotic cell cycle

NIC JONES

1. Introduction

Although most cellular genes are expressed constitutively throughout the cell cycle, a small number are expressed in a periodic fashion or in response to conditions that positively and negatively influence cell cycle progression. Such periodicity can be seen in all phases of the cycle and is common to all eukaryotes that have been studied. Genes that fall into this interesting class include proteins such as the cyclins that are directly implicated in the regulation of cell cycle progression, genes encoding products required for general DNA metabolism and synthesis, and genes that encode products that may regulate the initiation of DNA synthesis. The conserved nature of this pattern of expression, as well as the genes that fall into this class, suggests that periodicity in expression plays an important role in the control of proliferation; accordingly, understanding the mechanisms that underlie such periodicity will be important for understanding cell cycle control and how such control may be lost during oncogenesis.

In this chapter I will focus on one aspect of cell-cycle regulated transcription, namely, transcriptional activation that occurs in the late G1 phase of the cycle. Although a number of essential genes are periodically expressed this does not necessarily imply that in each case the periodicity is essential. In fact, it is likely that this is not the case and many or even most such genes could be expressed in a constitutive fashion without having any adverse effect on cell cycle progression. In some cases, however, the periodicity will be more critical and could be essential for maintaining the correct order of events during the cycle, or regulating the transition between different phases of the cycle. The clearest example of the importance of transcriptional regulation concerns the specific activation of gene expression that occurs in late G1 where it appears to be of major importance in a decision that every eukaryotic cell has to make: namely, whether to proceed into a new round of DNA synthesis and division or whether to leave the cycle and enter a resting state or a particular differentiation pathway. This decision is reached by the integration of various extracellular and intracellular signalling pathways triggered by positive and negative growth signals (reviewed in 1–3).

The molecular basis of this G1-specific transcription activation will be reviewed in both yeast and mammalian cells and, where appropriate, parallels between the two systems will be drawn. It is increasingly clear that the general features of the cell cycle are highly conserved and it can be expected therefore that mutual benefit will arise from the study of this activation process in such diverse cells.

As mammalian cells approach the end of the G1 phase a stage is reached called the 'restriction point', where the decision to either enter S phase or arrest in G1 is made. Once this stage is breached, the cells are committed to the next division cycle and are no longer sensitive to the withdrawal of growth-promoting factors or the presence of anti-proliferative factors. The cell cycle machinery has been engaged and cells automatically proceed through an ordered series of events culminating in mitosis. Following mitosis, the cells once again have an opportunity to evaluate their surroundings and decide whether to engage or disengage the cycle machinery. Similar decisions face the yeast cell at a point in the G1 period called START, which is functionally analogous to the mammalian restriction point (4). Prior to START, yeast cells can arrest their progression through G1 and enter a different developmental pathway involving sexual conjugation and meiotic division. Once cells have passed START, however, this developmental pathway is no longer an optional fate and they are committed to complete the mitotic cycle. A major feature of the restriction point and START is the specific transcriptional activation of a variety of genes.

2. START-specific transcription in budding yeast

At the G1 to S transition of the budding yeast *Saccharomyces cerevisiae*, there is an increase in the expression of at least 30 different genes (5). This increase is due to transcriptional activation and, in the cases where it has been examined, is mediated by one of two promoter elements known as the SCB or MCB elements. These are usually found in multiple copies in the promoters of genes expressed at this time and artificial promoters containing such elements can drive the expression of a reporter gene in a late G1-specific manner (6–8).

The SCB (Swi4/6 cell-cycle box) element has the consensus sequence CACGAAA and was first identified in the promoter of the *HO* gene (9), which encodes an endonuclease important for mating type switching (see Chapter 5). Subsequently, SCB elements have been found in a number of other promoters activated at late G1, including at least three different cyclin genes namely *CLN1*, *CLN2*, and *HCS26* (10). The MCB element on the other hand, controls the co-regulation of a large number of DNA synthesis genes, which include genes involved in the synthesis of DNA precursors as well as genes involved in the initiation and progression of DNA replication (5). Why so many genes required for DNA synthesis are periodically expressed in late G1 is not really clear since many of the gene products are stable and hence it is unlikely that they are limiting in the cell; some have been expressed in a constitutive manner without any adverse effect on cell cycle regulation. Of particular importance, however, are those genes encoding unstable products required for replication initiation; one or more of these are likely to be limiting and therefore could provide a biochemical link between START and the G1/S transition. The periodic expression of these genes would be expected to be important and could contribute to ensuring that replication occurs at the right time and only once per cycle. The MCB element has the consensus sequence ACGCGTNA, which contains the recognition site for the *Mlu* I restriction endonuclease (MCB refers to the term *Mlu* I cell cycle box).

The factor complexes that interact with both the SCB and MCB elements have been identified and the corresponding cDNAs isolated (reviewed in 11). In both cases the factors are composed of two subunits: a common subunit Swi6 complexed with either Swi4 or Mbp1 (12–17). The latter subunits have DNA-binding activity and therefore mediate the different DNA-binding specificities of the factors. The SBF (SCB binding factor) binds to the SCB and consists of a Swi6 and Swi4 heteromer, whereas the MBF (MCB binding factor) (also known as DSC1) binds to the MCB and consists of a heteromer between Swi6 and Mbp1. The two DNA—binding subunits can specifically bind to DNA in the absence of Swi6 although the binding is very weak (17, 18). Clearly, therefore, interaction with Swi6 augments binding. This may result from a Swi6-induced conformational change, although it is still formally possible that the augmentation results from Swi6 protein also interacting with DNA and thereby stabilizing the factor–DNA complex. The Swi6 protein alone, however, shows no DNA-binding activity (17, 18). Although the SBF normally activates SCB-containing promoters and the MBF activates MCB-containing promoters, some cross-regulation can occur. Thus, in the absence of Swi4, *CLN1* or *CLN2* expression can be mediated by the MBF, albeit less efficiently (10). Similarly *CLB5* can be regulated by the SBF in the absence of the MBF resulting from deletion of the *MBP1* gene (17, 19). Such cross-talk helps to explain the phenotypes obtained when the individual genes encoding these subunits are deleted. Neither deletion of *SWI4* nor *MBP1* is lethal, although in the former case the cells grow slowly and are larger (17, 20). A similar phenotype is observed upon deletion of the *SWI6* gene. In contrast to the viability of the single mutants, double mutants of *SWI4* and *MBP1* (17) or *SWI4* and *SWI6* (20) are inviable. The lethality appears to be due to a failure to express G1 cyclins since the double mutants can be partially rescued by expression of *CLN2* from a heterologous promoter. The phenotype of the double mutants underscores the essential nature of the transcriptional activation event that takes place at START; the double mutants fail to enter the cell cycle at START and are G1-arrested.

Alignment of the amino acid sequence of the Swi4 and Mbp1 proteins shows that they are highly related with extensive similarity in three regions (Fig. 1). One region is at the amino terminus (the first 112 amino acids of Mbp1 showing 50% identity to the corresponding region of Swi4), which has been shown to be the DNA-binding domain. The first 124 residues have been shown to be sufficient for DNA binding by Mbp1 (17), while the residues between positions 36 to 155 suffice for DNA binding by Swi4 (21). The carboxy-terminal region is also conserved and has been

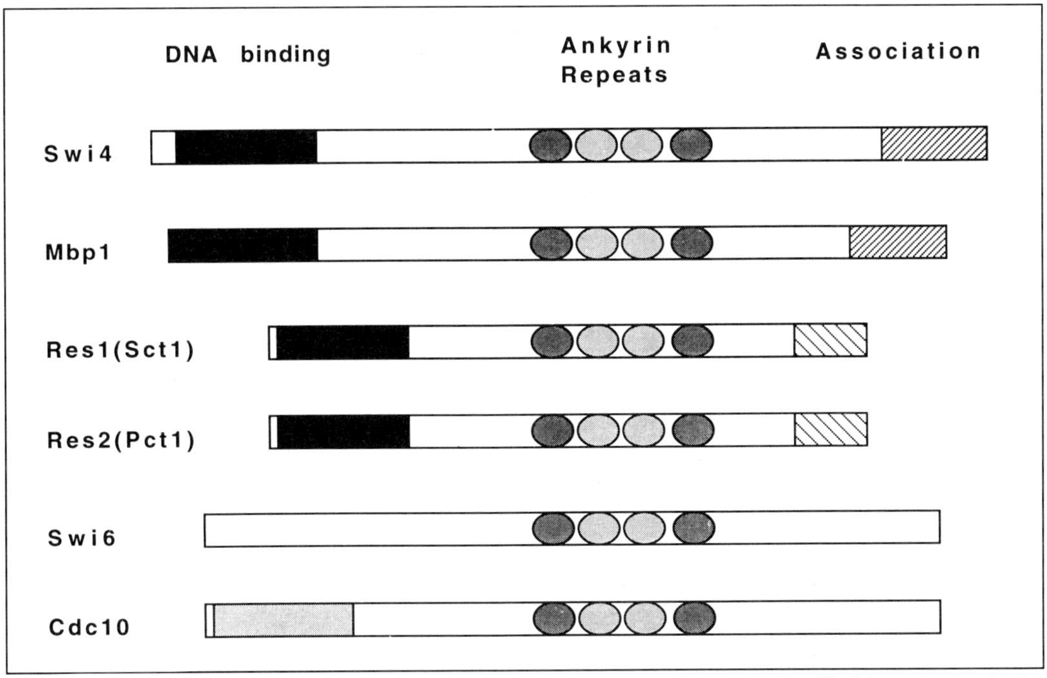

Fig. 1 Comparison of the SBF and MBF subunits in *S. cerevisiae* and *S. pombe*. The filled box represents the conserved DNA-binding domains; the shaded box in the Cdc10 protein represents the proposed vestigial binding domain. The ellipses represent the ankyrin repeats, both those that conform to the consensus well (black) and those that are degenerate (shaded). Homologies in the carboxy terminus involved in subunit interaction are also indicated.

shown to be important and sufficient for binding to the Swi6 protein (18, 22). This interaction also involves the carboxy-terminal region of Swi6 (22). The third region of similarity is centrally located and consists of four copies of a motif known as the ankyrin motif. This sequence motif has been found in many different regulatory proteins and has been implicated in protein–protein interactions (23). The two outermost repeats correspond well to the ankyrin repeat consensus, whereas the two inner repeats are degenerate (17, 23). Interestingly, these repeats are also conserved in the Swi6 protein in a similar location, suggesting the possibility that all three subunits originally derived from a common ancestor. The repeats are not directly required for DNA binding (17, 21) nor for subunit interaction (22) and yet they do appear to be important for function of the complex. Their role is therefore of major importance and, considering their function in other proteins, it is reasonable to expect that they are involved in specific protein–protein interactions. To date, however, no interactions involving these repeats have been described.

How the SBF and MBF complexes specifically activate transcription at START remains unknown. The activity of both complexes is dependent upon the *CDC28* gene, which encodes a cyclin-dependent kinase (cdk) essential for cell cycle control (24, 25). Its activity is required for progression through different transitions of the cycle including passage through START (26–28). The bulk of the evidence suggests

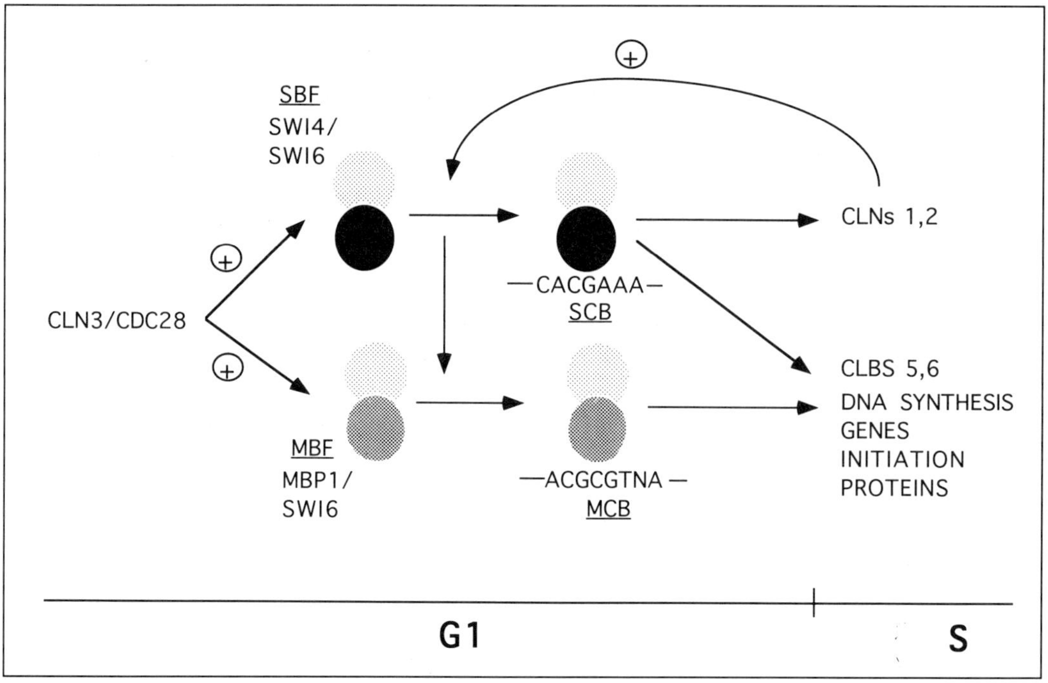

Fig. 2 A model depicting features of the start-specific transcription in *S. cerevisiae*. The SBF and MBF, with their particular subunit configurations, are shown together with the preferred consensus binding sites (SCB and MCB, respectively). Some of the groups of genes regulated by these complexes are indicated. A positive feedback loop is depicted whereby the activation of the *CLN1* and *CLN2* genes results in cyclins that can associate with the Cdc28 kinase and activate SBF and MBF complexes, thus giving rise to additional Cln1 and Cln2 expression.

that initial activation is due to Cdc28 complexed to the G1-specific cyclin Cln3 (29). Unlike the other two G1 cyclins, Cln1 and Cln2, Cln3 is not periodically expressed. It is absent in quiescent cells but is rapidly synthesized upon growth stimulation; its subsequent expression remains constant throughout the cell cycle. Cln1 and Cln2, on the other hand, are not present in G1 cells and are activated as a result of SBF activity at the G1/S transition (20). The level of Cln3 protein appears to determine the timing of START, which can only occur when cells reach a certain size (30). Thus, deletion of the *CLN3* gene results in an extended G1 period due to an increased size requirement at START, whereas the expression of mutant alleles of *CLN3*, which result in a more stable protein, decrease the size of cells at START. If Cln3–Cdc28 provides the initial trigger for SBF and MBF activation, it is rapidly and strongly reinforced by a positive feedback loop involving the Cln1 and Cln2 cyclins (24, 25). The expression of Cln1 and Cln2 is activated by SBF, whereupon they can complex with Cdc28 and activate further SBF or MBF activity. As a result of such a feedback loop, small changes in SBF/MBF activity are magnified and this process might contribute to the commitment that cells undergo when they pass through START (Fig. 2).

The key unanswered questions concern the mechanism of activation. How does Cln3 act as a sensor of cell size? How does Cdc28-associated kinase activity activate the SBF/MBF complexes? At present there are no real answers to these important questions. Although different factor subunits have been shown to be sensitive to direct phosphorylation by Cdc28, there is no evidence that this plays a regulatory role (K. Nasmyth and L. Breeden, personal communication). The major phosphorylation sites in Swi6 and Swi4 have been identified and mutated with no adverse effect on the timing or magnitude of activation. Thus, the involvement of Cdc28 may well be indirect in nature, possibly involving the modification of accessory proteins that interact with the complexes and mediate activation. An attractive possibility would involve the interaction with putative accessory proteins mediated by the ankyrin repeats of the subunits, since the repeats are required for activity and have been shown to be involved in protein–protein interactions in other proteins. The accessory proteins could consist of co-activator proteins necessary for transcription activation; in such a case, regulation at late G1 could be accomplished by limiting the interaction to this time of the cycle or regulating the activity of the co-activator by specific modification at this time of the cycle (perhaps by Cdc28-associated kinase activity). Another possible model stems from the observation that the Swi4 protein can directly interact with the mitotic cyclin Clb2 and its associated kinase activity, and it has been proposed that this interaction contributes to the down-regulation of the SBF complex following passage through the G1/S transition (31). This invokes the possibility that other cyclin–Cdc28 complexes might also interact with the SBF and MBF, and that, depending upon the nature of the associated kinase, activation or repression might result. For example, the associated kinase activity might target modification of components of the basic transcriptional machinery such as the large subunit of RNA polymerase II (see Chapter 1). It should be stressed, however, that, apart from the Clb2–Cdc28 interaction, no other direct protein–protein interactions involving the SBF or MBF complexes have been described. Presumably with the recent advances in methodology for detecting and isolating interacting proteins (such as the yeast two-hybrid screen), accessory proteins (if they exist) will be identified in the near future.

3. START-specific transcription in fission yeast

The fission yeast *Schizosaccharomyces pombe* is distantly related to budding yeast and, indeed, a comparison of the sequence of a number of common genes suggests that the two yeasts are as evolutionary distant from each other as either is to humans (32). The study of the cell cycle in both yeast systems has therefore been of tremendous value in illustrating the conservation of the cell cycle regulatory machinery and, as a consequence, provided important insights into many principles operating to regulate the cycle in mammalian cells. As is the case with the budding yeast, there is a commitment point (also known as START) prior to the G1/S transition and passage through this point is accompanied by a transcriptional activation programme (reviewed in 33). If this programme is blocked, by, for example, deletion of factors that mediate the activation (34–38), then cells arrest in G1. Less is known about the mechanics of this activation process than is the case for budding yeast. Nevertheless, from what has been discovered, it is perfectly clear that the process has been highly conserved; the DNA elements and the protein factors involved are highly related and in all likelihood, therefore, the mode of regulation is also similar.

Far fewer genes have been identified that are specifically activated at the G1/S transition. The reason for this is not clear but many of the genes encoding proteins required for general DNA metabolism are expressed in a non-periodic fashion. Perhaps it is the case that only the genes encoding limiting factors are expressed periodically in fission yeast. Four targets of this activation programme have so far been identified: *cdc22*$^{+}$, which encodes the regulatory subunit of ribonucleotide reductase (39, 40); *cdc18*$^{+}$, which is homologous to the budding yeast *CDC6* gene and whose product is likely to be involved in the initiation of DNA replication (41); *cdt1*$^{+}$, which also encodes a product required for DNA synthesis (42); and *cig2*$^{+}$, which encodes a B-type cyclin (43, 44). Both the *cdc22*$^{+}$ and *cdc18*$^{+}$ promoters contain MCB elements and, at least in the case of *cdc22*$^{+}$, they have been shown to be essential for START-specific activation.

Two MCB-binding complexes have been identified in *S pombe*, each consisting of two subunits (Fig. 3). One of the subunits is common to both complexes and is encoded by the *cdc10*$^{+}$ gene (45). The *cdc10*$^{+}$ gene is essential and required both for mitotic and pre-meiotic DNA synthesis (46). It appears to be the functional equivalent of the budding yeast Swi6 protein. The partner subunits of Cdc10, namely Res1 (Sct1) (37, 38) and Res2 (Pct1) (47, 48) provide DNA-binding specificity. Mutants of the *res1*$^{+}$ gene show severe growth abnormalities (37, 38); although they are viable at 30°C (37), they are very cold- and heat-sensitive resulting from failure to progress past START. In contrast, mutants of the *res2*$^{+}$ gene are viable (47, 48) and there is only a modest effect on mitotic division at low temperatures. It would appear, therefore, that it is the Res1–Cdc10 complex that plays the predominant role in allowing cells to progress into the mitotic S phase. Nevertheless, the Res2–Cdc10 complex is also active in mitotic cells and there is at least partial redundancy between Res1 and Res2 (47, 48). This is most clearly seen when both genes are deleted; the double mutation is lethal even at 30°C showing that when Res1 is absent, the Res2-associated activity becomes essential (47).

The most interesting diversity between the two complexes is evident when their respective requirements for pre-meiotic DNA synthesis and meiosis are tested. The *res2*$^{+}$ gene is essential for normal progression through the sexual differentiation pathway (47, 48). In wild-type cells, asci are developed that contain four ascospores, each containing a single nucleus. However, in *res2Δ* cells, the asci generally contain two or less spores, with a very significant number containing no spores at all. In addition, the distribution of the nuclei is abnormal, with some spores containing two nuclei and others none. In diploid cells, *res1*$^{+}$ is also required for pre-meiotic DNA synthesis (37). However, *S pombe* cells are normally found in the haploid state and in order to sporulate cells of opposite mating type arrest in G1 and conjugate before proceeding with the meiotic division pathway. Under these circumstances

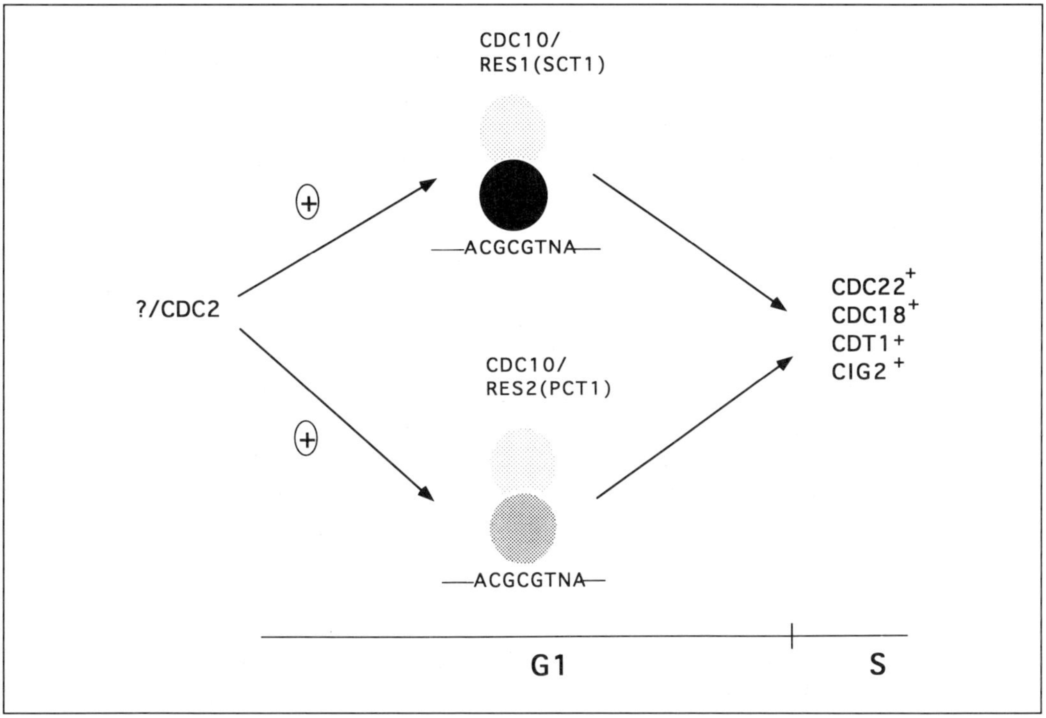

Fig. 3 A model depicting features of start-specific transcription in *S. pombe*. The two MBF complexes consisting of Cdc10–Res1(Sct1) and Cdc10–Res2(Pct1) subunits are indicated together with the known genes activated by one or other complex at the G1/S transition.

the *res1*$^{+}$ gene is not required. The most likely explanation for this apparent discrepancy stems from the finding that *res2*$^{+}$ expression is significantly increased upon conjugation and the resulting increase in Res2–Cdc10 activity is sufficient to overcome loss in *res1*$^{+}$ (47). It is clear, therefore, that both in the mitotic and meiotic pathways *res2*$^{+}$ can substitute for *res1*$^{+}$. The opposite is not the case, however; overexpression of *res1*$^{+}$ cannot overcome the meiotic defects of a *res2*Δ mutant (47; T. Takeda, personal communication). The two DNA-binding components, therefore, are only partially overlapping.

The Res1, Res2, and Cdc10 subunits are related to their budding yeast counterparts (Fig. 1). The Res1 and Res2 proteins are structurally similar to Swi4 and Mbp1, having a conserved amino-terminal DNA-binding domain, a centrally located ankyrin repeat region (again containing two good and two poor consensus repeats), and a carboxy-terminal domain involved in interaction with the Cdc10 subunit. The Cdc10 subunit also has the ankyrin repeats and, unlike Swi6, has the vestiges of a DNA-binding domain at its amino terminus. However, in the absence of Res1 or Res2, Cdc10 does not associate with DNA (48). The ankyrin repeats of Res2 are not required for either DNA binding or Cdc10 interaction (Y. Zhu, personal communication). Therefore, as is the case with the budding yeast factors, the role of the repeats in transcriptional regulation needs to be determined.

The activity of the fission yeast complexes is dependent upon the activity of *cdc2⁺* (49), the major (and possibly only) cyclin-dependent kinase (cdk) activity in *S pombe*. Cdc2 activity has been clearly shown to be required both at START, to allow G1/S transition, and at mitosis. It is possible that its only role at START is to activate transcription, although other roles are possible. However, the identity of the G1 cyclins involved together with *cdc2⁺* in mediating this activity is not known at present.

4. Late G1-specific transcription in mammalian cells

The conservation in the machinery that controls the specific transcription of genes at the G1/S transition in fission and budding yeast, poses the interesting and important question of whether similar complexes operate in higher eukaryotic cells. To date, however, no such complexes have been described although it must remain a possibility that they exist but have so far eluded detection. However, a family of transcription factors have been described that, although unrelated to the yeast factors, do show a number of functional similarities. The factors in question constitute the E2F family and a number of mammalian genes specifically expressed at the G1/S transition have binding sites for E2F in their transcriptional control sequences (reviewed in 50, 51). In some cases these binding sites have been shown to mediate their periodic expression.

An accumulating body of evidence suggests that E2F plays a major regulatory role in controlling the passage of mammalian cells through the restriction point and into the S phase (2, 3). The most convincing evidence is the demonstration that E2F activity is modulated by positive and negative regulators of cell cycle progression (see below). In addition, micro-injection of E2F protein into quiescent cells can induce their progression into S phase (52) and a dominant–negative E2F mutant has been shown to block the cell cycle progression of adenovirus EIA-induced quiescent cells (53). Clearly, therefore, the normal regulatory programme of the cell cycle can be subverted by the over-expression of E2F.

The E2F factor was first described as a factor required for the efficient transcription of an adenoviral gene (the E2A gene) during the early phase of infection (54, 55). The E2A promoter contains two binding sites for E2F that are inverted with respect to each other. This arrangement of sites was found to be important, since altering the spacing between the repeats or altering the orientation of either repeat resulted in significant loss of transcription. It is now known that this arrangement is essential for the formation of a stable, E2F-containing multimeric complex on the E2A promoter owing to the activity of a viral protein (the 19 kDa protein of the viral E4 gene) that directly interacts with E2F and promotes co-operative binding to DNA (56–59). In contrast, examination of E2F in uninfected cells led to the finding that it was often associated with the retinoblastoma tumour suppressor gene product (pRb) or the pRb-related proteins p107 and p130 (reviewed in 50, 51, 60, 61). The association with pRb was particularly exciting since considerable evidence existed to suggest that pRb could negatively control cell cycle progression, and that loss or mutation of the pRb gene was frequently found in tumour cells. This interest was reinforced by the finding that there was a good correlation between the regions of pRb altered in tumour cells and the regions required for E2F interaction (62, 63). These regions constitute a domain referred to as the pRb 'pocket', a common feature of pRb-related proteins. It has subsequently been shown that by interacting with pRb, the transcriptional activation potential of E2F is lost; this is also the case following interaction with p107 and p130 (50, 51). These results suggested, therefore, that the negative growth regulatory activity of the pocket-containing proteins was mediated through inactivation of E2F. This, in turn, would imply that E2F was critical for progression of cells from G1 phase into S phase. The many subsequent studies that have focused on the role of E2F in the cell support this possibility.

The E2F factor is not a single molecule but is rather a family of related proteins that together constitute the total E2F activity of the cell (50, 51). Four different E2F proteins have been described (E2F-1 to -4; 64–68) that share a number of conserved domains, including a centrally located domain involved in DNA binding and dimerization and a carboxy-terminal domain involved in transcriptional activation and interaction with pocket proteins. There is no significant similarity between the E2F proteins and the Swi family of proteins described in previously, even within their respective DNA-binding domains. This is somewhat surprising considering that the consensus binding site for E2F (CGCGAAA) has significant similarity to the SCB and MCB elements. *In vivo*, the different E2F factors bind to DNA as heterodimers with DP-1 (69–71), which is structurally quite different except for a small region within the DNA-binding/dimerization domain. The E2F proteins can bind to DNA as homodimers *in vitro* but there is no evidence that they do so *in vivo*; additionally, the heterodimer binds to DNA far more efficiently than the homodimer. There is also no evidence that the different E2F–DP-1 dimers have different DNA-binding specificities, although the possibility still exists. What is clear, however, is that at least some of the individual dimers have functional differences stemming from different profiles of protein–protein interactions. Thus, complexes containing E2F-1, -2, or -3 (72, 73) interact with pRb but not p107 or p130, whereas E2F-4-containing dimers interact with p107 but not pRb (67, 68). It may well be that yet another E2F member is involved in specific interactions with p130. The significance of these differences is not clear at present since the role of individual complexes in regulating cell cycle progression has not been elucidated. However, interaction with the pocket proteins has at least one consequence that is similar in all cases; the activity of the complex is negated or repressed.

The key to the regulation of E2F activity and hence the activation of transcription at the G1/S transition, is the regulation of its interaction with the pocket proteins. At mid to late G1, close to the restriction point, uncomplexed or free E2F can be detected. The accumulation of this form of E2F, which is the active form, is presumed to give rise to the stimulation of transcription of E2F-dependent genes (reviewed in 50, 51). How does free E2F arise? Much of the work relating to this question has come from the study of the E2F–pRb complex and a model has been developed based on the functional consequences of pRb phosphorylation. The pRb protein becomes hyperphosphorylated as cells progress to the late G1 phase of the cycle (74–77). It is suggested that this phosphorylation reverses the growth suppressive properties of pRb, since in its hyperphosphorylated form it can no longer associate with E2F (78, 79). This is likely to result in the accumulation of free E2F although it is not clear whether this accumulation arises from the failure of newly synthesized E2F to interact with phosphorylated pRb or the disruption of preformed E2F–pRb complexes by phosphorylation. There is no evidence for the latter possibility, and indeed E2F–pRb complexes can still be detected in the late G1 and S phases of the cycle.

There is some controversy as to how pRb hyperphosphorylation arises. It is clearly due to the activity of G1-specific cyclin-dependent kinases (cdks). Two different cdks that function during this period of the cycle have been described and are associated with either the D-type cyclins or with cyclin E (reviewed in 1, 2, 80). Three D-type cyclins have been identified in mammalian cells (D1, D2, and D3) that assemble into holoenzymes with either cdk4 or cdk6. Individual cell types do not necessarily express all three although the majority at least express D3 together with either D1 or D2. The D-type cyclins are not expressed in quiescent cells but are induced in mid G1 following the addition of growth factors or mitogens, whereupon they are subsequently expressed in a relatively non-periodic fashion. Cyclin E, however, which associates with cdk2, is expressed periodically, reaching a maximum at the G1/S transition. Both the D- and E-type cyclins appear to be important for cell cycle progression. Over-expression of cyclin D1 or E shortens the length of the G1 period, at least to a small extent (81, 82), and the micro-injection of antibodies specific to both cyclins blocks the progression of cells from G1 into S (81, 83). Which of these cdks gives rise to pRb hyperphosphorylation? The probable answer is that both are involved (reviewed in 1–3). Multiple sites on pRb are phosphorylated but only a subset of these are required to inactivate pRb function (84). Under normal circumstances, in order to get the full spectrum of phosphorylation, the combined action of both D- and E-type cyclin-directed cdks is likely to be necessary. This is not to say that under certain circumstances, for example when D-type activity is enhanced, hyperphosphorylation cannot result from D-type cyclin-directed cdks alone.

The model that has developed, therefore, is that after growth stimulation of quiescent cells, a programme of growth-dependent protein synthesis is initiated that includes the components of G1-specific cdks (Fig. 4). When this cdk activity reaches a threshold level, at the restriction point, pRb becomes hyperphosphorylated, leading to the accumulation of free E2F. This active form of E2F stimulates the expression of a number of genes required for S phase and the cells progress through the G1/S transition. Conditions that promote pRb hyperphosphorylation or block pRb hyperphosphorylation will control passage of cells through the restriction point. Whilst it is likely that this model encompasses some of the important aspects of the regulation of cell cycle progression through G1, additional layers of control are probably in operation. Very little is known or understood about the other pRb family members, p107 and p130. Upon over-expression, p107, like pRb,

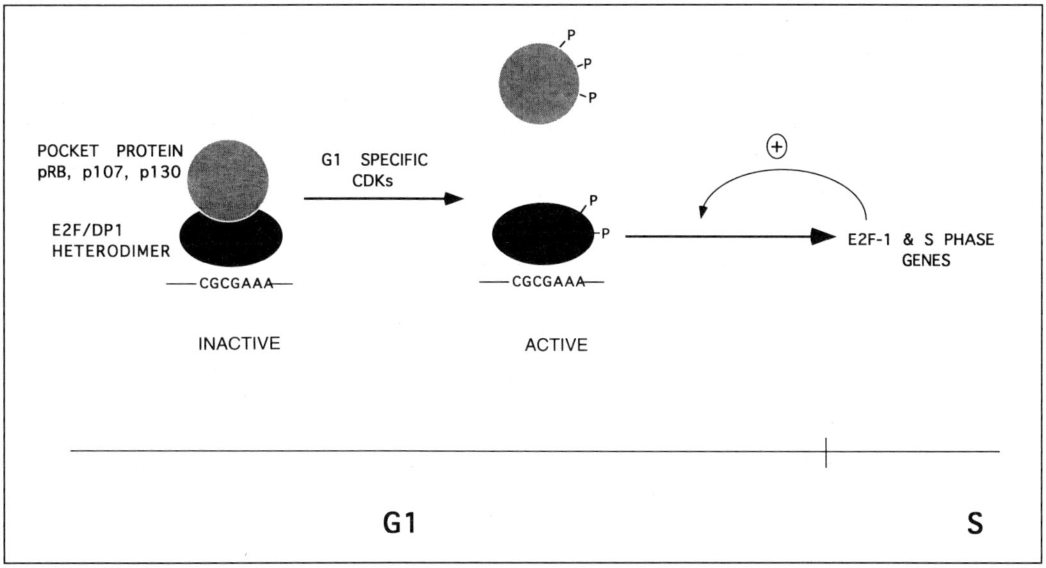

Fig. 4 A model depicting the features of E2F-mediated transcription activation in mammalian cells. The E2F–DP1 factor (black oval) is complexed to members of the pocket protein family, namely pRb, p107, and p130. These complexes are transcriptionally inactive. As cells progress through G1, both pRb (and possibly the other pocket proteins) and certain E2F family members (E2F-1) become hyperphosphorylated as a result of G1-specific cdk activity. The phosphorylated proteins cannot interact with each other. Uncomplexed or free E2F accumulates at late G1. This E2F is active and stimulates transcription of a number of genes whose products are required for S phase. A positive feedback loop is depicted whereby E2F positively activates E2F-1 transcription, thus contributing to the pool of free E2F at this period of the cell cycle.

can arrest cells (85). Potentially, therefore, it also serves to regulate progression through G1 and again is likely to be doing so by interacting with specific E2F complexes. Whether this interaction is regulated during the cell cycle, and what role p107 phosphorylation by G1-specific cdks might have, are unanswered questions. In addition, E2F-1 is a substrate for cyclin D-associated cdk activity (D. Mann, personal communication) and as a result of phosphorylation on two specific serine residues, can no longer bind to pRb (86). Thus, regulation could also be exercised through phosphorylation of the E2F-1 component of the E2F-1–pRb complex. Certain results with cyclin D mutants have suggested that cyclin D can promote S phase entry in the absence of pRb hyperphosphorylation (87); phosphorylation of E2F-1 under such circumstances would provide a possible explanation. Once E2F activation is initiated a positive feedback loop appears to operate to enhance the effect. This results from the transcriptional activation of the E2F-1 gene and, possibly, the E2F-2 and E2F-3 genes as well. In the case of the E2F-1 gene, the activation appears to be direct, since the promoter contains E2F binding sites (88, 89).

How do the pocket proteins negate E2F activity? The region that mediates the interaction is located in the carboxy terminus of E2F, embedded within the domain shown to be necessary for transcriptional activation (64–67). This suggests that activation is blocked by the pocket proteins physically preventing the activation domain from interacting with the basal transcription machinery. The activation domain interacts with TBP, the DNA-binding subunit of TFIID (90, and see Chapter 1). Mutations in the E2F carboxy terminus have been described that affect both pRb and TBP binding, suggesting that the two proteins recognize similar features of E2F. It would be expected, therefore, that TBP interaction would be prevented by pRb binding. However, there is also evidence that pRb can play a role in transcriptional repression beyond preventing E2F-mediated activation (91). Binding of the E2F–pRb complex to a promoter appears to result in active repression that can counteract the effect of other activator proteins bound to other sites in the promoter. The mechanism behind this repression is not known. This activity of the E2F–pRb complex suggests two different mechanisms by which promoters containing E2F sites can be regulated in a late G1-specific fashion. In one case, activation results from E2F activity, stimulating transcription at late G1 when it is freed from the negative influence of pRb or other pocket proteins. In the second case, activation is due to the activity of other DNA-binding transcription factors that cannot function outside late G1 due to the repressive effects of the E2F–pRb complex. Mutation of the E2F sites in these two cases would have a very different result. In the former case the promoter could no longer stimulate transcription at late G1 or at other times, while in the second case the promoter would be constitutively active. Examples of both types have been described (92–95).

The G1 phase of the mammalian cell cycle represents the major phase during which cell proliferation can be regulated (reviewed in 2). It is during this phase that growth stimulating and inhibiting activities exert their effect. The balance of these activities determines the level of G1-specific cdk activity and, as a result, the activity of the E2F factor. Mutation or over-expression of key players in this regulatory pathway characterizes a number of tumour cells and tumour cell lines. For example, the cyclin D1 gene is amplified and its mRNA over-expressed in several human tumours (reviewed in 96). It is anticipated that such over-expression would stimulate progression through G1 even in the face of anti-proliferative signals. Inactivation of both copies of the pRb gene gives rise to eye tumours as well as several other tumour types (reviewed in 97). Inactivation of MTS1 encoding the p16 inhibitor of cdk4 and cdk6 is common in cell lines (over 50%) and some tumours (98–102). A possible prediction from such observations, therefore, is that the E2F genes could also be targets for mutations leading to tumourogenesis. The mutations would have to be relatively subtle involving, for example, mutations that affect pocket protein binding but not transcription activation. Such mutants have been constructed *in vivo* showing that the possibility is a viable one. It will be interesting to see whether such mutations are indeed found in tumour cells.

It is not clear whether E2F represents the activity in mammalian cells equivalent to the Swi-like family of proteins in fission and budding yeast. It is possible that parallel systems exist, both controlling the transcription of genes in the late G1 phase. E2F and DP-1 homologues have been isolated from *Drosophila* (103) and at least E2F-specific binding activity has been described in yeast (104, 105). However, no yeast E2F-like gene has been found. The *Drosophila* E2F is of particular interest since it is similar to its mammalian homologue not only in the DNA-binding/dimerization region but also in the carboxy terminus, including the region that binds to pocket proteins. This suggests the possibility that pocket proteins may also exist in *Drosophila*. Despite the fact that E2F and the Swi-like proteins share no structural features, there are a number of parallels that can be drawn between the two transcriptional systems. As described above, the activity of the D-cyclins is important for the initial activation of E2F. In budding yeast, Cln3 performs such a function, being involved in the initial activation of the SBF and MBF complexes. A number of interesting comparisons are evident between the D-cyclins and Cln3. First, both are synthesized in a non-periodic fashion that distinguishes them from the majority of other cyclins involved in cell cycle control. Secondly, Cln3 is closer at the sequence level to D-cyclins than any of the other mammalian cyclins. Thirdly, they appear to share substrate specificities. The most startling example of this comes from a study of the pattern of pRb phosphorylation expressed in budding yeast (106). Surprisingly, the pattern of phosphorylation was found to be very similar to that seen in mammalian cells and was the result of both Cln3- and Cln-1- or -2-associated cdk activity. The loss of pRb hyperphosphorylation in a *CLN3Δ* mutant or a *CLN1ΔCLN2Δ* double mutant could be complemented by expression of cyclins D and E, respectively. Thus, with respect to pRb phosphorylation, cyclin D and Cln3 perform similar functions.

In addition to their regulation by G1-specific cdks, other similarities exist between E2F and the Swi-like proteins. Their respective binding sites are remarkably alike and the timing of their activation is similar, occurring as cells progress through START or the restriction point. Finally, there is some evidence that a fraction of the complexes can associate with cyclins and cdks. The E2F–p107 complex is often found to contain cdk2 together with either cyclin E or A (51, 52). The function of this multimeric complex is not understood. The SBF complex, as described above, can be found associated with Clb2–Cdc28, at least during part of the cell cycle.

5. Conclusions

There is mounting evidence that periodic transcription plays an important role in controlling the timing and order of cell cycle events. The conservation of the process in all eukaryotes emphasizes this importance. Many of the regulatory factors, such as the cyclins, are only expressed during particular periods of the cycle when they are functionally required. Surprisingly little is known about the underlying mechanisms that regulate cyclin expression and maintain their order of expression. It will be particularly interesting to discover whether the same mechanisms are used in all eukaryotes. Rather more is known about the regulation of transcription that takes place just prior to the G1/S transition. The reason for this is understandable; the transition from G1 into S represents a major regulatory step of the cycle, particularly in higher eukaryotes, and, as such, it has commanded a considerable amount of attention. What has emerged is that the transcriptional activation of a number of genes at this step is crucial for the transition to take place. As outlined in this chapter, many of the components involved in such activation have now been identified in a range of different eukaryotic systems. Over the next few years it will become clearer how these components interact and regulation takes place. This will be a major step towards understanding how the cell cycle is regulated.

References

1. Hunter, T. and Pines, J. (1994) Cyclins and cancer II: cyclin D and CDK inhibitors come of age. *Cell*, **79**, 573.
2. Sherr, C. J. (1994) G1 phase progression: cycling on cue. *Cell*, **79**, 551.
3. Peeper, D. S., van der Eb, A.J., and Zantena, A. (1994) The G1/S cell-cycle checkpoint in eukaryotic cells. *Biochim. Biophys. Acta*, **1198**, 215.
4. Pringle, J. R. and Hartwell, L. H. (1981) The *Saccharomyces cerevisiae* cell cycle. In *The molecular biology of the yeast* Saccharomyces*: life cycle and inheritance* (ed. J. N. Strathern, E. W. Jones, and J. R. Broach), p. 97. Cold Spring Harbor Laboratory Press, CSH, New York.
5. Merrill, G. F., Morgan, B. A., Lowndes, N. F., and Johnston, L. H. (1992) DNA synthesis control in yeast: an evolutionarily conserved mechanism for regulating DNA synthesis genes. *BioEssays*, **14**, 823.
6. Breeden, L. and Nasmyth, K. (1987) Cell cycle control of the yeast *HO* gene: *cis*- and *trans*-acting regulators. *Cell*, **48**, 389.
7. McIntosh, E. M., Atkinson, T., Storms, R. K., and Smith, M. (1991) Characterization of a short, *cis*-acting DNA sequence which conveys cell cycle stage-dependent transcription in *Saccharomyces cerevisiae*. *Mol. Cell. Biol.*, **11**, 329.
8. Lowndes, N. F., Johnson, A. L., and Johnston, L. H. (1991) Coordination of expression of DNA synthesis genes in budding yeast by a cell-cycle regulated *trans* factor. *Nature*, **350**, 247.
9. Nasmyth, K. (1986) A repetitive DNA sequence that confers cell cycle START (CDC28)-dependent transcription of the *HO* gene in yeast. *Cell*, **42**, 225.
10. Ogas, J., Andrews, B. J., and Herskowitz, I. (1991) Transcriptional activation of *CLN1*, *CLN2*, and a putative new G1 cyclin (*HCS26*) by SWI4, a positive regulator of G1-specific transcription. *Cell*, **66**, 1015.
11. McKinney, J. and Cross, F. (1992) A switch-hitter at the start of the cell cycle. *Curr. Biol.*, **2**, 421.
12. Andrews, B. J. and Herskowitz, I. (1989) The yeast SWI4 protein contains a motif present in developmental regulators and is part of a complex involved in cell-cycle dependent transcription. *Nature*, **342**, 830.
13. Andrews, B. J. and Herskowitz, I. (1989) Identification of a DNA binding factor involved in cell cycle control of the yeast *HO* gene. *Cell*, **57**, 21.
14. Lowndes, N. F., Johnson, A. L., and Johnston, L. H. (1991) Coordination of expression of DNA synthesis genes in budding yeast by a cell-cycle regulated transfactor. *Nature*, **350**, 247.
15. Lowndes, N., Johnson, A. L., Breeden, L., and Johnston, L. H. (1992) SWI6 protein is required for transcription of the periodically expressed DNA synthesis genes in budding yeast. *Nature*, **357**, 505.
16. Dirick, L., Moll, T., Auer, H., and Nasmyth, K. (1992) A central role for SWI6 in modulated cell cycle Start-specific transcription in yeast. *Nature*, **357**, 508.

17. Koch, C., Moll, T., Neuberg, M., Ahorn, H., and Nasmyth, K. (1993) A role for the transcription factors Mbp1 and Swi4 in progression from G1 to S phase. *Science*, **261**, 1551.
18. Sidorova, I. and Breeden, L. (1993) Analysis of the SWI4/SWI6 protein complex, which directs G1/S-specific transcription in *Saccharomyces cerevisiae*. *Mol. Cell. Biol.*, **13**, 1069.
19. Epstein, C. B. and Cross, F. R. (1992) CLB5: a novel B cyclin from budding yeast with a role in S phase. *Genes Dev.*, **6**, 1695.
20. Nasmyth, K. and Dirick, L. (1991) The role of SWI4 and SWI6 in the activity of G1 cyclins in yeast. *Cell*, **66**, 995.
21. Primig, M., Sockanathan, S., Auer, H., and Nasmyth, K. (1992) Anatomy of a transcription factor important for the start of the cell cycle in *Saccharomyces cerevisiae*. *Nature*, **358**, 593.
22. Andrews, B. J. and Moore, L. A. (1992) Interaction of the yeast Swi4 and Swi6 cell cycle regulatory proteins *in vitro*. *Proc. Natl Acad. Sci. USA*, **89**, 11852.
23. Bork, P. (1993) Hundreds of ankyrin-like repeats in functionally diverse proteins: mobile modules that cross phyla horizontally. *Proteins: Struct. Function, Genet.*, **17**, 363.
24. Cross, F. R. and Tinkelenberg, A. H. (1991) A potential positive feed-back loop controlling *CLN1* and *CLN2* gene expression at the START of the yeast cell cycle. *Cell*, **65**, 875.
25. Dirick, L. and Nasmyth, K. (1991) Positive feedback in the activation of G1 cyclins in yeast. *Nature*, **351**, 754.
26. Piggott, J. R., Rai, R., and Carter, B. L. A. (1982) A bifunctional gene product involved in two phases of the yeast cell cycle. *Nature*, **298**, 391.
27. Reed, S. I. and Wittenberg, C. (1990) Mitotic role for the Cdc28 protein kinase of *Saccharomyces cerevisiae*. *Proc. Natl Acad. Sci. USA*, **87**, 5697.
28. Surana, U., Robitsch, H., Price, C., Schuster, T., Fitch, I., Futcher, A. B., and Nasmyth, K. (1991) The role of cdc28 and cyclins during mitosis in the budding yeast *Saccharomyces cerevisiae*. *Cell*, **65**, 145.
29. Tyers, M., Tokiwa, G., and Futcher, B. (1993) Comparison of the *Saccharomyces cerevisiae* G1 cyclins: Cln3 may be an upstream activator of Cln1, Cln2 and other cyclins. *EMBO J.*, **12**, 1955.
30. Nash, R., Tokiwa, G., Anand, S., Erickson, K., and Futcher, A. B. (1988) The *WHI1*$^+$ gene of *Saccharomyces cerevisiae* tethers cell division to cell size and is a cyclin homolog. *EMBO J.*, **7**, 4335.
31. Amon, A., Tyers, M., Futcher, B., and Nasmyth, K. (1993) Mechanisms that help the yeast cell cycle clock tick: G2 cyclins transcriptionally activate G2 cyclins and repress G1 cyclins. *Cell*, **74**, 993.
32. Sipiczki, M. (1989) Taxonomy and phylogenesis. In *Molecular biology of the fission yeast* (ed. A. Nasim, B. F. Johnson, and P. Young), p. 431. Academic Press, New York.
33. Forsburg, S. L. and Nurse, P. (1991) Cell cycle regulation in the yeast *Saccharomyces cerevisiae* and *Schizosaccharomyces pombe*. *Annu. Rev. Cell Biol.*, **7**, 227.
34. Beach, D., Piper, M., and Nurse, P. (1982) Construction of a *Schizosaccharomyces pombe* gene bank in a yeast bacterial shuttle vector and its use to isolate genes by complementation. *Mol. Gen. Genet.*, **187**, 326.
35. Nurse, P. and Bissett, Y. (1981) Gene required in G1 for commitment to cell cycle and in G2 for control of mitosis in fission yeast. *Nature*, **292**, 558.
36. Aves, S. J., Durkacz, B. W., Carr, A., and Nurse, P. (1985) Cloning, sequencing and transcriptional control of the *Schizosaccharomyces pombe cdc10* start gene. *EMBO J.*, **4**, 457.
37. Tanaka, K., Okazaki, K., Okazaki, N., Ueda, T., Sugiyama, A., Nojima, H., and

Okayama, H. (1992) A new cdc gene required for S phase of *Schizosaccharomyces pombe* encodes a protein similar to the *cdc10* and SWI4 gene products. *EMBO J.*, **11**, 4923.

38. Caligiuri, M. and Beach, D. (1993) *Sct1* functions in partnership with *cdc10* in a transcription complex that activates cell cycle START and inhibits differentiation. *Cell*, **72**, 607.
39. Gordon, C. and Fantes, P. (1986) The *cdc22* gene of *Schizosaccharomyces pombe* encodes a cell cycle regulated transcript. *EMBO J.*, **5**, 2981.
40. Fernandez-Sarabia, M. J., McInery, C., Harris, P., Gordon, C., and Fantes, P. (1993) The cell cycle genes *cdc22* and *suc22* of the fission yeast *Schizosaccharomyces pombe* encode the large and small subunits of ribonucleotide reductase. *Mol. Gen. Genet.*, **238**, 241.
41. Kelly, T. J., Martin, G. S., Forsburg, S. L., Stephen, R. J., Russo, A., and Nurse, P. (1993) S-phase is coupled to start and mitosis in fission yeast by the $cdc18^+$ gene product. *Cell*, **74**, 371.
42. Hofmann, J. F. X. and Beach, D. (1994) *cdt1* is an essential target of the Cdc10/Sct1 transcription factor: requirement for DNA replication and inhibition of mitosis. *EMBO J.*, **13**, 425.
43. Bueno, A. and Russell, P. (1992) Dual functions of *CDC6*: a yeast protein required for DNA replication also inhibits nuclear division. *EMBO J.*, **11**, 2167.
44. Obara-Ishihara, T. and Okayama, H. (1994) A B-type cyclin negatively regulates conjugation via interacting with cell cycle 'start' genes in fission yeast. *EMBO J.*, **13**, 1863.
45. Lowndes, N. F., McInerny, C. J., Johnson, A. L., Fantes, P. A., and Johnston, L. H. (1992) Control of DNA synthesis genes in fission yeast by the cell-cycle gene $cdc10^+$. *Nature*, **355**, 449.
46. Grallert, B. and Sipiczki, M. (1991) Common genes and pathways in the regulation of the mitotic and meiotic cell cycles of *Schizosaccharomyces pombe*. *Curr. Genet.*, **20**, 199.
47. Miyamoto, M., Tanaka, K., and Okayama, H. (1994) $res2^+$, a new member of the $cdc10^+$/*SWI4* family, controls the start of mitotic and meiotic cycles in fission yeast. *EMBO J.*, **13**, 1873.
48. Zhu, Y., Takeda, T., Nasmyth, K., and Jones, N. (1994) $pct1^+$, which encodes a new DNA-binding partner of $p85^{cdc10}$, is required for meiosis in the fission yeast *Schizosaccharomyces pombe*. *Genes Dev.*, **8**, 885.
49. Broek, D., Bartlett, R., Crawford, K., and Nurse, P. (1991) Involvement of $p34^{cdc2}$ in establishing the dependency of S phase on mitosis. *Nature*, **349**, 388.
50. LaThangue, N. B. (1994) DRTF1/E2F: an expanding family of heterodimeric transcription factors implicated in cell-cycle control. *Trends Biochem. Sci.*, **19**, 108.
51. Lam, E. W. F. and LaThangue, N. B. (1994) DP and E2F proteins: coordinating transcription with cell cycle progression. *Curr. Opin. Cell Biol.*, **6**, 859.
52. Johnson, D. G., Schwarz, J. K., Cress, W. D., and Nevins, J. R. (1993) Expression of E2F1 induces quiescent cells to enter S phase. *Nature*, **365**, 349.
53. Dobrowolski, S. F., Stacey, D. W., Harter, M. L., Stine, J. T., and Hiebert, S. W. (1994) An E2F dominant negative mutant blocks E1A induced cell cycle progression. *Oncogene*, **9**, 2605.
54. Kovesdi, I., Reichel, R., and Nevins, J. R. (1986) Identification of a cellular transcription factor involved in E1A trans-activation. *Cell*, **45**, 219.
55. Yee, A. S., Raychaudhuri, P., Jokoi, L., and Nevins, J. R. (1987) Promoter interaction of E1A-inducible factor E2F and its potential role in the formation of a multicomponent complex. *EMBO J.*, **6**, 2061.
56. Hardy, S. and Shenk, T. (1989) E2F from adenovirus-infected cells binds cooperatively

to DNA containing two properly orientated and spaced recognition sites. *Mol. Cell. Biol.*, **9**, 4495.

57. Hardy, S., Engel, D. A., and Shenk, T. (1989) An adenovirus early region 4 gene product is required for induction of the infection-specific form of cellular E2F activity. *Genes Dev.*, **3**, 1062.
58. Huang, M. M. and Hearing, P. (1989) Adenovirus early region 4 encodes two gene products with redundant effects in lytic infection. *J. Virol.*, **63**, 2605.
59. Reichel, R., Neill, S. D., Kovesdi, I., Simon, M. C., Raychaudhuri, P., and Nevins, J. R. (1989) The adenovirus E4 gene, in addition to the E1A gene, is important for the trans-activation of E2 transcription and for E2F activation. *J. Virol.*, **63**, 3643.
60. Nevins, J. R. (1992) E2F: a link between the Rb tumor suppressor protein and viral oncoproteins. *Science*, **258**, 424.
61. Helin, K. and Harlow, E. (1993) The retinoblastoma protein as a transcriptional repressor. *Trends Cell Biol.*, **3**, 43.
62. Zamanian, M. and LaThangue, N. B. (1992) Adenovirus E1A prevents the retinoblastoma gene product from repressing the activity of a cellular transcription factor. *EMBO J.*, **11**, 2603.
63. Hiebert, S. W., Chellappan, S. P., Horowitz, J. M., and Nevins, J. R. (1992) The interaction of pRb with E2F coincides with an inhibition of the transcriptional activity of E2F. *Genes Dev.*, **6**, 177.
64. Helin, K., Lees, J. A., Vidal, M., Dyson, N., Harlow, E., and Fattaey, A. (1992) A cDNA encoding a pRB-binding protein with properties of the transcription factor E2F. *Cell*, **70**, 337.
65. Kaelin, W. G., jun., Krek, W., Selers, W. R., DeCaprio, J. A., Ajchenbaum, F., Fuchs, C. S., Chittenden, T., Li, Y., Farnham, P. J., Blanar, M. A., Livingston, D. M., and Flemington, E. K. (1992) Expression cloning of a cDNA encoding a retinoblastoma-binding protein with E2F-like properties. *Cell*, **70**, 351.
66. Shan, B., Zhu, X., Chen, P.-L., Durfee, T., Yang, Y., Sharp, D., and Lee, W.-H. (1992) Molecular cloning of cellular genes encoding retinoblastoma-associated proteins: identification of a gene with properties of the transcription factor E2F. *Mol. Cell. Biol.*, **12**, 5620.
67. Ginsberg, D., Vairo, G., Chittenden, T., Xiao, Z.-H., Xu, G., Wydner, K. L., DeCaprio, J. A., Lawrence, J. B., and Livingston, D. M. (1994) E2F-4, a new member of the E2F transcription factor family, interacts with p107. *Genes Dev.*, **8**, 2665.
68. Beijersbergen, R. L., Kerkhoven, R. M., Zhu, L., Cartée, L., Voorhoeve, P. M., and Bernards, R. (1994) E2F-4, a new member of the E2F gene family, has oncogenic activity and associates with p107, *in vivo*. *Genes Dev.*, **8**, 2680.
69. Bandara, L. R., Buck, V. M., Zamanian, M., Johnston, L. H., and LaThangue, N. B. (1993) Functional synergy between DP-1 and E2F-1 in the cell cycle regulating transcription factor DRTF1/E2F. *EMBO J.*, **12**, 4317.
70. Helin, K., Wu, C., Fattaey, A. R., Lees, J. A., Dynlacht, B. D., Ngwu, C., and Harlow, E. (1993) Heterodimerization of the transcription factors E2F-1 and DP-1 leads to cooperative *trans*-activation. *Genes Dev.*, **7**, 1850.
71. Krek, W., Livingston, D. M., and Shirodkar, S. (1993) Binding to DNA and the retinoblastoma gene product promoted by complex formation of different E2F family members. *Science*, **262**, 1557.
72. Lees, J. A., Saito, M., Vidal, M., Valentine, M., Look, T., Harlow, E., Dyson, N., and Helin, K. (1993) The retinoblastoma protein binds to a family of E2F transcription factors. *Mol. Cell. Biol.*, **13**, 7813.

73. Ivey-Hoyle, M., Conroy, R., Huber, H. E., Goodhart, P. J., Oliff, A., and Heimbrook, D. C. (1993) Cloning and characterization of E2F-2, a novel protein with the biochemical properties of the transcription factor E2F. *Mol. Cell. Biol.*, **13**, 7802.
74. Buchkovich, K., Duffy, L. A., and Harlow, E. (1989) The retinoblastoma protein is phosphorylated during specific phases of the cell cycle. *Cell*, **58**, 1097.
75. Chen, P., Scully, P., Shew, J., Wang, J. Y. J., and Lee, W. (1989) Phosphorylation of the retinoblastoma gene product is modulated during the cell cycle and cellular differentiation. *Cell*, **58**, 1193.
76. DeCaprio, J. A., Ludlow, J. W., Lynch, D., Furukawa, Y., Griffin, J., Piwnica-Worms, H., Huang, C.-M., and Livingston, D. M. (1989) The product of the retinoblastoma susceptibility gene has properties of a cell cycle regulatory element. *Cell*, **58**, 1085.
77. Mihara, K., Cao, X., Yen, A., Chandler, S., Driscoll, B., Murphree, A. L., T'Ang, A., and Fung, Y. (1989) Cell cycle-dependent regulation of phosphorylation of the human retinoblastoma gene product. *Science*, **246**, 1300.
78. Chellappan, S. P., Hiebert, S., Mudryj, M., Horowitz, J. M., and Nevins, J. R. (1991) The E2F transcription factor is a cellular target for the Rb protein. *Cell*, **65**, 1053.
79. Schwarz, J. K., Devoto, S. H., Smith, E. J., Chellappan, S. P., Jakoi, L., and Nevins, J. R. (1993) Interactions of the p107 and Rb proteins with E2F during the cell proliferation response. *EMBO J.*, **12**, 1013.
80. Draetta, G. (1994) Mammalian G1 cyclins. *Curr. Opin. Cell Biol.*, **6**, 842.
81. Quelle, D., Ashmun, R., Shurtleff, S., Kato, J., Bar-Sagi, D., Roussel, M., and Sherr, C. (1993) Overexpression of mouse D-type cyclins accelerates G1 phase in rodent fibroblasts. *Genes Dev.*, **7**, 1559.
82. Ohtsubo, M. and Roberts, J. (1993) Cyclin-dependent regulation of G1 in mammalian fibroblasts. *Science*, **259**, 1908.
83. Baldin, V., Lukas, J., Marcote, M. J., Pagano, M., and Draetta, G. (1993) Cyclin D1 is a nuclear protein required for cell cycle progression in G1. *Genes Dev.*, **7**, 812.
84. Mittnacht, S., Lees, J. A., Desai, D., Harlow, E., Morgan, D.O., and Weinberg, R. A. (1994) Distinct sub-populations of the retinoblastoma protein show a distinct pattern of phosphorylation. *EMBO J.*, **13**, 118.
85. Zhu, L., Van den Heuvel, S., Helin, K., Fattaey, A., Ewen, M., Livingston, D., Dyson, N., and Harlow, E. (1993) Inhibition of cell proliferation by p107, a relative of the retinoblastoma protein. *Genes Dev.*, **7**, 1111.
86. Fagan, R., Flint, K. J., and Jones, N. (1994) Phosphorylation of E2F-1 modulates its interaction with the retinoblastoma gene product and the adenoviral E4 19 kDa protein. *Cell*, **78**, 799.
87. Dowdy, S. F., Hinds, P. W., Louie, K., Reed, S. I., Arnold, A., and Weinberg, R. A. (1993) Physical interaction of the retinoblastoma protein with human D cyclins. *Cell*, **73**, 499.
88. Johnson, D. G., Ohtani, K., and Nevins, J. R. (1994) Autoregulatory control of E2F1 expression in response to positive and negative regulators of cell cycle progression. *Genes Dev.*, **8**, 1514.
89. Hsiao, K.-M., McMahon, S. L., and Farnham, P. J. (1994) Multiple DNA elements are required for the growth regulation of the mouse E2F1 promoter. *Genes Dev.*, **8**, 1526.
90. Hagemeier, C., Cook, A., and Kouzarides, T. (1993) The retinoblastoma protein binds E2F residues required for activation *in vivo* and TBP binding *in vitro*. *Nucl. Acids Res.*, **21**, 4998.
91. Weintraub, S. J., Prater, C. A., and Dean, D. C. (1992) Retinoblastoma protein switches the E2F site from a positive to a negative element. *Nature*, **358**, 259.

92. Slansky, J. E., Li, Y., Kaelin, W. G., and Farnham, P. J. (1993) A protein synthesis dependent increase in E2F1 mRNA correlates with growth regulation of the dihydrofolate reductase promoter. *Mol. Cell. Biol.*, **13**, 1610.
93. Blake, M. and Azizkhan, J. C. (1989) Transcription factor E2F is required for efficient overexpression of the hamster dihydrofolate reductase gene *in vitro* and *in vivo*. *Mol. Cell. Biol.*, **9**, 4944.
94. Dalton, S. (1992) Cell cycle regulation of the human *cdc2* gene. *EMBO J.*, **11**, 1797.
95. Means, A. L., Slansky, J. E., McMahon, S. L., Knuth, M. W., and Farnham, P. J. (1992) The HIP1 binding site is required for growth regulation of the dihydrofolate reductase gene promoter. *Mol. Cell. Biol.*, **12**, 1054.
96. Motokura, T. and Arnold, A. (1993) Cyclin D and oncogenesis. *Curr. Opin. Genet. Dev.*, **3**, 5.
97. Hinds, P. W. (1995) The retinoblastoma tumour suppressor protein. *Curr. Opin. Genet. Dev.*, **5**, 79.
98. Kamb, A., Gruis, N. A., Weaver-Feldhaus, J., Liu, Q., Harshman, K., Tavtigan, S. V., Stockert, E., Day, R. S. I., Johnson, B. E., and Skolnick, M. H. (1994) A cell cycle regulator potentially involved in the genesis of many tumor types. *Science*, **264**, 436.
99. Nobori, T., Miura, K., Wu, D. J., Lois, A., Takabayashi, K., and Carson, D. A. (1994) Deletion of the cyclin-dependent kinase-4 inhibitor gene in multiple human cancers. *Nature*, **368**, 753.
100. Hussussian, C. J., Struewing, J. P. M, Goldstein, A. M., Higgins, P. A., Ally, D. S., Sheahan, M. D., Clark, W. H., jun., Tucker, M. A., and Dracopoli, N. C. (1994) Germline p16 mutations in familial melanoma. *Nature Genet.*, **8**, 15.
101. Kamb, A., Shattuck-Eidens, D., Beles, R., Liu, Q., Gruis, N. A., Ding, W., Hussey, C., Tran, T., Miki, Y., and Weaver-Feldhaus, J. (1994) Analysis of the p16 gene (CDKN2) as a candidate for the chromosome 9p melanoma susceptibility locus. *Nature Genet.*, **8**, 22.
102. Caldas, C., Hanh, S. A., DaCosta, L. T., Redston, M. S., Schutte, M., Seymour, A. B., Weinstein, C. L., Hruban, R. H., Yeo, C. J., and Kern, S. E. (1994) Frequent somatic mutations and homozygous deletions of the p16 (MTS1) gene in pancreatic adenocarcinoma. *Nature Genet.*, **8**, 27.
103. Ohtari, K. and Nevins, J. R. (1994) Functional properties of a *Drosophila* homolog of the E2F1 transcription factor. *Mol. Cell. Biol.*, **14**, 1603.
104. Mai, B. and Lipp, M. (1993) Identification of a protein from *Saccharomyces cerevisiae* with E2F-like DNA-binding and transactivating properties. *FEBS Lett.*, **321**, 153.
105. Malhorta, P., Manohar, C. F., Swaminathan, S., Toyana, R., Dhar, R., Reichel, R., and Thimmapaya, B. (1993) E2F site activates transcription in fission yeast *Schizosaccharomyces pombe* and binds to a 30-KDa transcription factor. *J. Biol. Chem.*, **268**, 20392.
106. Hatakeyama, M., Brill, J. A., Fink, G. A., and Weinburg, R. A. (1994) Collaboration of G1 cyclins in the functional inactivation of the retinoblastoma protein. *Genes Dev.*, **8**, 1759.

9 | Control of transcriptional elongation by the human immunodeficiency virus Tat protein

JONATHAN KARN, MARK J. CHURCHER, KAROLA RITTNER, NICHOLAS J. KEEN, and MICHAEL J. GAIT

1. Introduction

AIDS is induced by the human immunodeficiency virus (HIV), a retrovirus with an infectious cycle similar to the oncogenic retroviruses. Immediately after infection, a double-stranded DNA copy of the RNA genome of HIV is synthesized, translocated to the nucleus, and integrated into the host genome. The integrated pro-viral form of HIV carries the long terminal repeat (LTR) sequences at both ends of the genome. The 5′ LTR acts as an inducible promoter, while the 3′ LTR supplies a polyadenylation signal.

One of the most surprising results to emerge from the intensive research into HIV molecular biology that was prompted by the AIDS crisis was the discovery that the regulation of HIV gene expression is achieved by the binding of viral regulatory proteins to specific RNA target sequences rather than to DNA targets. In this chapter, we describe how one of these RNA-binding proteins, the viral *trans*-activator protein Tat, is thought to regulate transcription through a novel elongation mechanism.

2. The Tat–TAR interaction

The first evidence that gene expression in HIV requires virally encoded *trans*-acting factors came from experiments in 1985 by Haseltine, Sodroski, and colleagues in which it was noted that transcription of reporter genes placed under the control of the viral LTR was stimulated 200- to 300-fold in cells that had been previously infected by HIV (1, 2). The *trans*-acting factor, now known to be encoded by the HIV-1 Tat gene, is a small, basic nuclear protein. As shown in Fig. 1, the Tat gene contains two exons that partially overlap the *rev* and *env* genes. Depending on the viral isolate, Tat is between 86 and 130 amino acids in length. The variable length

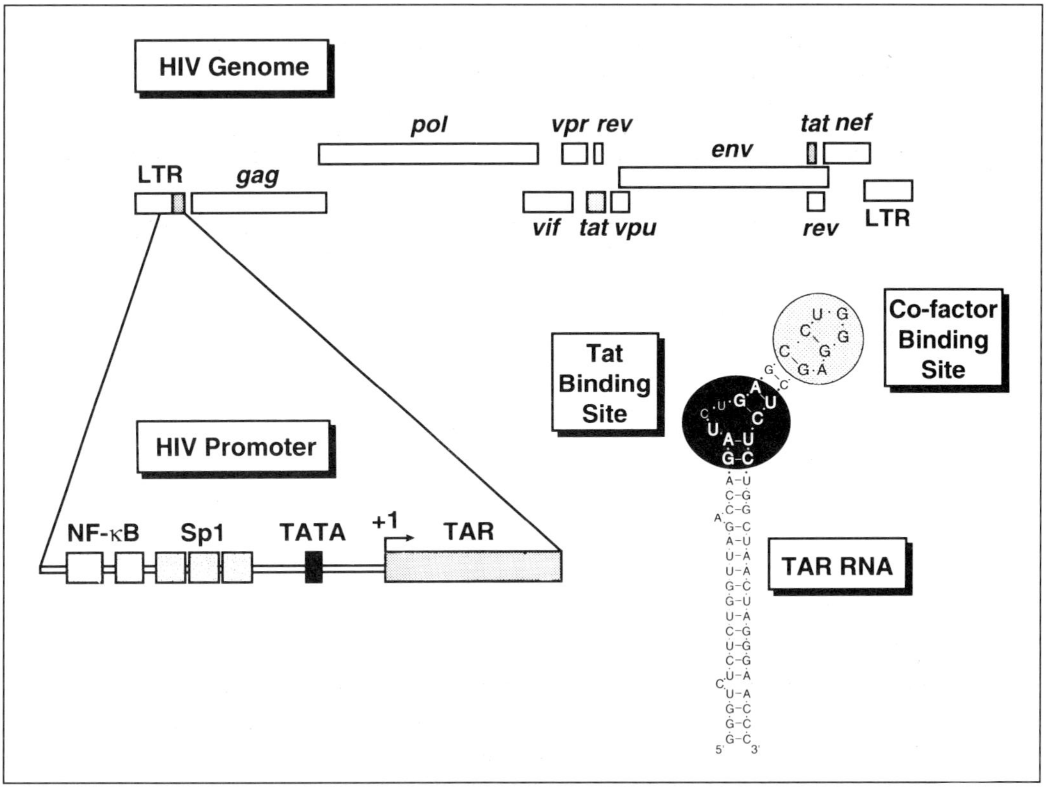

Fig. 1 Structure of the HIV genome (top), the viral promoter (bottom left), and TAR RNA (bottom right). The *tat* gene is encoded by two exons that are highlighted on the HIV genome map. The viral promoter has a structure typical of promoters activated by RNA polymerase II. Immediately upstream of the TATA-box are two tandem NF-κB binding sites and three tandem Sp1 binding sites. Immediately downstream of the start of transcription is the *trans*-activation response region (TAR). TAR encodes an RNA that can fold into the stem–loop structure shown to the right. Tat is able to bind directly to the highlighted region near to apex of the stem containing a U-rich bulge. Critical residues for Tat binding are shown in white. A cellular co-factor is believed to bind to the apical loop structure.

sequences are always in the second exon. The first exon of the Tat gene encodes a 72-amino acid protein that is functional (3).

2.1 TAR is functional as an RNA transcript

Deletion analysis of the viral LTR showed that Tat activity also requires the *trans*-activation-responsive region (TAR), a regulatory element located downstream of the initiation site for transcription (1–12). Because of its position in the HIV genome, each viral mRNA carries a copy of TAR at its 5′-end.

The location of TAR within a transcribed region was surprising and raised the possibility that Tat could be acting through an RNA element rather than a DNA element. By contrast to enhancer elements, the TAR element is only functional when it is placed downstream of the start of HIV transcription and in the correct orientation (3, 8–11, 13).

The RNA transcript of TAR is able to form a stable stem–loop structure 59 nucleotides long (Fig. 1; refs 3, 14–18). Point mutations that disrupt base-pairing in the upper TAR RNA stem invariably abolish Tat-stimulated transcription (7, 11, 12, 19–23).

Berkhout *et al.* (10) have shown that the TAR RNA sequence must be correctly folded during transcription in order for *trans*-activation to occur. In this experiment, an antisense sequence, which was designed to compete for the formation of the TAR RNA hairpin–loop structure, was inserted either upstream or downstream of TAR. When the antisense sequence was placed on the 5′-side of TAR, TAR RNA was unable to form a normal stem–loop structure and, as a result, the Tat response was inhibited. By contrast, incorporation of the antisense sequence on the 3′-side of TAR allowed normal *trans*-activation, since the TAR RNA present on nascent transcripts was able to adopt its normal structure prior to the transcription of the downstream antisense sequences.

2.2 Tat is an RNA-binding protein

The genetic evidence described above strongly suggested that Tat recognizes TAR RNA by direct binding. In 1989 we provided the first demonstration that recombinant Tat expressed in *Escherichia coli* could bind specifically to the stem of the TAR RNA (24).

Figure 2 shows an example of a competition binding experiment similar to the experiments we used to define the Tat binding site on TAR RNA (19, 22, 24). In the absence of competitor RNA, recombinant Tat protein formed a one-to-one complex with TAR RNA with a binding constant of approximately 3 nM. Unlabelled TAR RNA acts as an effective competitor with a $D_{1/2}$ (the concentration of competitor required to reduce binding by 50%) of approximately 3 nM. By contrast, TAR RNAs carrying mutations in the binding site, such as the mU23 (U_{23} to C) and mGC (G_{26}:C_{39} to C:G) mutations were poor competitors for protein binding and there was virtually no loss of radioactivity from the Tat–TAR complex in the presence of 20 nM competitor RNA ($D_{1/2} > 20$ nM).

2.3 Principles of RNA recognition by Tat

Extensive mutagenesis studies (12, 15, 16, 19, 22, 24), as well as chemical probing and peptide binding studies (17, 18, 25–31) have now defined the key elements required for TAR recognition.

There are numerous points of contact between Tat and TAR RNA that involve functional groups on bases near the bulge as well as adjacent phosphate residues. Although the bulge can be varied in size between 2 to 4 residues, the first residue in the bulge must always be a uridine (U_{23}) to allow high-affinity Tat binding (12, 15, 22, 26). The other residues in the bulge, C_{24} and U_{25}, appear to act predominantly as spacers and may be replaced by other nucleotides, or even by non-nucleotide

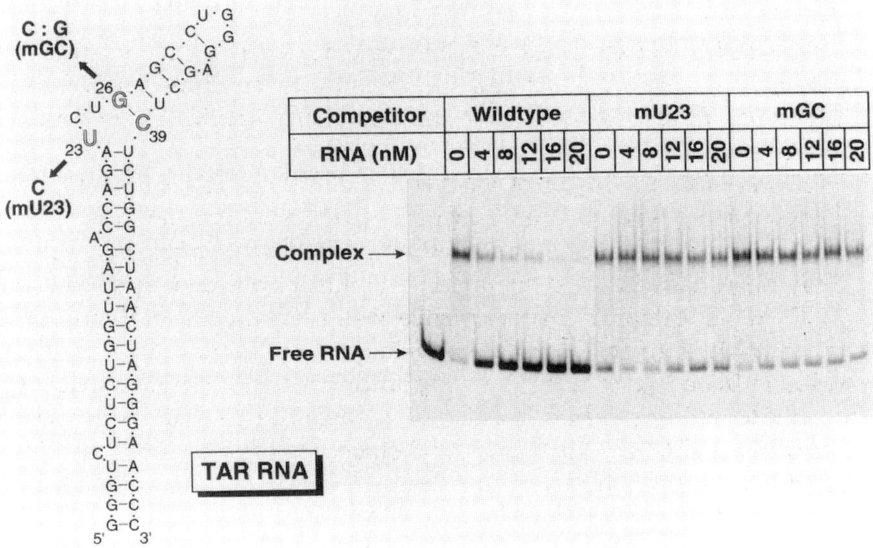

Fig. 2 Tat binds specifically to TAR RNA *in vitro*. (Left): sequence of TAR RNA and mutations that affect Tat binding. Bases in the stem are numbered from the start of the TAR RNA sequence, at the initiation site for transcription. (Right): gel mobility shift assays for Tat binding to TAR RNA. Free RNA was fractionated from the Tat–TAR RNA complex by electrophoresis on non-denaturing polyacrylamide gels. Complexes were formed between 2 nM ^{35}S-labelled TAR RNA and 5 nM Tat protein in the presence of 0 to 20 nM unlabelled competitor TAR RNA. TAR RNAs carrying the mU23 (U_{23} to C) and mGC (G_{26}:C_{39} to C:G) mutations competed with more than 10-fold lower affinity than the wild-type sequence.

linkers (19, 28). Tat recognition also requires two base pairs in the stem above the U-rich bulge, G_{26}:C_{39} and A_{27}:U_{38} (15, 19). Two base pairs below the bulge, A_{22}:U_{40} and G_{21}:C_{41}, also make significant contributions to Tat binding. Critical phosphate contacts involve phosphates P21, P22, and P40, which are located below the bulge on both strands (19, 25, 29, 30, 32).

Nuclease and chemical probing experiments (3, 14–18) suggest that the bulged residues bend the TAR RNA stem and widen the major groove. Consistent with these results, recent NMR data have suggested that TAR RNA is flexible, but can be stabilized in an 'open' configuration with a widened major groove after binding to basic peptides, or even to the amino acid derivative argininamide (33, 34).

The studies of the interactions of Tat with RNA outlined above have provided new insights into the chemistry of nucleic acid recognition (35). There are many examples of RNA-binding proteins that recognize bases displayed in single-stranded loop and bulge structures (36). Tat was one of the first examples of an RNA-binding protein that recognizes functional groups on Watson–Crick base pairs exposed within the major groove of a distorted double-stranded RNA (17). This type of RNA recognition is also used by the HIV-1 Rev protein (35), and is likely to extend to many other protein–RNA interactions.

2.4 RNA binding is required for Tat activity

In normal circumstances Tat is brought in close proximity to the transcribing polymerase because of its ability to bind to TAR RNA. There is a direct correlation between Tat binding to TAR RNA and *trans*-activation. Mutations in TAR that show reduced binding to Tat protein also produce a corresponding reduction in the levels of Tat-stimulated transcription *in vivo* (12, 15, 19, 22) and *in vitro* (37–40).

Experiments using hybrid proteins have also provided evidence that TAR RNA functions primarily as a 'loading site' that brings Tat in proximity to the transcription machinery (41–45). Fusion proteins containing sequences from Tat and bacteriophage R17 coat protein can stimulate transcription from HIV LTRs when the TAR RNA sequence is replaced entirely by an RNA stem–loop structure carrying R17 operator sequence (42). As in the case of Tat recognition of TAR RNA, binding of the fusion protein to its RNA target appears to be direct, since mutations in the R17 RNA operator sequence that reduce its affinity to coat protein produced a corresponding decrease in *trans*-activation by the Tat–R17 fusion protein *in vivo*. Similar results are obtained when the HIV–Rev protein is used as a fusion partner and an RNA binding site derived from the Rev response element (RRE) is used to replace TAR (41, 44, 45).

Studies of the fusion proteins have also been useful in the mapping of regions of Tat that are required to activate transcription, but are not required for RNA binding. The results show that most of the N-terminal 48 amino acid residues of Tat are necessary for transcriptional activity, whereas residues from the C-terminus, including the 'core' and basic regions of Tat are required for RNA binding (43–45). These observations are in agreement with peptide binding studies that show that sequences from the C-terminus of Tat are able to bind TAR RNA (12, 18, 19, 26, 31, 46).

3. Tat stimulates transcriptional elongation

The HIV LTR characteristically gives rise to two populations of transcripts: short non-polyadenylated RNAs terminating at, or near, the end of the TAR stem loop, and longer polyadenylated mRNAs (47). In the absence of Tat, the short transcripts predominate, whereas in the presence of Tat there is a dramatic increase in the levels of long transcripts (47–50). Nuclear run-on experiments have shown that these short transcripts arise because of a strong transcriptional polarity. In the absence of Tat, the majority of the transcriptional complexes formed at the HIV promoter stall or disengage near TAR, but Tat is able to increase the density of RNA polymerases found downstream of the promoter (47, 51–53).

3.1 Does Tat act as a generalized elongation factor?

The observation that Tat can alter the elongation capacity of RNA polymerases (reviewed above) has led a number of groups to suggest that Tat either acts as a specific elongation factor or helps recruit a specific cellular elongation factor (22,

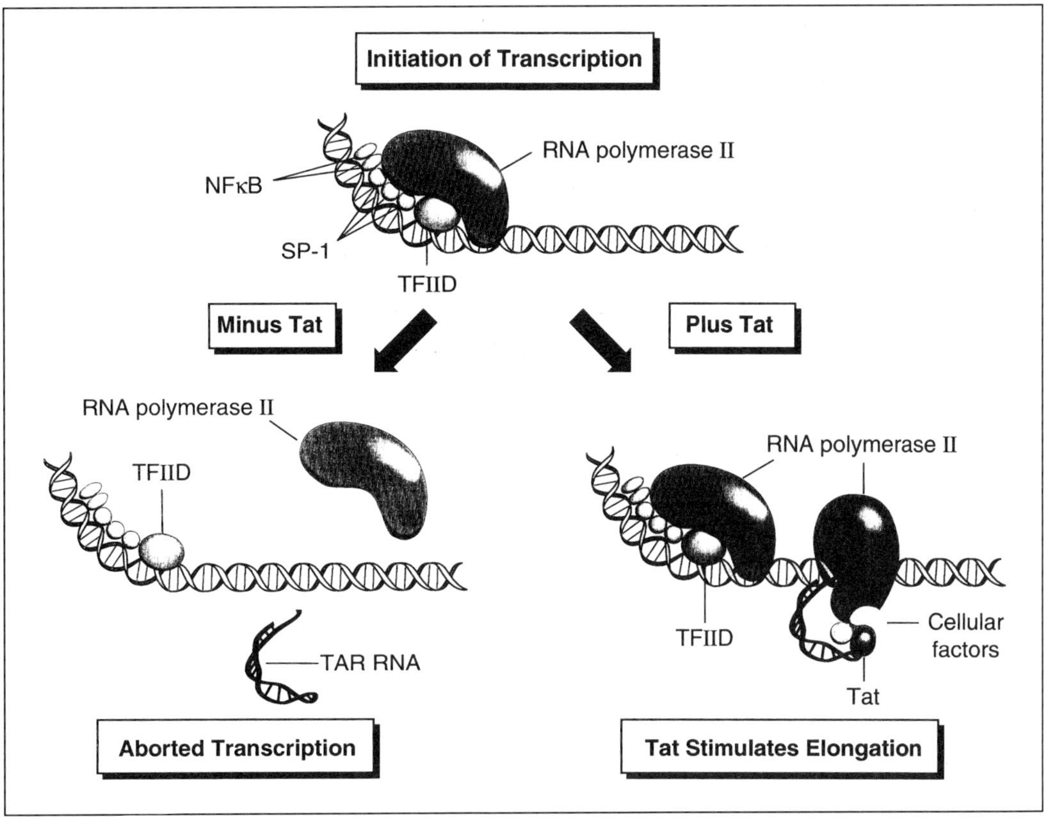

Fig. 3 Model for the control of transcription elongation by Tat. In a newly infected cell, the binding of cellular transcription factors to the viral LTR DNA sequences stimulates a basal level of transcription. In the absence of Tat the polymerase is unstable and disengages from the template at random sites downstream of TAR. As a result, the majority of the transcripts are short and non-polyadenylated. When Tat is present in the cell, it can bind to nascent TAR RNA sequences and form a complex with RNA polymerase, and cellular co-factors. This new complex is capable of efficient elongation and leads to the production of high levels of the HIV mRNAs.

37–40, 47, 53, 54). According to the proposal illustrated in Fig. 3, the RNA polymerase engaged by the HIV core promoter is believed to be intrinsically unstable. Soon after transcription through TAR, the polymerase either pauses or falls off the template. If Tat is present in the cell, it can associate with TAR RNA and the transcribing RNA polymerase. It seems likely that this reaction involves not only the binding of Tat to TAR RNA but also the recruitment of cellular co-factors, most likely including the TAR RNA loop-binding proteins and cellular elongation factors (38). The modified transcription complex is then able to transcribe the remainder of the HIV genome efficiently.

The anti-termination mechanism envisioned in Fig. 3 is closely analogous to the mechanism used by the bacteriophage λN protein (22, 39, 47, 55). However, in the λN system anti-termination takes place at a specific ρ-dependent termination site, whereas in the HIV system, TAR does not appear to be a unique site of termination.

Mutations in TAR that abolish *trans*-activation, or deletions of TAR, do not result in constitutively high levels of LTR expression (3, 6, 11, 20). It therefore seems likely that Tat acts as a more general elongation factor for RNA polymerase II that can overcome blocks to elongation at a variety of distal sites, including TAR (22, 37, 39, 40, 50, 53, 54).

3.2 *Trans*-activation *in vitro*

Detailed analysis of the *trans*-activation mechanism is now possible because of the development of efficient cell-free transcription systems that respond to Tat (37–40, 54). In standard cell-free systems, the HIV LTR behaves as a highly active promoter even in the absence of Tat. However, this high level of Tat-independent transcription can be reduced by a number of different methods. The original protocols exploited the observation that Tat-independent transcription *in vitro* is comparatively short-lived, and after 30 minutes, most transcription in the reaction becomes Tat dependent (37, 40, 54). Alternatively, extracts that have been pre-incubated in the presence of ATP (40, 54), or prepared in the presence of sodium citrate (38), or 100 μM $ZnSO_4$ (39), show reduced levels of Tat-independent transcription and can be used to detect Tat-dependent transcription.

3.3 Tat promotes elongation through distal terminator sites

The experiment shown in Fig. 4 (39) demonstrates that the Tat-stimulated RNA polymerase has an intrinsic anti-termination activity, consistent with the mechanism outlined in Fig. 3. A synthetic terminator sequence (τ) consisting of a stable RNA stem–loop structure followed by a tract of nine uridine residues was placed approximately 200 nucleotides downstream of the start of transcription. The presence of τ caused approximately 30% of the transcribing polymerases to disengage prematurely from the template. Addition of recombinant Tat protein to the reaction stimulated the production of run-off product (ρ) by more than 25-fold.

Stimulation of transcription by Tat in the cell-free system requires a functional TAR element. As shown in Fig. 4, the template carrying the ΔUCU mutation, which removes the bulge sequence from TAR, was stimulated less than three-fold by the addition of Tat. Studies using an extensive series of templates carrying mutations in TAR have shown that Tat-dependent *trans*-activation *in vitro* has the same sequence requirements seen *in vivo* (39). For example, LTRs carrying the mGC mutation, which only binds Tat non-specifically, respond only weakly to Tat in the cell-free system (Fig. 4).

3.4 Initiation is not increased by Tat in the cell-free systems

Several recent experiments using the cell-free systems suggest that Tat acts in these systems exclusively at the level of elongation (37–40, 54). The relative effects of Tat on initiation and elongation are demonstrated by RNase protection experiments shown in Fig. 5. In these assays the distribution of transcripts that extend to various

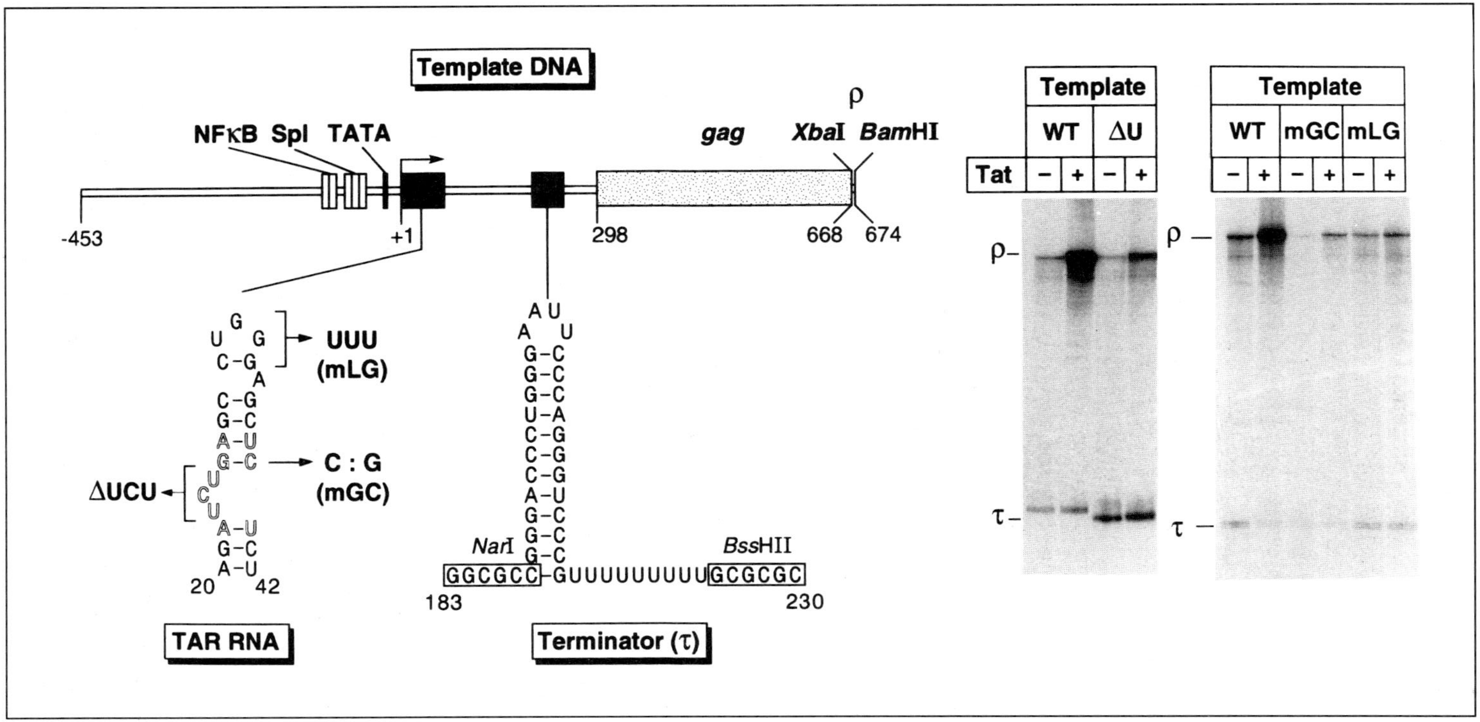

Fig. 4 Tat promotes read-through of distal terminators *in vitro*. The structure of template DNAs is shown on the top left. Each plasmid carried synthetic terminators (τ) inserted approximately 230 nucleotides downstream of the start of transcription and were linearized by cleavage at the *Bam*H1 site (or an adjacent *XbaI* site) located approximately 650 nucleotides downstream of the start of transcription. Transcription reactions were performed essentially as described (37, 39, 54) in the presence of 0 (–) or 200 ng (+) recombinant Tat protein. Right panels : templates carried an intact LTR mutation in the TAR element. The ΔU (ΔU_{23-25}) and mGC (G_{26}:C_{39} to C:G) mutations reduce Tat binding by more than 10-fold, whereas the mLG (G_{32-34} to UUU) mutation does not affect Tat binding. Addition of Tat protein increased the synthesis of the run-off product (ρ) approximately 25-fold from the wild-type template, but stimulated transcription by less than three-fold from the template carrying mutated TAR sequences. For each template, the level of transcription products ending at the terminator (τ) remained relatively constant in the presence or absence of Tat.

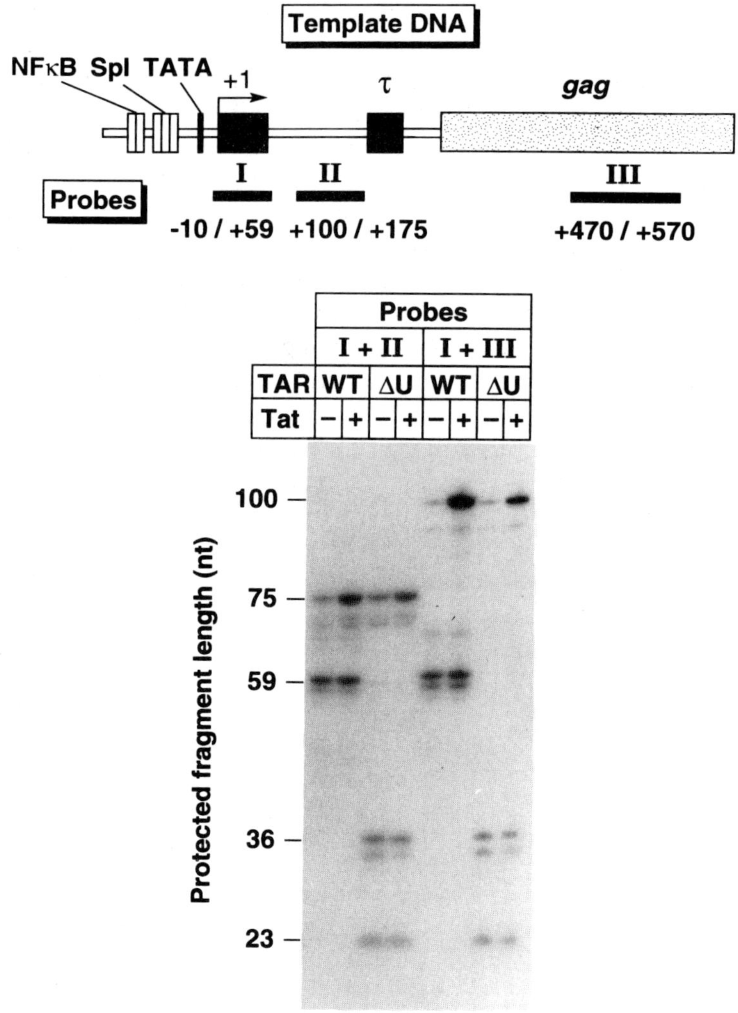

Fig. 5 Tat stimulates elongation but not initiation. Templates carrying either a wild-type (WT) TAR sequence or TAR carrying the ΔU (ΔU_{23-25}) mutation were transcribed *in vitro* and the products were analysed by RNase protection assays. The probes were ^{32}P-labelled antisense transcripts that were complementary to HIV sequences –10 to +59 (Probe I), +100 to +175 (Probe II), and +470 to +570 (Probe III). The probes were hybridized to the unlabelled products of the cell-free transcription reactions, digested with T1 ribonuclease, and the protected fragments fractionated by polyacrylamide gel electrophoresis. Because of the deletion in the ΔU template, Probe I is cleaved into two products of 36 nucleotides and 23 nucleotides. Addition of Tat stimulates transcription through the Probe II and Probe III regions of the template, but initiation, as measured by hybridization to Probe I appears to be constant. Note that the ΔU mutation substantially reduces the response to Tat.

points downstream of the start of the transcription are measured by hybridizing the transcribed RNAs to a series of ^{32}P-labelled complementary probes. Probe I overlaps the start of transcription and TAR provides a measure of the total initiation events. In the presence or absence of Tat the levels of transcripts that can hybridize to Probe I are constant, suggesting that initiation rates are constant. By contrast, in the presence of Tat there is a significant increase in the levels of transcripts that extend to the more distal Probes II and III. Addition of Tat stimulates an approximately five-fold increase in the number of transcripts that extend through the Probe II region. Tat simultaneously stimulates a 20-fold increase in the number of transcripts hybridizing to Probe III, which is located downstream of the terminator sequence. Thus, the data suggest that most of the polymerases that initiate transcription in the absence of Tat disengage from the template before reaching the terminator. Because Tat is able to stimulate elongation through the terminator, it produces a dramatic increase in the number of transcripts detected by Probe III. The results of these experiments are consistent with the observation of Marciniak *et al.* (54) that relatively few transcripts from the LTR are able to extend more than 200 nucleotides in the absence of Tat, whereas in the presence of Tat, the majority of the transcripts initiated from the LTR are able to extend until the end of the template.

Control experiments using LTRs carrying the ΔUCU mutation show that the ability of Tat to stimulate elongation requires a functional TAR sequence (Fig. 5). Because of the deletion in TAR, transcripts hybridizing to Probe I are cleaved into two fragments of 23 and 36 nucleotides. Constant levels of these transcripts are detected in both the presence and absence of Tat using Probe I and at levels comparable to those seen using the wild-type templates. This suggests that initiation is unaffected by the ΔUCU mutation. As expected, the ability of Tat to stimulate transcription through the Probe II and III regions is very much reduced when the ΔUCU mutation is present.

3.5 Tat-modified RNA polymerases have altered enzymological properties

In addition to pausing at RNA stem–loop structures, elongation by RNA polymerase can be blocked by DNA-binding proteins, such as the *lac* repressor (56–58). In the experiment shown in Fig. 6, a *lac* operator site was inserted downstream of the HIV LTR. When purified *lac* repressor was added to transcription reactions using these templates a significant fraction of the RNA polymerases were arrested at this site. The block imposed by the *lac* repressor protein can be used to trap polymerases that have initiated transcription in either the presence or absence of Tat. Unlike polymerases arrested by terminator sequences, the majority of the transcription complexes pausing at the *lac* repressor do not disengage from the template (56–58). These properties have allowed us to perform an experiment (Fig. 7) to measure the elongation capacities of Tat-positive and Tat-negative transcription complexes in the absence of new initiation events.

Transcription reactions were performed in the presence of *lac* repressor protein

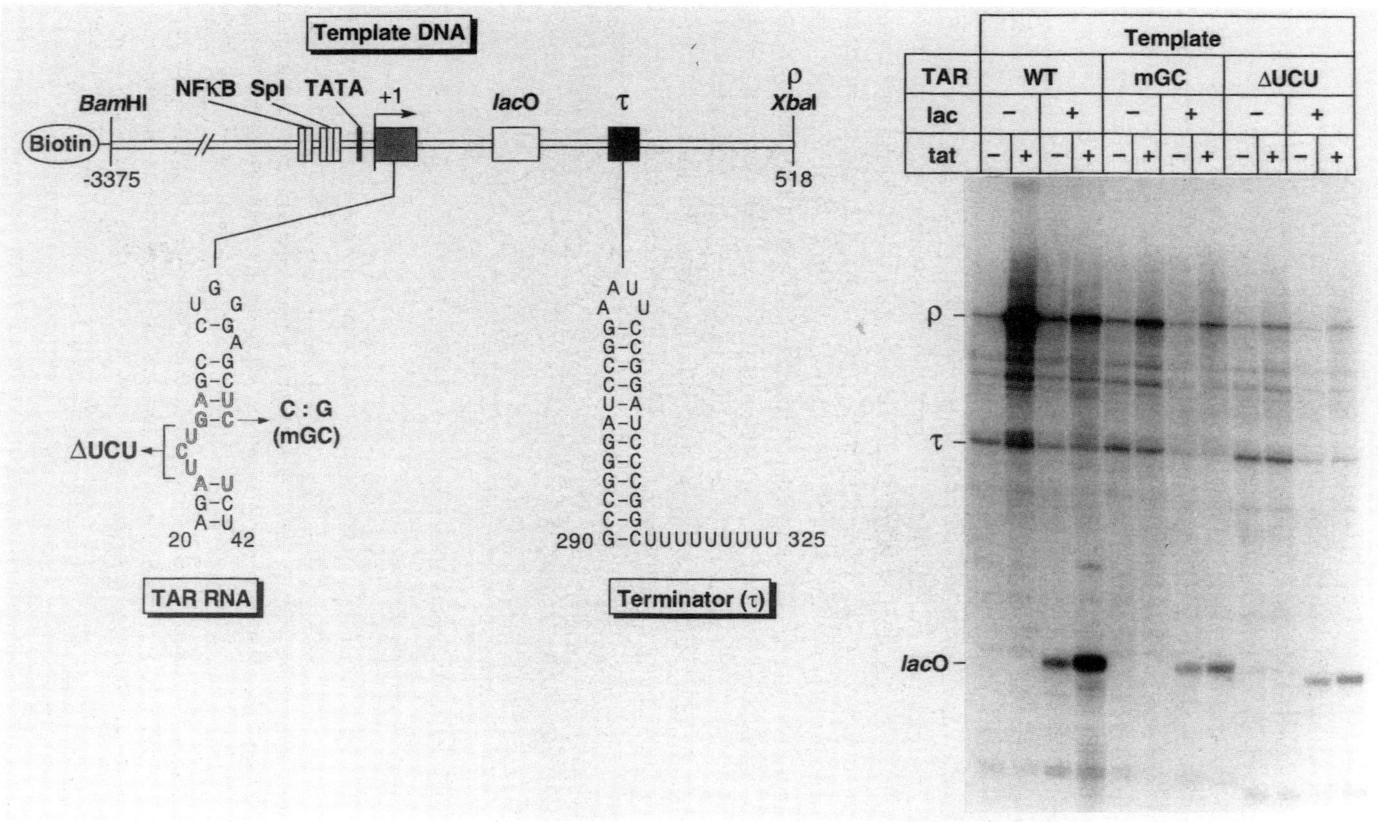

Fig. 6 Elongating RNA polymerase can be arrested by *lac* repressor. Transcription reactions were performed as described in Fig. 3, but using templates that include a *lac* operator (*lac*O) site cloned between TAR and the terminator sequence (τ). Addition of *lac* repressor (0.5 μg/ml) caused the transcribing RNA polymerases to arrest at the *lac*O site.

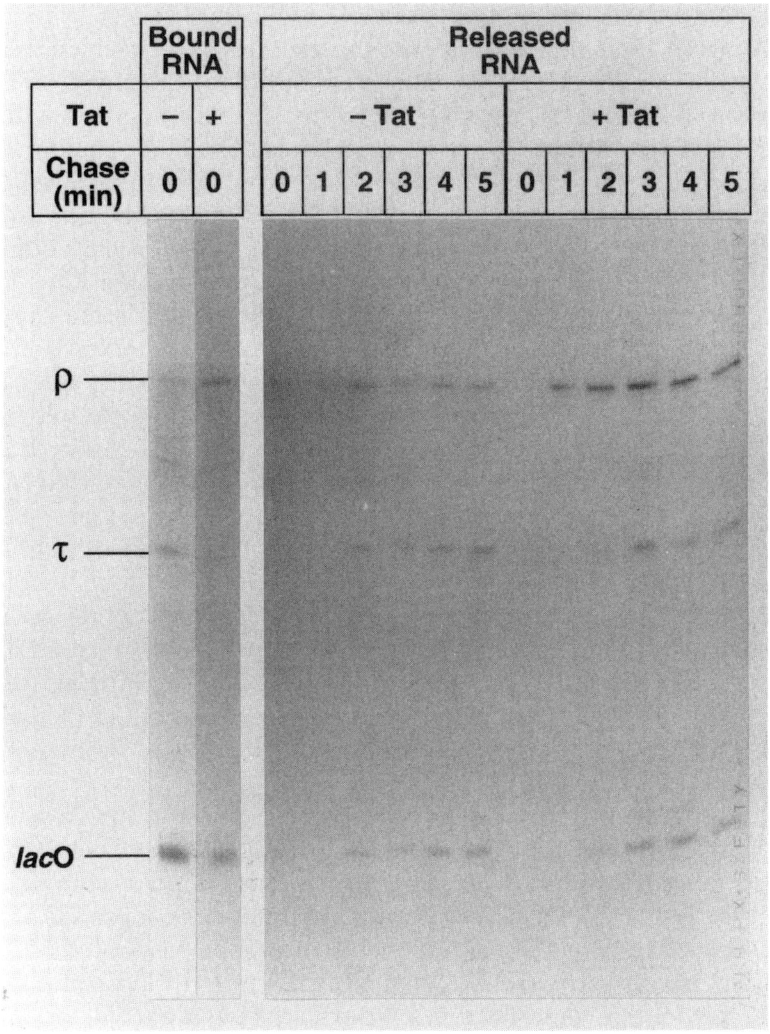

Fig. 7 Tat can stimulate read-through of *lac* repressor. Transcription reactions were performed as described in Fig. 6, using a biotinylated template DNA carrying a wild-type TAR sequence. The transcription reactions were incubated in the presence of 0 (–), or 200 ng (+) recombinant Tat protein. Following the transcription reaction the templates were bound to streptavidin-coated magnetic beads and extensively washed. The bound fractions carry RNA polymerases stalled at the *lac*O site together with labelled nascent RNA transcripts. In order to obtain equivalent levels of bound transcription reactions approximately three times more transcription mix from the minus Tat reactions was bound to the beads. The bound transcription complexes were incubated with 50 μM NTPs and the release of RNA from the templates was monitored at 1, 2, 3, 4, and 5 minutes. Note that the rate of full-length RNA release is significantly greater from transcription complexes that were exposed to Tat.

and biotinylated template DNAs. The newly synthesized RNA was labelled with radioactive nucleotides, and the arrested complexes, together with the nascent RNA, were then purified by binding to Streptavidin-coated magnetic beads. As shown in Fig. 7, most of the labelled RNA that is retained on the template DNAs bound to the beads corresponds to the transcripts associated with RNA polymerase, which is stalled by the *lac* repressor. In order to obtain equivalent amounts of transcription complexes from both reactions, approximately three times more sample from the Tat-minus reactions were bound to the beads than from the Tat-plus reactions. The arrested transcription complexes were then restarted by addition of unlabelled NTPs. As transcription proceeds, pre-labelled RNA is released from the immobilized templates. Transcription complexes that were exposed to Tat release significantly more run-off RNA than transcription complexes in the absence of Tat. Furthermore, the kinetics of RNA release are significantly different. By one minute after nucleotide addition the Tat-plus transcription complexes have completed release of run-off RNA, but there was no significant release of RNA from the Tat-minus transcription complexes. Approximately 50% of the RNA released from the Tat-minus transcription reactions corresponds to transcripts ending at the artificial terminator. By contrast, there is significantly more read-through of this sequence in the presence of Tat.

The experiment shown in Fig. 7 was performed while *lac* repressor was still bound to the templates. When the *lac* repressor is released by addition of the inducer isopropyl-thio-β-D-galactoside (IPTG), the kinetics of transcription are faster, but it is still possible to demonstrate differences in the behaviour of the Tat-associated transcription complexes and the complexes formed in the absence of Tat (data not shown).

Further evidence that Tat-modified transcription complexes have distinct properties compared with the unmodified transcription complexes comes from studies of transcription inhibitors. The Tat-modified transcription complex can be selectively inhibited both by DRB *in vitro* (54) and by inhibitors of protein kinases *in vivo* (59).

It seems likely that elongation factors normally associated with RNA polymerase are also associated with the Tat-modified transcription complexes. Kato *et al.* (38) have shown that the addition of purified elongation factor, TFIIF, increases the basal level of transcription from the viral LTR and largely overcomes the requirement for Tat. By contrast, addition of a second elongation factor, TFIIS, can boost transcription from the viral LTR in a co-operative manner with Tat.

3.6 TAR is functional at a distance

If Tat binds to the TAR RNA that is produced by an elongating RNA polymerase, as postulated by the mechanism depicted in Fig. 3, then it should also be possible to introduce Tat when TAR is placed considerably downstream of the promoter. In an early experiment, Selby et al. (11) found that maximal promoter activity required TAR to be located immediately downstream of the start of transcription. When TAR was displaced by 88 nucleotides, Tat-stimulated transcriptional activity was reduced by over 80%. This experiment has been interpreted to mean that TAR needs to be near the start of transcription in order to allow Tat to interact with the upstream elements in the promoter (60). However, there are two alternative explanations for the observations of Selby *et al.* (11). First, because transcription is severely attenuated in the absence of Tat, the majority of initiated polymerases may be prevented from reaching the displaced TAR and recruiting Tat. The contribution of the displaced TAR sequence can be measured by introducing mutations into it, but this important control was not included. Secondly, the templates constructed by Selby et al. (11) carry linkers inserted at +8. These insertions could have disrupted transcription initiation by interfering with the HIV INR (see Section 4).

In order to re-examine whether Tat could be introduced at a distance, under conditions where the sequences near the start of HIV transcription were unaltered, we prepared templates carrying duplicated TAR elements (Fig. 8). The first TAR carried the mGC mutation and is therefore unable to bind Tat efficiently. The second TAR carried either an intact Tat binding site, or, as a control, the mGC mutation in the Tat binding site. As shown in Fig. 8 (bottom left), when the first TAR is inactivated by the mGC mutation but a wild-type sequence is present in the second TAR, Tat is able to efficiently stimulate transcription. The double mutation has negligible activity. Because transcription is attenuated, the amount of run-off transcripts produced from templates where only TAR-2 is active is approximately 50% that obtained from templates where TAR-1 is active.

3.7 TAR loop sequences are required for Tat activity

Sequences in the apical loop of TAR RNA are not required for Tat binding (12, 22, 24). However, mutations in the TAR RNA loop strongly inhibit *trans*-activation (7, 11, 20–23). For example, templates with mutations in the loop domain of TAR (G_{32-34} to UUU, mLGC) were unresponsive to Tat in the cell-free system (Fig. 4; 39). One simple explanation for the behaviour of these loop mutations is that cellular RNA-binding proteins participate in *trans*-activation by binding to the loop sequences (22, 61, 62). Consistent with this idea, experiments using duplicated TAR sequences show that active TAR elements must carry both an intact loop sequence and an intact Tat binding site. When TAR-1 is inactivated by the mLG mutation in the loop, a Tat response can still be demonstrated by introducing an intact TAR element downstream (Fig. 8, bottom right). By contrast, templates carrying mutations in the loop sequences of both TAR-1 and TAR-2 are unable to respond to Tat, even though both these TAR sequences could, in principle, bind Tat.

To test whether Tat binding mediated by TAR-1 can be rescued by co-factor binding at a loop sequence provided by TAR-2, the mGC mutation was introduced into TAR-2. Since both sites must be present on the same TAR, it seems likely that co-operative interactions between the loop factor and Tat are required in order to obtain an efficient Tat response. Similarly, the insertion of extra Watson–Crick base pairs into the upper stem of TAR RNA alters the spacing between the Tat binding site and the loop and inhibits Tat-stimulated transcription (19, 61). It should also be

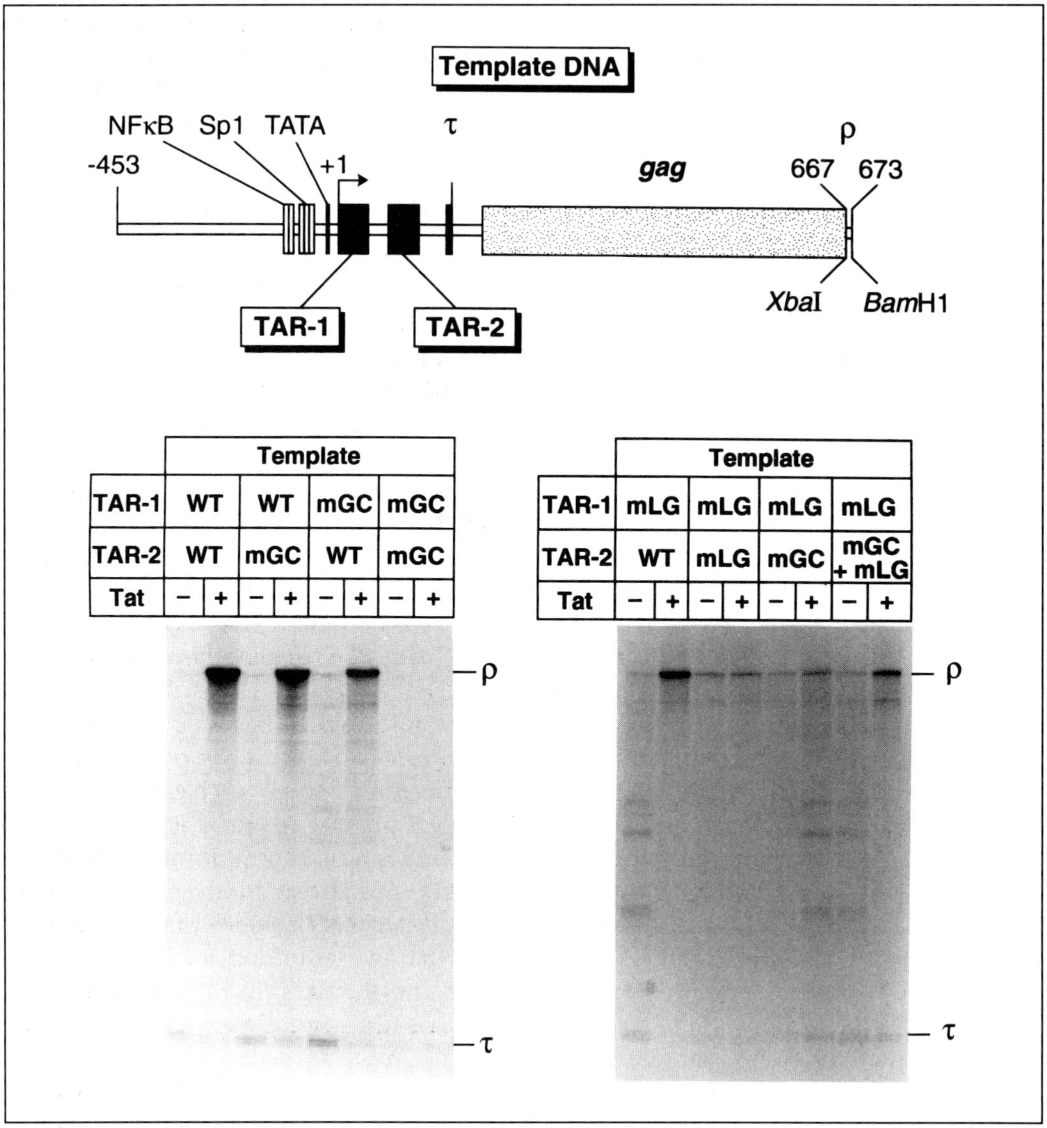

Fig. 8 TAR is functional at a distance. Template DNAs were constructed that carried a duplicated TAR (TAR-2) element inserted 100 nucleotides downstream of TAR (TAR-1). The effects of mutations in both elements were analysed in transcription reactions performed as described in the legend to Fig. 4. Left panel : templates carrying TAR duplications are able to respond to Tat even when the first TAR is inactivated by mutation. A template carrying the mGC (G_{26}:C_{39} to C:G) mutation in TAR 1 and a wild-type sequence in TAR 2 produced approximately 10-fold more runoff product (ρ) after addition of Tat. When both TAR 1 and TAR 2 were inactivated by the mGC mutation, there was no significant production of the run-off product. Right panel : TAR activity requires both a functional Tat-binding site and loop element. Mutations in the loop sequence (mLG) of TAR-1 can be rescued by a downstream wild-type TAR-2 sequence. However, if TAR-2 carries the mGC mutation no rescue is observed.

noted that this experiment excludes the possibility that the loop sequence is acting as a DNA element, since in that circumstance an intact loop sequence provided by TAR-2 should allow rescue of Tat activity by TAR-1, which carries an intact Tat binding site.

3.8 Recognition of TAR by cellular co-factors

Additional evidence supporting the view that cellular co-factors that recognize the TAR RNA loop sequences are also essential for Tat activity comes from experiments using TAR RNA as a competitive inhibitor. An excess of TAR RNA can effectively inhibit transcription from the HIV LTR both *in vivo* (63–66) and *in vitro* (37, 67). This activity requires both an intact Tat binding site and an intact loop sequence, implying that multiple protein binding sites on TAR are recognized by cellular proteins.

A third experiment, suggesting that Tat requires a positive cellular co-factor recognizing the TAR loop sequence, comes from observations about the species tropism of Tat. The Tat protein of HIV-1 is highly active in human cells, but shows significantly reduced activity in rodent cells (68, 69). DNA transfection experiments show that the block to *trans*-activation can be largely relieved by factor(s) expressed on human chromosome 12. Significantly, the residual *trans*-activation that is seen in cell lines lacking chromosome 12 is independent of mutations in the TAR RNA loop (69). Because of these considerations, there has been a considerable effort to identify these possible co-factors for Tat, which bind to the TAR loop sequence (67, 70, 71) or to the TAR RNA stem (72–75). Unfortunately, none of the proteins found so far has all of the properties that would be expected for a Tat co-factor. In particular, the genetic evidence suggests that the Tat co-factors should be able to associate with the TAR RNA loop in a sequence-dependent fashion, form a ternary complex with Tat, and show reduced activities in extracts obtained from cells that lack chromosome 12. For example, although Marciniak *et al.* (70) reported that a protein of approximately 68 kDa could be identified by UV-cross-linking to the TAR RNA loop, this protein is not a strong candidate for a Tat co-factor since it also binds with comparatively high affinity to TAR RNAs carrying mutations in the TAR loop that are known to significantly reduce *trans*-activation.

Perhaps the best candidate for a cellular co-factor for Tat comes from the laboratories of R. Gaynor and K. Jones (67, 71). They have independently identified another cellular protein, with a size of approximately 185 kDa, called TRP-185 or TRP-1, that participates in TAR RNA binding and can be UV-cross-linked to TAR RNA. Efficient complex formation requires a lower molecular weight cellular co-factor and has a strict requirement for the intact sequences of the TAR loop. One puzzling feature of TRP-185 activity is that it also requires an intact bulge sequence for binding. Since Tat also recognizes this region of TAR it is able to compete with TRP-185 for TAR binding. Although this observation does not by itself rule out a role for TRP-185 in the *trans*-activation mechanism, it is inconsistent with the prediction that a cellular co-factor for Tat should be able to form a stable ternary complex together with Tat and TAR. A demonstration of whether TRP-185, or any other candidate co-factor for Tat, is required for the *trans*-activation mechanism will have to await reconstitution of the system *in vitro* using purified components.

4. Promoter requirements for *trans*-activation

An important feature of the *trans*-activation mechanism envisioned in Fig. 3 is that transcription from the HIV LTR is initiated by a polymerase with poor elongation capacity. In common with most eukaryotic promoters transcribed by RNA polymerase II, the HIV LTR includes multiple upstream DNA regulatory elements that serve as binding sites for cellular transcription factors (Fig. 9).

4.1 Structure of the HIV promoter

Although the complete HIV LTR is approximately 600 nucleotides in length, the core promoter, which includes all the elements that are necessary for transcription initiation *in vitro*, extends from only –78 to approximately +26. The core promoter includes three tandem Sp1 binding sites (76–78), a TatA element (20, 79–82), and an initiator (Inr) sequence (80, 83).

Sp1 functions as an upstream activator. Deletion of one or more of the Sp1 sites significantly reduces Tat-dependent transcription from the viral LTR (23, 77, 84, 85). The TATA element is also required for high level expression from the viral LTR (20, 79–82). Converting the HIV-1 TATA and flanking sequences into those found in several other viral and cellular promoters reduces both basal and Tat-stimulated transcription to varying extents (81, 82, 86).

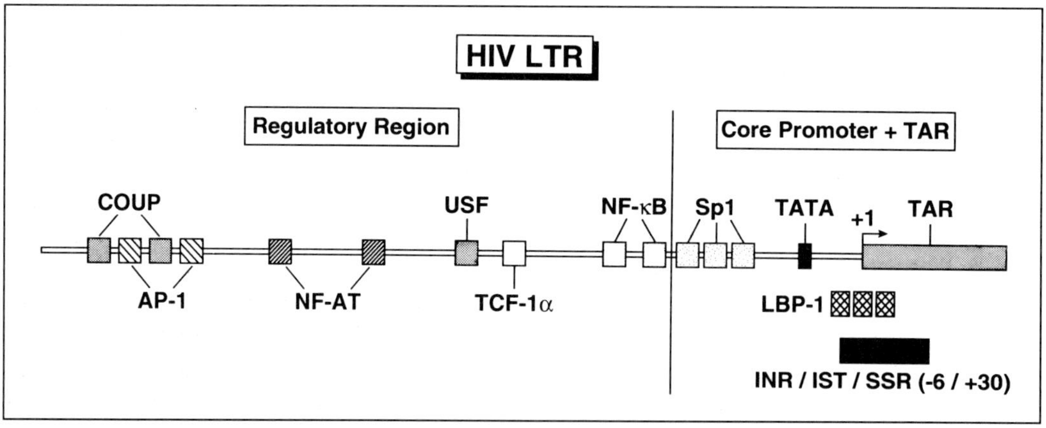

Fig. 9 Control elements in the HIV LTR. The HIV promoter can be divided into the 'core' region, which contains three tandem Sp1 sites, a TATA element, and an initiator (Inr) element. The Inr sequence, which has also been called an inducer of short transcripts (IST) (50, 88) or start site region (SSR) (83), extends from –6 to approximately +30 and therefore partially overlaps the TAR element. The HIV regulatory region carries a series of binding sites for DNA transcription factors. Depending on the cell type, these factors can modulate HIV transcriptional activity. However, in contrast to the core promoter elements, the virus is still able to grow when these elements are deleted.

The initiator (Inr) element was originally defined as a motif surrounding the RNA initiation site in the TATA-less promoter of the terminal deoxynucleotidyl-transferase (TdT) gene (87), but these types of elements can also stimulate transcription from promoters that contain TATA elements (see Chapter 1). A sequence very similar to the TdT and adenovirus major late promoter (Ad-MLP) Inr sequences is located near the start of transcription of the HIV LTR and includes sequences between –5 and +30 (80, 83). Mutations in the HIV Inr can dramatically reduce initiation rates (50, 80, 83, 88).

Further evidence that the HIV Inr is a key promoter element stems from the observation that a number of cellular DNA-binding proteins, which can bind to HIV sequences that overlap the initiator element, can inhibit transcription from the LTR. These proteins include the leader binding protein (LBP-1 or UBP-1) (20, 89–91) and YY1, a protein that binds to the polypyrimidine tracts found in the Inr element (92).

4.2 Regulatory promoter elements and the control of basal transcription from the HIV LTR

The HIV growth cycle may also include a latent stage where viral gene expression is silent. Little or no virus production is detected in infected but non-cycling T cells obtained from the peripheral blood of patients, but stimulation of these cells with mitogens dramatically increases virus production (93). At least six cellular transcription factors, AP-1, COUP, NF-AT, TCF-1a, USF, and NF-κB are known to bind to sites located upstream of the core promoter (Fig. 9). These elements can regulate HIV expression in both a positive and negative fashion in specific cell types (59, 77, 94–100). It seems likely that these cellular DNA-binding transcription factors can act as 'molecular switches' that allow HIV to make the transition from a latent state to lytic growth. The rate of virus growth is further moderated by the regulatory proteins. If the LTR activity is too weak to allow the production of necessary levels of Tat, then virus replication will be aborted.

The best-studied element in the HIV regulatory region is the 'enhancer region', which contains two NF-κB binding motifs (99). These elements are crucial for Tat-independent (basal) expression from the viral LTR in unstimulated lymphoid cells (99, 101). Addition of phorbol esters induces NF-κB DNA-binding activity and stimulates transcription from the LTR (99). Similarly, basal expression from the HIV LTR in undifferentiated monocytes or macrophages is highly dependent on the 'enhancer' sequences, but this requirement is lost after the cells are induced to differentiate (102, 103). A number of different cellular proteins can bind to NF-κB motifs (97), but the best-studied group are members of the Rel family (reviewed in Chapter 4) and these appear able to bind to the NF-κB sequences in the HIV LTR in a co-operative manner (98, 101).

However, it is an over-simplification to assume that all activation of basal transcription is invariably mediated through the NF-κB sites. Phorbol esters can induce transcription from viral LTRs from which the κB sites have been deleted, through a mechanism that appears to involve core promoter elements (59, 100). Furthermore, in contrast to the elements of the core promoter, mutation of the NF-κB sites results in only a modest inhibition of virus growth in most cell lines (85).

4.3 Control of initiation rates *in vitro* by core promoter elements

In the cell-free system, each of the core elements also appears to make a significant contribution to initiation and to optimal Tat responses. Figure 10 demonstrates the transcriptional activity of template DNAs carrying 3, 2, 1, or no Sp1 sites in the cell-free system. In this experiment, Tat responses were measured by RNase protection assays using a probe overlapping the start site of transcription to measure initiation rates, and a probe downstream of the terminator to measure the effects of Tat on elongation. As the Sp1 sites were removed, there was a progressive decrease in initiation rates. For each template, initiation (measured by hybridization to Probe I) was constant in the presence or absence of Tat and Tat was able to stimulate the production of full-length RNA. However, as the Sp1 sites are removed, the Tat responses decrease progressively. The loss of the Tat responses appears to be more rapid than would be expected from the changes in initiation rates measured by Probe I. For example, the SP3 template shows a four-fold greater Tat response than the SP-2 template, yet the levels of initiation by these two promoters are comparable. Similarly, the SP-0 template shows almost no Tat response, but its level of initiation is similar to the Sp1 template, which responds well to Tat.

The HIV Inr includes an unusual start site consisting of three G residues. As shown in Fig. 11, mutations in the start site sequence reduce initiation rates dramatically. The presence of at least one G residue appears to be important for efficient initiation. Templates carrying the TGG and TTG start sites have higher activity than templates that carry ACT, which is present in the adenovirus Inr, or ATT, which is present in the TdT Inr. Templates carrying a TTT start site are nearly inactive. Each of these mutations produces loss of Tat responsiveness by the promoter.

The experiment shown in Fig. 12 demonstrates that the HIV start site can support initiation in the absence of a functional TATA sequence, as expected for a sequence with Inr activity. A series of templates was constructed that carried the wild-type HIV Inr sequence, or mutant Inr sequences carrying the TTT start site mutation, or an ATT start site sequence. Each of these Inr sequences was combined with either a normal HIV TATA box sequence or with an inactive TATA box sequence (A at position –26 replaced by a G residue). The templates were then tested by run-off transcription assays in the cell-free system (Fig. 12). As expected, mutations in either the HIV TATA-box or the Inr sequences severely inhibited transcription initiation. However, the two sequences are synergistic; double mutations (KTX-17) are much more severely impaired than the single mutations in either the TATA box (–TATA) or the Inr element (KTX-14).

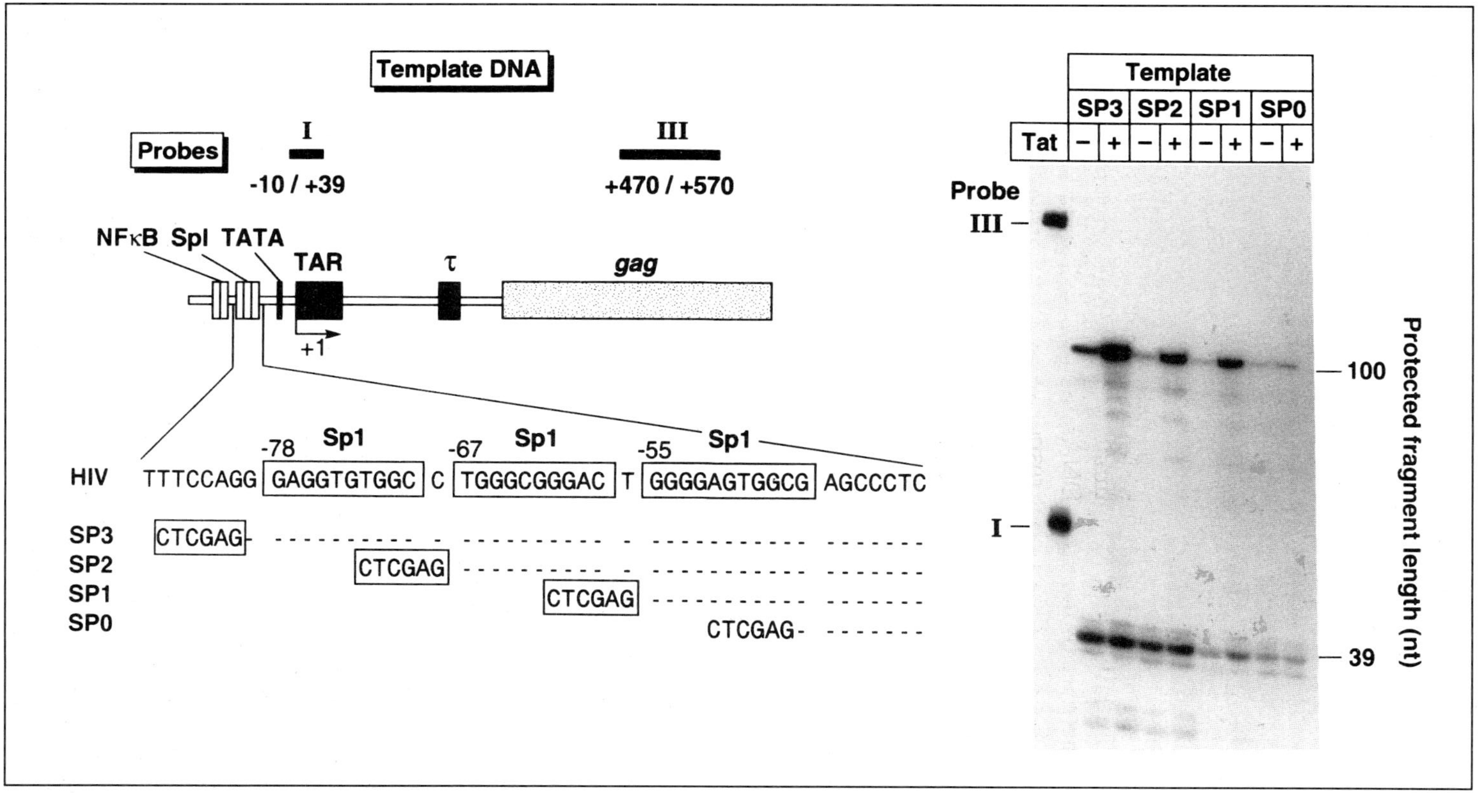

Fig. 10 Efficient initiation and Tat responses require a full complement of the Sp1 upstream elements. Templates carried either the SP-3, SP-2, SP-1, or SP-0 deletions. Transcription reactions were performed and analysed by RNase protection assays as described in the legend to Fig. 5. Note that in this experiment Probe I is slightly smaller than the Probe used in Fig. 5 and extends from –10 to +39. Tat was able to stimulate transcription through the Probe III region; however, the magnitude of Tat responses decreased as the Sp1 sites were deleted. The SP-3 template showed a 20-fold response, the SP-2 template showed a 10-fold response, the SP-1 template showed a five-fold response, and the SP-0 template showed a less than two-fold response. Initiation, as measured by hybridization to Probe I, also decreased as the Sp1 sites were removed. However, the change in initiation rates is insufficient to account for the changes in the Tat responses.

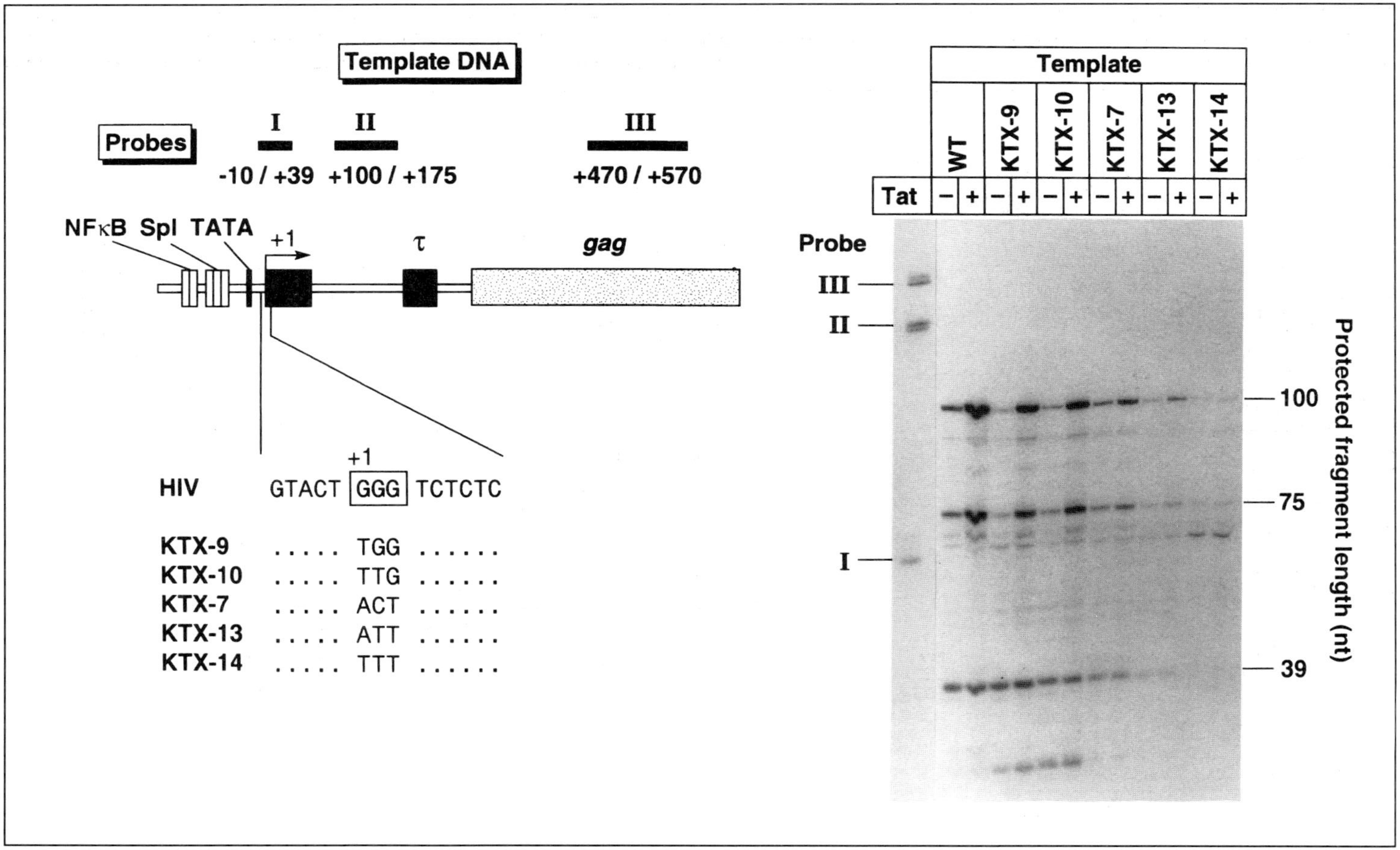

Fig. 11 Efficient initiation requires an intact Inr element. Templates carried a series of mutations at the transcription start site. Transcription reactions were performed and analysed by RNase protection assays as described in the legend to Fig. 5. Mutations in the start site sequence behave similarly to mutations in the Sp1 sites (Fig. 10). In each case the rate of initiation, as measured by hybridization to Probe I, is inhibited, and there is a concomitant decline in the Tat responses. For example, the KTX-7 template shows less than a two-fold response to Tat. The preferred transcription start site carries at least one G residue. Templates such as KTX-14, which carry no purines at the start site, are almost completely inactive for initiation.

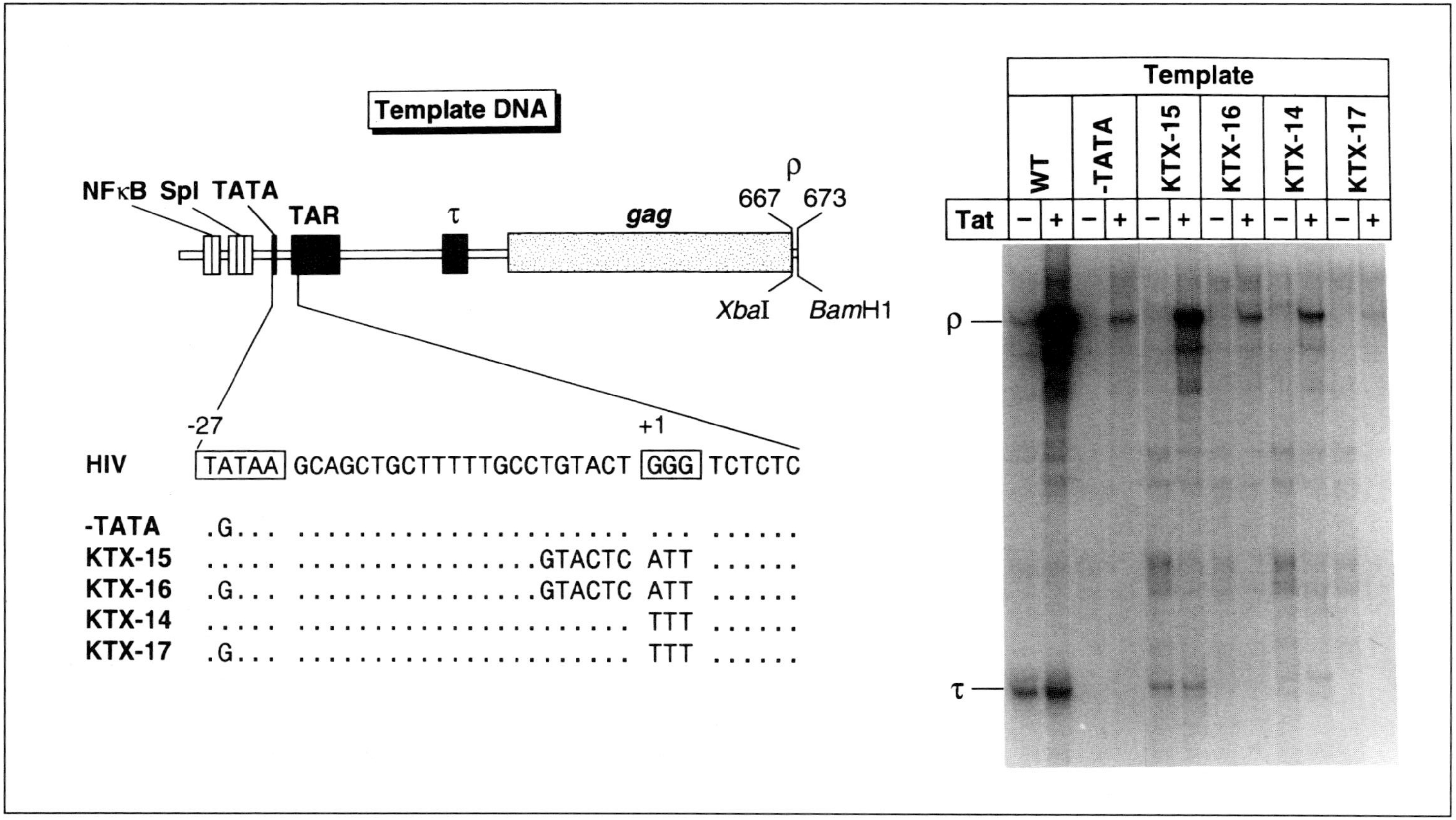

Fig. 12 The TATA box and Inr element act synergistically to stimulate transcription initiation from the HIV LTR. The effects of mutations in both elements were analysed in transcription reactions performed as described in the legend to Fig. 4. Mutations in either the TATA element (–TATA) or the Inr element severely inhibited transcription. However, the double mutation (KTX-17) is much more severely impaired than single mutations in the TATA box (–TATA) or Inr element (KTX-14). Note that the proportion of transcripts ending at the terminator sequence (τ) is severely restricted by TATA box or Inr mutations. This suggests that removal of these elements permits initiation by a polymerase that is more highly processive than the polymerase normally engaged by the HIV LTR.

4.4 Selection of Tat-responsive polymerase by core promoter elements

The data described above suggest strongly that each of the three elements in the HIV core promoter, the Sp1 sites, the TATA box, and the Inr, participate co-operatively in configuring the promoter for optimal *trans*-activation. These results are in agreement with *in vivo* studies that show that mutations that remove Sp1 sites or interfere with the TATA element produce more severe reductions in the Tat-activated response than in the basal, Tat-independent response (60, 81, 82, 84, 86). The spacing between Sp1 and the TATA box is also critical. When the two elements are separated, Tat-dependent transcription is selectively inhibited (104).

Experiments using hybrid promoters have shown that Sp1 can be replaced by other types of activator elements, including sequences derived from the adenovirus major late promoter, the U2 snRNA promoter, as well as the α- and β-globin promoters from chicken, rabbit, and human (3, 38, 40, 50, 60, 105). Similarly, artificial promoters that carry GAL4 sites in place of the Sp1 sites of HIV can be activated by a GAL4–VP16 hybrid protein and produce a strong Tat response (106). However, the Tat response does require initiation by RNA polymerase II since Pol-III-dependent promoters, such as the snRNA U6 promoter, are unable to respond to Tat (50).

The most indispensable promoter element for *trans*-activation appears to be the Inr. When an Inr and TAR were introduced downstream of heterologous promoters, initiation was stimulated and a strong Tat response was observed (50, 88). However, when the HIV Inr was replaced by a comparable region from the start of the β-globin gene, initiation was reduced and the Tat response was abolished (50).

It is tempting to speculate that all three elements from the HIV core promoter participate co-operatively in the binding of the initiation factor TFIID. TFIID is a multi-subunit complex that contains the TATA-binding protein, TBP, and co-factors called TAFs (107–117; reviewed in Chapter 1). The TAFs can make direct contact with acidic activator proteins such as Sp1. In addition, it has been recently shown that the Inr of the adenovirus major late promoter is recognized directly by TAF150, an essential component of the TFIID complex, which also recognizes the TATA box (116, see Chapter 1). Thus the mutational data are consistent with the view that selection of Tat-responsive polymerases by the HIV promoter involves the recruitment of a specific TFIID complex that can recognize preferentially the HIV Inr sequence co-operatively in the presence of Sp1 and the HIV TATA region.

4.5 The 'inducer of short transcripts (IST)'

Transcription from the HIV LTR characteristically produces a population of short transcripts ending near TAR. Hernandez and co-workers have shown the short transcripts are produced whenever the HIV TAR region (–6 to +59) is placed downstream of a heterologous promoter (50, 88). These observations suggested that this region of the promoter includes a novel DNA element that acts as an inducer of short transcripts, or IST element (50, 88). However, more recent data suggest that the IST element is not a distinct element. Rather, as described above, the Inr element stimulates initiation of transcription by a poorly processive polymerase. When this enzyme transcribes through the TAR region it tends to disengage from the template and give rise to short transcripts (39, 47–50, 88, 118). In support of this view, it is important to note that each of the mutations reported by Hernandez and co-workers to disrupt the IST element also affect the Inr and reduce initiation (50, 88).

The termination activity of TAR is due to its secondary structure. When alternative stem–loop structures are introduced in place of TAR, the short transcripts always end at the base of the stem (48, 88). Similarly, in cell-free systems the major termination sites, like TAR, correspond to regions of the transcripts that can fold into stable RNA hairpin loop structures (39, 48). The primary termination products downstream of TAR have heterogeneous 3′ ends, which are then trimmed by nucleases to give rise to short transcripts that end at the base of the TAR stem (39).

5. Can Tat stimulate initiation as well as elongation?

There has been considerable debate about whether Tat acts exclusively to stimulate transcriptional elongation, as described above, or whether Tat can also stimulate transcriptional initiation. In an influential paper, Mathews and co-workers reported that in addition to stimulating elongation, Tat could stimulate initiation two- to five-fold (51). However, later analysis suggested that any effect that Tat may have on initiation is variable and that the effects of Tat on elongation can be demonstrated in the absence of changes in the initiation rate (50, 52, 106). Furthermore, as discussed above, studies using cell-free systems demonstrate that Tat appears to act exclusively to stimulate elongation, at least *in vitro* (37–40, 54).

5.1 Tethering of Tat to DNA elements

Perhaps the strongest evidence that Tat could be acting at the level of initiation comes from the observation that Tat can stimulate transcription when it is tethered to DNA-binding sites positioned upstream of the HIV-1 promoter. In these experiments Tat was fused to DNA-binding domains derived from a variety of transcription factors including c-Jun, GAL4, LexA, and BPV-1 E2 (60, 78, 84, 105, 106, 119, 120). As in the case of activation via RNA binding, mutations in each of the critical residues in the N-terminal region of Tat inhibits Tat-stimulated transcription (84).

It should be stressed that Tat is a very inefficient activator of transcription when it is introduced to the promoter via DNA. Ghosh *et al.* (106) have recently shown that Tat requires at least five tandem binding sites to produce a measurable effect on transcription, whereas, by contrast, the herpes simplex virus *trans*-activator protein VP16, which is believed to stimulate initiation, was highly active when only a single binding site was present at the promoter. Similarly Kamine *et al.* (78, 119, 120) have shown that a GAL4–Tat fusion protein is unable to efficiently activate transcription on its own and requires the presence of at least one functional Sp1 site. Furthermore, when Tat was introduced to the promoter via TAR RNA it produced effects that were synergistic with the effects of VP16 introduced via DNA. By contrast, in a reciprocal experiment, VP16 was not synergistic with Tat when it was introduced via TAR RNA. Synergistic effects between Tat and proteins that stimulate initiation have also been observed using NF-κB and E1A (52, 121). Thus, it seems likely that Tat acts at a different step of the transcription process than the acidic *trans*-activators and by a distinct mechanism of action.

5.2 Binding of Tat to cellular transcription factors

A number of studies have also suggested that Tat might be able to interact directly with components of the basal transcription apparatus. For example, Jeang *et al.* (122) have reported that purified Tat protein binds directly to the cellular transcription factor Sp1. The Oct and NF-κB factors can also bind to Tat affinity columns with somewhat lower affinities than Sp1. In a similar experiment, Kashanchi *et al.* (115) have found that Tat can bind directly to the TATA-binding protein, TBP. In both of these experiments, the binding interaction involves the 'minimal' region of the Tat protein, which includes the essential residue, lysine-41.

It is not yet known whether similar binding reactions occur *in vivo*. Binding of Tat to the transcription factors appears to be relatively weak and requires a large excess of Tat protein. A second problem, is that it is not yet known whether these protein–protein interactions can take place in the presence of the nucleic acid sequences such as TAR RNA and the DNA-binding sites for the Sp1 and TBP proteins that normally bind these proteins. Tat has also been reported to bind to a number of still undefined cellular proteins (123, 124). At present it is not possible to determine whether any of these interactions is physiologically important, although it seems unlikely that each of these binding reactions could occur simultaneously.

5.3 Models of Tat activity linking control of initiation and elongation

An alternative view of Tat function to the mechanism outlined in Fig. 3 is that Tat might stimulate initiation of an incoming polymerase with altered elongation properties (55, 60, 78, 84, 125, 126). At present these models are difficult to discount, although they appear to be mechanistically more complicated than the elongation model outlined in Fig. 3. Because Tat is introduced to the HIV promoter via its interactions with TAR RNA, one implication of any initiation model is that Tat does not act on the first round of transcription. This could only occur if the RNA polymerase synthesizing TAR remains in close proximity to the promoter and Tat is free to 'reach back' and make contact with the next initiation complex. However, this is inconsistent with observations that Tat is functional in cell-free systems under pulse–chase conditions where re-initiation of transcription is limited (38, 39, 54). A second postulate of the initiation model is that there are two classes of RNA polymerases with different elongation capacity that pre-exist in the cells. It is not obvious how Tat could distinguish between these two types of polymerases. By contrast, one clear prediction from the elongation model outlined in Fig. 3 is that Tat directly associates with transcription complexes that have progressed through the TAR region.

At present, there seems little reason to evoke specialized mechanisms to explain the reported effects of Tat on initiation reviewed above. For example, it seems likely that an RNA polymerase stalled near the promoter at TAR would block subsequent rounds of initiation. If Tat is able to release the stalled RNA polymerase then it could appear to simultaneously stimulate initiation. Similarly, if Tat is able to bind directly to RNA polymerase, or a closely associated elongation factor such as TFIIS or TFIIF, then it seems likely that this interaction could give rise to a weak initiation activity when Tat is tethered to DNA.

6. Conclusions

Studies of the HIV Tat protein have revealed a unique example of a viral protein that regulates transcriptional elongation. In this chapter we have reviewed the considerable progress that has been made in defining the mechanism of action of the HIV *trans*-activator protein. However, it remains a challenge for the future to devise biochemical experiments that will allow the identification of the novel components of Tat-modified transcription complexes. In particular, the cellular co-factors that interact with the TAR loop element and the elongation factors that modify the Tat-activated transcription complexes need to be better defined, and the system needs to be reconstituted using purified components. There is still more work to be done on this fascinating problem!

Acknowledgements

We thank our colleagues at LMB, in particular, Drs J. Butler, A. Klug, and G. Varani for helpful discussions and the MRC AIDS-Directed Programme for support.

References

1. Sodroski, J., Patarca, R., Rosen, C., Wong-Staal, F., and Haseltine, W. A. (1985) Location of the *trans*-acting region on the genome of human T-cell lymphotropic virus type III. *Science*, **229**, 74.
2. Sodroski, J. G., Rosen, C. A., Wong-Staal, F., Salahuddin, S. Z., Popovic, M., Arya, S., Gallo, R. C., and Haseltine, W. A. (1985) *Trans*-acting transcriptional regulation of human T-cell leukemia virus type III long terminal repeat. *Science*, **227**, 171.
3. Muesing, M. A., Smith, D. H., and Capon, D. J. (1987) Regulation of mRNA accumulation by a human immunodeficiency virus *trans*-activator protein. *Cell*, **48**, 691.
4. Rosen, C. A., Sodroski, J. G., and Haseltine, W. A. (1985) The location of *cis*-acting regulatory sequences in the human T cell lymphotropic virus type III (HTLV-III/LAV) long terminal repeat. *Cell*, **41**, 813.

5. Arya, S. K., Guo, C., Josephs, S. F., and Wong-Staal, F. (1985) *Trans*-activator gene of human T-lymphotropic virus type III (HTLV-III). *Science*, **229**, 69.
6. Cullen, B. R. (1986) *Trans*-activation of human immunodeficiency virus occurs via a bimodal mechanism. *Cell*, **46**, 973.
7. Feng, S. and Holland, E. C. (1988) HIV-1 Tat *trans*-activation requires the loop sequence within TAR. *Nature*, **334**, 165.
8. Hauber, J. and Cullen, B. R. (1988) Mutational analysis of the *trans*-activation responsive region of human immunodeficiency virus type 1 long terminal repeat. *J. Virol.*, **62**, 673.
9. Jakobovits, A., Smith, D. H., Jakobovits, E. B., and Capon, D. J. (1988) A discrete element 3′ of human immunodeficiency virus 1 (HIV-1) and HIV-2 mRNA initiation sites mediates transcriptional ativation by an HIV *trans*-activator. *Mol. Cell. Biol.*, **8**, 2555.
10. Berkhout, B., Silverman, R. H., and Jeang, K.-T. (1989) Tat *trans*-activates the human immunodeficiency virus through a nascent RNA target. *Cell*, **59**, 273.
11. Selby, M. J., Bain, E. S., Luciw, P., and Peterlin, B. M. (1989) Structure, sequence and position of the stem-loop in TAR determine transcriptional elongation by Tat through the HIV-1 long terminal repeat. *Genes Dev.*, **3**, 547.
12. Roy, S., Delling, U., Chen, C.-H., Rosen, C. A., and Sonenberg, N. (1990) A bulge structure in HIV-1 TAR RNA is required for Tat binding and Tat-mediated *trans*-activation. *Genes Dev.*, **4**, 1365.
13. Peterlin, B. M., Luciw, P. A., Barr, P. J., and Walker, M. D. (1986) Elevated levels of mRNA can account for the *trans*-activation of human immunodeficiency virus. *Proc. Natl Acad. Sci. USA*, **83**, 9734.
14. Colvin, R. A. and Garcia-Blanco, M. A. (1992) Unusual structure of the human immunodeficiency virus type 1 *trans*-activation response element. *J. Virol.*, **66**, 930.
15. Delling, U., Reid, L. S., Barnett, R. W., Ma, M. Y.-X., Climie, S., Sumner-Smith, M., and Sonenberg, N. (1992) Conserved nucleotides in the TAR RNA stem of human immunodeficiency virus type 1 are critical for *Tat* binding and *trans*-activation: model for TAR RNA tertiary structure. *J. Virol.*, **66**, 3018.
16. Harper, J. W. and Logsdon, N. J. (1991) Refolded HIV-1 Tat protein protects both bulge and loop nucleotides in TAR RNA from ribonucleolytic cleavage. *Biochemistry*, **30**, 8060.
17. Weeks, K. M. and Crothers, D. M. (1991) RNA recognition by Tat-derived peptides : interaction in the major groove? *Cell*, **66**, 577.
18. Weeks, K. M., Ampe, C., Schultz, S. C., Steitz, T. A., and Crothers, D. M. (1990) Fragments of the HIV-1 Tat protein specifically bind TAR RNA. *Science*, **249**, 1281.
19. Churcher, M., Lamont, C., Dingwall, C., Green, S. M., Lowe, A. D., Butler, P. J. G., Gait, M. J., and Karn, J. (1993) High affinity binding of TAR RNA by the human immunodeficiency virus Tat protein requires amino acid residues flanking the basic domain and base pairs in the RNA stem. *J. Mol. Biol.*, **230**, 90.
20. Garcia, J. A., Harrich, D., Soultanakis, E., Wu, F., Mitsuyasu, R., and Gaynor, R. B. (1989) Human immunodeficiency virus type 1 LTR TatA and TAR region sequences required for transcriptional regulation. *EMBO J.*, **8**, 765.
21. Roy, S., Parkin, N. T., Rosen, C. A., Itovitch, J., and Sonenberg, N. (1990) Structural requirements for *trans*-activation of human immunodeficiency virus type 1 long terminal repeat-directed gene expression by Tat: importance of base-pairing, loop sequence and bulges in the Tat-responsive sequence. *J. Virol.*, **64**, 1402.
22. Dingwall, C., Ernberg, I., Gait, M. J., Green, S. M., Heaphy, S., Karn, J., Lowe, A. D., Singh, M., and Skinner, M. A. (1990) HIV-1 Tat protein stimulates transcription by binding to a U-rich bulge in the stem of the TAR RNA structure. *EMBO J.*, **9**, 4145.

23. Berkhout, B. and Jeang, K.-T. (1989) *Trans*-activation of human immunodeficiency virus type 1 is sequence specific for both the single-stranded bulge and loop of the *trans*-acting responsive hairpin: a quantitative analysis. *J. Virol.*, **63**, 5501.
24. Dingwall, C., Ernberg, I., Gait, M. J., Green, S. M., Heaphy, S., Karn, J., Lowe, A. D., Singh, M., Skinner, M. A., and Valerio, R. (1989) Human immunodeficiency virus 1 Tat protein binds *trans*-activation-responsive region (TAR) RNA *in vitro*. *Proc. Natl Acad. Sci. USA*, **86**, 6925.
25. Calnan, B. J., Tidor, B., Biancalana, S., Hudson, D., and Frankel, A. D. (1991) Arginine-mediated RNA recognition: the arginine fork. *Science*, **252**, 1167.
26. Cordingley, M. G., LaFemina, R. L., Callahan, P. L., Condra, J. H., Sardana, V. V., Graham, D. J., Nguyen, T. M., LeGrow, K., Gotlib, L., Schlabach, A. J., and Colonno, R. J. (1990) Sequence-specific interaction of Tat and Tat peptides with the TAR sequence of HIV-1 *in vitro*. *Proc. Natl Acad. Sci. USA*, **87**, 8985.
27. Delling, U., Roy, S., Sumner-Smith, M., Barnett, R., Reid, L., Rosen, C. A., and Sonenberg, N. (1991) The number of positively charged amino acids in the basic domain of Tat is critical for *trans*-activation and complex formation with TAR RNA. *Proc. Natl Acad. Sci. USA*, **88**, 6234.
28. Sumner-Smith, M., Roy, S., Barnett, R., Reid, L. S., Kuperman, R., Delling, U., and Sonenberg, N. (1991) Critical chemical features in *trans*-acting-responsive RNA are required for interaction of human immunodeficiency virus type 1 Tat protein. *J. Virol.*, **65**, 5196.
29. Tao, J. and Frankel, A. D. (1992) Specific binding of arginine to TAR RNA. *Proc. Natl Acad. Sci. USA*, **89**, 2723.
30. Hamy, F., Asseline, U., Grasby, J., Iwai, S., Pritchard, C., Slim, G., Butler, P. J. G., Karn, J., and Gait, M. J. (1993) Hydrogen-bonding contacts in the major groove are required for human immunodeficiency virus type-1 Tat protein recognition of TAR RNA. *J. Mol. Biol.*, **230**, 111.
31. Weeks, K. M. and Crothers, D. M. (1992) RNA binding assays for Tat-derived peptides : implications for specificity. *Biochemistry*, **31**, 10281.
32. Pritchard, C. E., Grasby, J. A., Hamy, F., Zachareck, A. M., Singh, M., Karn, J., and Gait, M. J. (1994) Methylphosphonate mapping of phosphate contacts critical for RNA recognition by the human immunodeficiency virus Tat and Rev proteins. *Nucl. Acids Res.*, **22**, 2592.
33. Puglisi, J. D., Tan, R., Calnan, B. J., Frankel, A. D., and Williamson, J. R. (1992) Conformation of the TAR RNA-arginine complex by NMR spectroscopy. *Science*, **257**, 76.
34. Puglisi, J. D., Chen, L., Frankel, A. D., and Williamson, J. R. (1993) Role of RNA structure in arginine recognition of TAR RNA. *Proc. Natl Acad. Sci. USA*, **90**, 3680.
35. Gait, M. J. and Karn, J. (1993) RNA recognition by the human immunodeficiency virus Tat and Rev proteins. *Trends Biochem. Sci.*, **18**, 255.
36. Nagai, K. (1992) RNA-protein interactions. *Curr. Opin. Struct. Biol.*, **2**, 131.
37. Marciniak, R. A., Calnan, B. J., Frankel, A. D., and Sharp, P. A. (1990) HIV-1 Tat protein *trans*-activates transcription *in vitro*. *Cell*, **63**, 791.
38. Kato, H., Sumimoto, H., Pognonec, P., Chen, C.-H., Rosen, C. A., and Roeder, R. G. (1992) HIV-1 Tat acts as a processivity factor *in vitro* in conjunction with cellular elongation factors. *Genes Dev.*, **6**, 655.
39. Graeble, M. A., Churcher, M. J., Lowe, A. D., Gait, M. J., and Karn, J. (1993) Human immunodeficiency virus type 1 transactivator protein Tat, stimulates transcriptional read-through of distal terminator sequences *in vitro*. *Proc. Natl Acad. Sci. USA*, **90**, 6184.

40. Laspia, M. F., Wendel, P., and Mathews, M. B. (1993) HIV-1 Tat overcomes inefficient transcriptional elongation *in vitro*. *J. Mol. Biol.*, **232**, 732.
41. Southgate, C., Zapp, M. L., and Green, M. R. (1990) Activation of transcription by HIV-1 Tat protein tethered to nascent RNA through another protein. *Nature*, **345**, 640.
42. Selby, M. J. and Peterlin, B. M. (1990) *Trans*-activation by HIV-1 Tat via a heterologous RNA binding protein. *Cell*, **62**, 769.
43. Derse, D., Carvalho, M., Carroll, R., and Peterlin, B. M. (1991) A minimal lentivirus Tat. *J. Virol.*, **65**, 7012.
44. Tiley, L. S., Madore, S. J., Malim, M. H., and Cullen, B. R. (1992) The vp16 transcription activation domain is functional when targeted to a promoter-proximal RNA sequence. *Genes Dev.*, **6**, 2077.
45. Madore, S. J. and Cullen, B. R. (1993) Genetic analysis of the cofactor requirement for human immunodeficiency virus type 1 Tat function. *J. Virol.*, **67**, 3703.
46. Calnan, B. J., Biancalana, S., Hudson, D., and Frankel, A. D. (1991) Analysis of arginine-rich peptides from the HIV Tat protein reveals unusual features of RNA-protein recognition. *Genes Dev.*, **5**, 201.
47. Kao, S.-Y., Calman, A. F., Luciw, P. A., and Peterlin, B. M. (1987) Anti-termination of transcription within the long terminal repeat of HIV-1 by *tat* gene product. *Nature*, **330**, 489.
48. Toohey, M. G. and Jones, K. A. (1989) *In vitro* formation of short RNA polymerase II transcripts that terminate within the HIV-1 and HIV-2 promoter-proximal downstream regions. *Genes Dev.*, **3**, 265.
49. Kessler, M. and Mathews, M. B. (1992) Premature termination and processing of human immunodeficiency virus type 1-promoted transcripts. *J. Virol.*, **66**, 4488.
50. Ratnasabapathy, R., Sheldon, M., Johal, L., and Hernandez, N. (1990) The HIV-1 long terminal repeat contains an unusual element that induces the synthesis of short RNAs from various mRNA and snRNA promoters. *Genes Dev.*, **4**, 2061.
51. Laspia, M. F., Rice, A. P., and Mathews, M. B. (1989) HIV-1 Tat protein increases transcriptional initiation and stabilizes elongation. *Cell*, **59**, 283.
52. Laspia, M. F., Rice, A. P., and Mathews, M. B. (1990) Synergy between HIV-1 Tat and adenovirus E1a is principally due to stabilization of transcriptional elongation. *Genes Dev.*, **4**, 2397.
53. Feinberg, M. B., Baltimore, D., and Frankel, A. D. (1991) The role of Tat in the human immunodeficiency virus life cycle indicates a primary effect on transcriptional elongation. *Proc. Natl Acad. Sci. USA*, **88**, 4045.
54. Marciniak, R. A. and Sharp, P. A. (1991) HIV-1 Tat protein promotes formation of more-processive elongation complexes. *EMBO J.*, **10**, 4189.
55. Greenblatt, J., Nodwell, J. R., and Mason, S. W. (1993) Transcriptional antitermination. *Nature*, **364**, 401.
56. Reines, D. and Mote, J. J. (1993) Elongation factor SII-dependent transcription by RNA polymerase II through a sequence-specific DNA-binding protein. *Proc. Natl Acad. Sci. USA*, **90**, 1917.
57. Deuschle, U., Hipskind, R. A., and Bujard, H. (1990) RNA polymerase II transcription blocked by *Escherichia coli lac* repressor. *Science*, **248**, 480.
58. Kuhn, A., Bartsch, I., and Grummt, I. (1990) Specific interaction of the murine transcription termination factor TTFI with class I RNA polymerases. *Nature*, **344**, 559.
59. Jakobovits, A., Rosenthal, A., and Capon, D. J. (1990) *Trans*-activation of HIV-1 LTR-directed gene expression by Tat requires protein kinase C. *EMBO J.*, **9**, 1165.

60. Berkhout, B., Gatignol, A., Rabson, A. B., and Jeang, K.-T. (1990) TAR-independent activation of the HIV-1 LTR: evidence that Tat requires specific regions of the promoter. *Cell*, **62**, 757.
61. Berkhout, B. and Jeang, K.-T. (1991) Detailed mutational analysis of TAR RNA: critical spacing between the bulge and loop recognition domains. *Nucl. Acids Res.*, **19**, 6169.
62. Cullen, B. R. (1990) The HIV Tat protein: An RNA sequence-specific processivity factor? *Cell*, **63**, 655.
63. Lisziewicz, J., Rappaport, J., and Dhar, R. (1991) Tat-regulated production of multimerized TAR RNA inhibits HIV-1 gene expression. *New Biologist*, **3**, 82.
64. Graham, G. J. and Maio, J. J. (1990) RNA transcripts of the human immunodeficiency virus *trans*-activation response element can inhibit action of the viral *trans*-activator. *Proc. Natl Acad. Sci. USA*, **87**, 5817.
65. Sullenger, B. A., Gallardo, H. F., Ungers, G. E., and Gilboa, E. (1990) Overexpression of TAR sequences renders cells resistant to human immunodeficiency virus replication. *Cell*, **63**, 601.
66. Sullenger, B. A., Gallardo, H. F., Ungers, G. E., and Gilboa, E. (1991) Analysis of *trans*-acting response decoy RNA-mediated inhibition of human immunodeficiency virus type 1 *trans*-activation. *J. Virol.*, **65**, 6811.
67. Sheline, C. T., Milocco, L. H., and Jones, K. A. (1991) Two distinct nuclear transcription factors recognize loop and bulge residues of the HIV-1 TAR RNA hairpin. *Genes Dev.*, **5**, 2508.
68. Newstein, M., Stanbridge, E. J., Casey, G., and Shank, P. R. (1990) Human chromosome 12 encodes a species-specific factor which increases human immunodeficiency virus type 1 Tat-mediated *trans*-activation in rodent cells. *J. Virol.*, **64**, 4565.
69. Hart, C. E., Galphin, J. C., Westhafer, M. A., and Schochetman, G. (1993) TAR loop-dependent human immunodeficiency virus *trans*-activation requires factors encoded on human chromosome 12. *J. Virol.*, **67**, 5020.
70. Marciniak, R. A., Garcia-Blanco, M. A., and Sharp, P. A. (1990) Identification and characterization of a HeLa nuclear protein that specifically binds to the *trans*-activation response (TAR) element of human immunodeficiency virus. *Proc. Natl Acad. Sci. USA*, **87**, 3624.
71. Wu, F., Garcia, J., Sigman, D., and Gaynor, R. (1991) Tat regulates binding of the human immunodeficiency virus *trans*-activating region RNA loop-binding protein TRP-185. *Genes Dev.*, **5**, 2128.
72. Han, X.-M., Laras, A., Rounseville, M. P., Kumar, A., and Shank, P. R. (1992) Human immunodeficiency virus type 1 Tat-mediated *trans*-activation correlates with the phosphorylation of a cellular TAR RNA stem-binding factor. *J. Virol.*, **66**, 4065.
73. Rounseville, M. P. and Kumar, A. (1992) Binding of a host cell nuclear protein to the stem region of human immunodeficiency virus type 1 *trans*-activation-responsive RNA. *J. Virol.*, **66**, 1688.
74. Gatignol, A., Kumar, A., Rabson, A., and Jaeng, K.-T. (1989) Identification of cellular proteins that bind to the human immunodeficiency virus type 1 *trans*-activation responsive TAR element RNA. *Proc. Natl Acad. Sci. USA*, **86**, 7828.
75. Gatignol, A., Buckler-White, A., Berkhout, B., and Jeang, K.-T. (1991) Characterization of a human TAR RNA-binding protein that activates the HIV-1 LTR. *Science*, **251**, 1597.
76. Jones, K., Kadonaga, J., Luciw, P., and Tjian, R. (1986) Activation of the AIDS retrovirus promoter by the cellular transcription factor, Sp1. *Science*, **232**, 755.
77. Harrich, D., Garcia, J., Mitsuyasu, R., and Gaynor, R. (1990) TAR independent activa-

tion of the human immunodeficiency virus in phorbol ester stimulated T lymphocytes. *EMBO J.*, **9**, 4417.

78. Kamine, J., Subramanian, T., and Chinnadurai, G. (1991) Sp1-dependent activation of a synthetic promoter by human immunodeficiency virus type 1 Tat protein. *Proc. Natl Acad. Sci. USA*, **88**, 8510.
79. Kliewer, S., Garcia, J., Pearson, L., Soultanakis, E., Dasgupta, A., and Gaynor, R. (1989) Multiple transcriptional regulatory domains in the human immunodeficiency virus type 1 long terminal repeat are involved in basal and E1A/E1B-induced promoter activity. *J. Virol.*, **63**, 4616.
80. Jones, K. A., Luciw, P. A., and Duchange, N. (1988) Structural arrangements of transcription control domains within the 5′-untranslated leader regions of the HIV-1 and HIV-2 promoters. *Genes Dev.*, **2**, 1101.
81. Berkhout, B. and Jeang, K.-T. (1992) Functional roles for the TatA promoter and enhancers in basal and Tat-induced expression of the human immunodeficiency virus type 1 long terminal repeat. *J. Virol.*, **66**, 139.
82. Olsen, H. S. and Rosen, C. A. (1992) Contribution of the TatA motif to Tat-mediated transcriptional activation of the human immunodeficiency virus gene expression. *J. Virol.*, **66**, 5594.
83. Zenzie-Gregory, B., Sheridan, P., Jones, K. A., and Smale, S. T. (1993) HIV-1 core promoter lacks a simple initiator element but contains bipartite activator at the transcription start site. *J. Biol. Chem.*, **268**, 15823.
84. Southgate, C. D. and Green, M. R. (1991) The HIV-1 Tat protein activates transcription from an upstream DNA-binding site: implications for Tat function. *Genes Dev.*, **5**, 2496.
85. Ross, E. K., Buckler-White, A. J., Rabson, A. B., Englund, G., and Martin, M. A. (1991) Contribution of NF-κB and Sp1 binding motifs to the replicative capacity of human immunodeficiency virus type 1: distinct patterns of viral growth are determined by T-cell types. *J. Virol.*, **65**, 4350.
86. Lu, X., Welsh, T. M., and Peterlin, B. M. (1993) The human immunodeficiency virus type 1 long terminal repeat specifies two different transcription complexes, only one of which is regulated by Tat. *J. Virol.*, **67**, 1752.
87. Smale, S. T. and Baltimore, D. (1989) The 'initiator' as a transcription control element. *Cell*, **57**, 103.
88. Sheldon, M., Ratnasabapathy, R., and Hernandez, N. (1993) Characterization of the inducer of short transcripts, a human immunodeficiency virus type 1 transcriptional element that activates the synthesis of short RNAs. *Mol. Cell. Biol.*, **13**, 1251.
89. Jones, N. C., Rigby, P. W. J., and Ziff, E. B. (1988) *Trans*-acting protein factors and the regulation of eukaryotic transcription: lessons from studies of DNA tumor viruses. *Genes Dev.*, **2**, 267.
90. Wu, F. K., Garcia, J. A., Harrich, D., and Gaynor, R. B. (1988) Purification of the human immunodeficiency virus type 1 enhancer and TAR binding proteins EBP-1 and UBP-1. *EMBO J.*, **7**, 2117.
91. Kato, H., Horikoshi, M., and Roeder, R. G. (1991) Repression of HIV-1 transcription by a cellular protein. *Science*, **251**, 1476.
92. Margolis, D. M., Somasundaran, M., and Green, M. R. (1994) Human transcription factor YY1 represses human immunodeficiency virus type 1 transcription and virion production. *J. Virol.*, **68**, 905.
93. Fauci, A. S. (1988) The human immunodeficiency virus: infectivity and mechanisms of pathogenesis. *Science*, **239**, 617.

94. Gaynor, R. B. (1992) Cellular transcription factors involved in the regulation of HIV-1 gene expression. *AIDS*, **6**, 347.
95. Garcia, J. A., Wu, F. K., Mitsuyasu, R., and Gaynor, R. B. (1987) Interactions of cellular proteins involved in the transcriptional regulation of human immunodeficiency virus. *EMBO J.*, **6**, 3761.
96. Li, Y., Ross, J., Scheppler, J. A., and Franza, B. R. J. (1991) An *in vitro* transcription analysis of early responses of the human immunodeficiency virus type 1 long terminal repeat to different transcriptional activators. *Mol. Cell. Biol.*, **11**, 1883.
97. Phares, W., Franza, B. R. J., and Herr, W. (1992) The κB enhancer motifs in human immunodeficiency virus type 1 and simian virus 40 recognize different binding activities in human Jurkat and H9 T cells : evidence for NF-κB-independent activation of the κB motif. *J. Virol.*, **66**, 7490.
98. Liu, J., Perkins, N. D., Schmid, R. M., and Nabel, G. J. (1992) Specific NF-κB subunits act in concert with Tat to stimulate human immunodeficiency virus type 1 transcription. *J. Virol.*, **66**, 3883.
99. Nabel, G. and Baltimore, D. A. (1987) An inducible transcription factor activates expression of human immunodeficiency virus in T cells. *Nature*, **326**, 711.
100. Sakaguchi, M., Zenzie-Gregory, B., Groopman, J. E., Smale, S. T., and Kim, S. (1991) Alternative pathway for induction of human immunodeficiency virus gene expression: involvement of the general transcription machinery. *J. Virol.*, **65**, 5448.
101. Perkins, N. D., Edwards, N. L., Duckett, C. S., Agranoff, A. B., Schmid, R. M., and Nabel, G. J. (1993) A cooperative interaction between NF-κB and Sp1 is required for HIV-1 enhancer activation. *EMBO J.*, **12**, 3551.
102. Griffin, G. E., Leung, K., Folks, T. M., Kunkel, S., and Nabel, G. J. (1989) Activation of HIV gene expression during monocyte differentiation by induction of NF-κB. *Nature*, **339**, 70.
103. Moses, A. V., Ibanez, C., Gaynor, R., Ghazal, P., and Nelson, J. A. (1994) Differential role of long terminal repeat control elements for the regulation of basal and Tat-mediated transcription of the human immunodeficiency virus in stimulated and unstimulated primary human macrophages. *J. Virol.*, **68**, 298.
104. Huang, L.-M. and Jeang, K.-T. (1993) Increased spacing between Sp1 and TatAA renders human immunodeficiency virus type 1 replication defective: implications for Tat function. *J. Virol.*, **67**, 6937.
105. Han, P., Brown, R., and Barsoum, J. (1991) *Trans*-activation of heterologous promoters by HIV-1 Tat. *Nucl. Acids Res.*, **19**, 7225.
106. Ghosh, S., Selby, M. J., and Peterlin, B. M. (1993) Synergism between Tat and vp16 in *trans*-activation of HIV-1 LTR. *J. Mol. Biol.*, **234**, 610.
107. Takada, R., Nakatani, Y., Hoffmann, A., Kokubo, T., Hasegawa, S., Roeder, R. G., and Horikoshi, M. (1992) Identification of human TFIID components and direct interaction between a 250-kDa polypeptide and the TatA box-binding protein (TFIIDt). *Proc. Natl Acad. Sci. USA*, **89**, 11809.
108. Brou, C., Chaudhary, S., Davidson, I., Lutz, Y., Wu, J., Egly, J.-M., Tora, L., and Chambon, P. (1993) Distinct TFIID complexes mediate the effect of different transcriptional activators. *EMBO J.*, **12**, 489.
109. Timmers, H. T. M., Meyers, R. E., and Sharp, P. A. (1992) Composition of transcription factor B-TFIID. *Proc. Nat. Acad. Sci. USA*, **89**, 8140.
110. Timmers, H. T. M. and Sharp, P. A. (1991) The mammalian TFIID protein is present in two functionally distinct complexes. *Genes Dev.*, **5**, 1946.

111. Pugh, B. F. and Tjian, R. (1991) Transcription from a TatA-less promoter requires a multisubunit TFIID complex. *Genes Dev.*, **5**, 1935.
112. Pugh, B. F. and Tjian, R. (1992) Diverse transcriptional functions of the multisubunit eukaryotic TFIID complex. *J. Biol. Chem.*, **267**, 679.
113. Bungert, J., Kober, I., During, F., and Seifart, K. H. (1992) Transcription factor eUSF is an essential component of isolated transcription complexes on the duck histone H5 gene and it mediates the interaction of TFIID with a TatA-deficient promoter. *J. Mol. Biol.*, **223**, 885.
114. Emanuel, P. A. and Gilmour, D. S. (1993) Transcription factor TFIID recognizes DNA sequences downstream of the TatA element in the *Hsp70* heat shock gene. *Proc. Natl Acad. Sci. USA*, **90**, 8449.
115. Kashanchi, F., Piras, G., Radonovich, M. F., Duvall, J. F., Fattaey, A., Chiang, C.-M., Roeder, R. G., and Brady, J. N. (1994) Direct interaction of human TFIID with the HIV-1 transactivator Tat. *Nature*, **367**, 295.
116. Verrijzer, C. P., Yokomori, K., Chen, J.-L., and Tjian, R. (1994) *Drosophila* $TAF_{II}150$: Similarity to yeast gene TSM-1 and specific binding to core promoter DNA. *Science*, **264**, 933.
117. Kashanchi, F., Shibata, R., Ross, E. K., Brady, J. N., and Martin, M. A. (1994) Second-site long terminal repeat (LTR) revertants of replication-defective human immunodeficiency virus: effects of revertant TatA box motifs on virus infectivity, LTR-directed expresssion, *in vitro* RNA synthesis, and binding of basal transcription factors TFIID and TFIIA. *J. Virol.*, **68**, 3298.
118. Bengal, E. and Aloni, Y. (1991) Transcriptional elongation by purified RNA polymerase II is blocked at the *trans*-activation-responsive region of human immunodeficiency virus type 1 *in vitro*. *J. Virol.*, **65**, 4910.
119. Kamine, J. and Chinnadurai, G. (1992) Synergistic activation of human immunodeficiency virus type 1 promoter by the viral Tat protein and cellular transcription factor Sp1. *J. Virol.*, **66**, 3932.
120. Kamine, J., Subramanian, T., and Chinnadurai, G. (1993) Activation of a heterologous promoter by human immunodeficiency virus type 1 Tat protein requires Sp1 and is distinct from the mode of activation by acidic transcriptional activators. *J. Virol.*, **67**, 6828.
121. Liu, J., Perkins, N. D., Schmid, R. M., and Nabel, G. J. (1992) Specific NF-κB subunits act in concert with Tat to stimulate human immunodeficiency virus type 1 transcription. *J. Virol.*, **66**, 3883.
122. Jeang, K.-T., Chun, R., Lin, N. H., Gatignol, A., Glabe, C. G., and Fan, H. (1993) *In vitro* and *in vivo* binding of human immunodeficiency virus type 1 Tat protein and Sp1 transcription factor. *J. Virol.*, **67**, 6224.
123. Desai, K., Loewenstein, P. M., and Green, M. (1991) Isolation of a cellular protein that binds to the human immunodeficiency virus Tat protein and can potentiate *trans*-activation of the viral promoter. *Proc. Natl Acad. Sci. USA*, **88**, 8875.
124. Nelbock, P., Dillon, P. J., Perkins, A., and Rosen, C. A. (1990) A cDNA for a protein that interacts with the human immunodeficiency virus Tat *trans*-activator. *Science*, **248**, 1650.
125. Sharp, P. A. and Marciniak, R. A. (1989) HIV TAR: an RNA enhancer? *Cell*, **59**, 229.
126. Cullen, B. R. (1993) Does HIV-1 Tat induce a change in viral initiation rights? *Cell*, **73**, 417.

Index

Notes: (1). Entries sometimes appear twice to differentiate between a gene (italic) and its product (not italic) (2). References to Figures and Tables are indicated in bold.